Genetics, Cytogenetics and Breeding of Crop Plants

Pulses and Oilseeds

Genetics, Cytogenetics and Breeding of Crop Plants

(Volume 1)

Pulses and Oilseeds

Editors
P.N. Bahl
P.M. Salimath

SCIENCE PUBLISHERS, INC.

Science Publishers. Inc.
10 Water Street, # 310
Lebanon, New Hampshire 03766
United States of America

ISBN 1-886106-59-2

Foreword

Since the rediscovery of Mendelian laws of inheritance, the pace of development in the science of genetics has been amazing. The application of principles of genetics to crop improvement during the last three or four decades has substantially contributed to improvements in quality and quantity of food, feed, and fibre products. During this period, the National Agricultural Research Systems (NARS) of the developing countries, particularly those engaged in crop improvement programmes in South and Southeast Asia, made major strides with visible impact on crop yields. This era also witnessed the emergence of a number of International Institutes for Crops Improvement in the tropical countries. These International Research Centres have complemented the breeding efforts of NARS in elevating the yield potential of crops.

Though the principles of plant breeding are universal, its practical applications vary greatly in details. Moreover, the strategies, approaches, and techniques of genetic improvement in the crop species have been refined over the years, based on our understanding of the latest developments in related sciences, particularly genetics and cytogenetics of the respective crop species. This two-volume work, which includes information on all these aspects for each crop, can be regarded as highly valuable. The editors deserve to be congratulated for their efforts. They have done well in focusing on important selected crop species of pulses and oilseeds (volume I) and cereals and commercial crops (volume II) that serve as standard case studies. I hope students, researchers, teachers, and professional plant breeders will find this book more valuable considering the experience and expertise of the contributors of each chapter. It should prove useful to everyone interested in the latest developments in the field.

(R.S. Paroda)
Director General
Indian Council of Agricultural Research

Preface

Genetics, a relatively young branch of science, has had a spectacular growth since the rediscovery of mendelian laws of inheritance in the beginning of the twentieth century. Genetics and cytogenetics are the branches of science to which generally plant breeding appeals. the application of principles of genetics and cytogenetics has brought about significant production and productivity advances in crop plants world-wide, particularly during the last three or four decades. However, The Search For Enhanced germplasm is a neverending process, particularly in view of the constant demand for higher productivity, better quality, increased resistance to pests and diseases, and better tolerance/resistance to abiotic stresses. The coordinated and interdisciplinary approach to crop improvement has paid rich dividends in many countries, resulting in the quantum jumps in yield levels of many crops. However, the easy options are getting exhausted with time and there is no alternative for effective crop improvement strategies in meeting future challenges.

We believe that a clear understanding of the principles of genetics and cytogenetics of a crop species is essential in planning innovative breeding approaches. In fact, the detailed genetic and cytogenetic information of a crop species would greatly help us to understand the rationale behind breeding approaches employed for its genetic improvement. In the past, perhaps, no conscious effort has been made to review the literature on genetics, cytogenetics, and breeding of crop plants in a single volume. In this book attempt has been made to cover the most recent advances made in these areas.

Volume I of the book deals with pulse and oilseed crops, which constitute an important source of food, vegetable oils, protein, animal feed, and industrial raw materials. Crop species of this group have been chosen deliberately and dealt with in this volume keeping in view their international and regional importance. We have yet to witness breaking of the yield barriers in many of these crops, as happened in case of cereals.

Crop improvement activities normally revolve around specific crops and the importance of crop species to be improved upon differs with the environment. Moreover, the application of general principles of plant breeding to specific crop improvement, which in essence forms 'practice', very much varies from one crop species to another. Unlike the subject of plant breeding principles, which can very well be documented by an experienced plant breeder, the approaches or strategies, which differ so much from crop to crop, can hardly be presented by a single author with any competence. For this reason, the successful and experienced plant breeders from National Agricultural Research Systems and International Crop Improvement Institutes were asked to share their experiences and those of their colleagues. This book is published for the benefit of students, practising plant breeders, researchers, and teachers. It is our hope that it will serve them as a text-book and reference.

We are sincerely grateful to all the contributors from Indian Agricultural Universities and Institutes. Our special thanks are due to contributors from foreign countries and International Research Centres who readily agreed to share their experience of global and inter-regional programmes.

The Editors

Contents

1. CHICKPEA

P.M. Salimath

Professor and Head, Dept. of Genetics and Plant breeding, College of Agriculture, UAS, Dharwad (Karnataka)

P.N. Bahl

Former Deputy Director General (Crop Sciences), Indian Council of Agricultural Research (ICAR), New Delhi

K.B. Singh

Principal Chickpea Breeder, ICRISAT-ICARDA Chickpea Project, International Center for Agricultural Research in the Dry Areas (ICARDA), P.O. Box 5466, Aleppo, Syria

Introduction

Chickpea (*Cicer arietinum* L.) is the third-most important pulse crop in the world, after dry beans (*Phaseolus vulgaris* L.) and dry peas (*Pisum sativum* L.). The total world production of chickpea (average of 1990–92) was 7.1 million t produced from 10.1 million ha (FAO Production Yearbook, 1992).

Table 1
Triennium average of area, production and yield of chickpea in India from 1952–54 to 1991–93

Period	Area (mha)	Production (mt)	Yield (kg/ha)
1952-54	7.35	4.14	561
1955-57	9.57	5.76	602
1958-60	9.83	5.84	593
1961-63	9.35	5.80	621
1964-66	8.75	4.83	353
1967-69	7.79	4.63	594
1970-72	7.83	5.28	673
1973-75	7.26	4.22	583
1976-78	8.09	5.57	688
1979-81	7.09	4.48	628
1982-84	7.48	4.89	656
1985-87	7.23	4.96	684
1988-90	6.35	4.33	678
1991-93	6.52	4.61	708

Asia alone produces 92 per cent, Africa 4 per cent, and the Americas 3 per cent of total world production of chickpea (Singh, 1990). In Asia, India is the largest producer of chickpea, contributing over 70 per cent of the world production. A comparison of the data for 1934–38 with those for 1961–65 shows a 50 per cent increase in production in both Asia and the world. However, since 1965, there has been a slight reduction in production due to decline in area sown (Saxena, 1990).

In India, chickpea is grown on an area of about 6.5 million ha with a production of about 4.5 million t and yield close to 700 kg/ha (average of 1988–1993). The trend in area, production, and yield in India is shown in Table 1 and Fig. 1. The total area under chickpea rose from 7.35 million

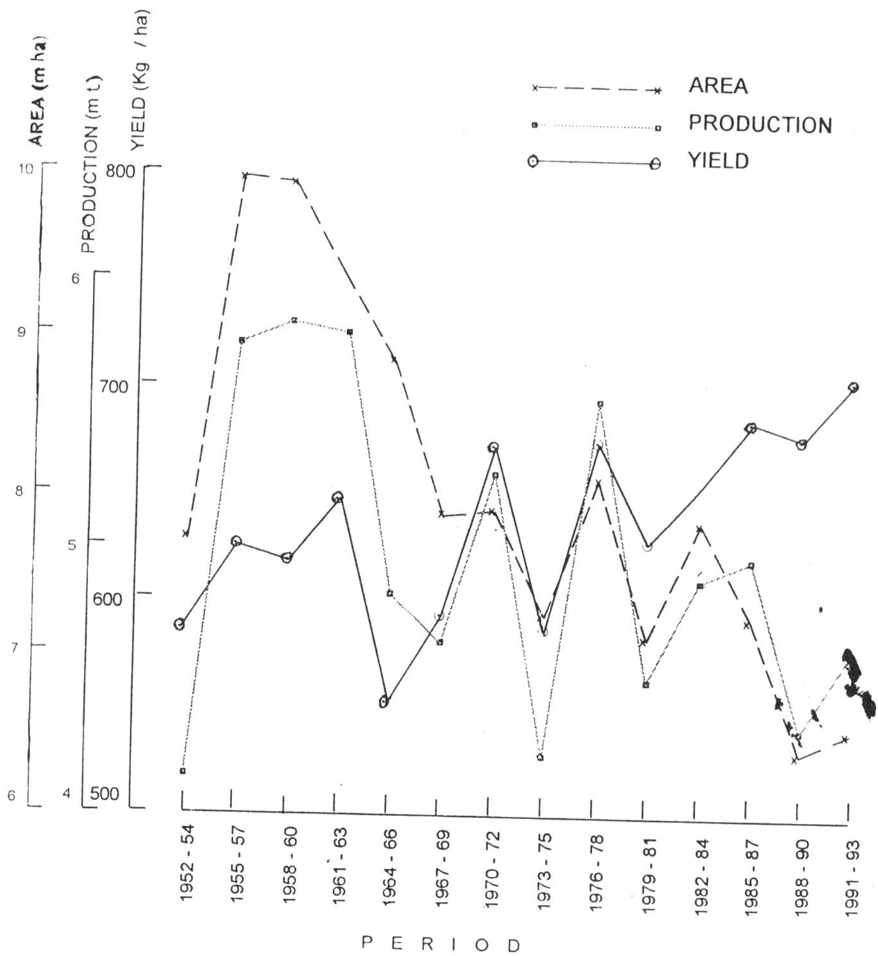

Fig. 1 Area, producution and yield of chickpea (1952–54 to 1991–93) in India

ha in 1952–54 to 9.83 million ha in 1958–60. Thereafter, it has gradually declined, reaching 6.52 million ha in 1991–93. With the advent of high yielding cultivars of cereals, particularly wheat and rice, and increased irrigation facilities, chickpea areas were diverted to some of these crops. Consequently, the production has also declined, especially after 1961–63. On the other hand, the yield per hectare has shown a gradual increase over the years, reaching the highest of 708 kg/ha during 1991–93.

In India, chickpea cultivation can be divided into different agroclimatic zones (Singh, 1980). The area under cultivation in different zones is as follows: 20–25 per cent in northwest plains, 20–25 per cent in north-central plains, 15–20 per cent in Bundelkhand highlands, 15–25 per cent in central plateau and plains and 5–10 per cent in eastern and peninsular zone. Rajasthan, Madhya Pradesh, Uttar Pradesh, and Haryana are the major chickpea producers and together contribute 80 per cent in area and 84 per cent in production in India (Bahl and Baldev, 1981). Other important chickpea-growing states are Punjab, Maharashtra, Bihar, Karnataka, West Bengal, Gujarat, and Andhra Pradesh. The productivity goes down as the cultivation moves from north to south. This can be partly attributed to the shorter growing season in southern India. On an average, crop matures from 90 days in Karnataka to 160 days in Punjab.

Chickpea is predominantly consumed in the form of whole grains or *dhal*. However, a sizeable portion of chickpea is converted into flour and is used in the preparation of a variety of snack foods, sweets and condiments. Chickpea is relatively free from antinutritional factors, has a high protein digestibility, and is rich in phosphorus and calcium compared to other pulses (Saxena, 1990). Chickpea straw has a forage value comparable to other straws commonly used for livestock feed. Because of these diversified uses and its ability to grow better with low inputs under harsh edaphic and arid environments than many crops, it is an important component in the cropping system of subsistence farmers in the Indian subcontinent, West Asia, and North Africa. Some of these attributes, together with its ability to meet more than 70 per cent of its nitrogen requirement from symbiotic nitrogen fixation (Saxena, 1988), make chickpea a potential crop for the alternative agriculture that is now attracting considerable attention in the industrialized world, especially in Australia.

In the past 30 years, there has been a many-fold increase in cereal production. On the other hand, there has been only a nominal improvement in pulse production, with the exception of pea. Many factors have been responsible for this and some related to genetic improvement of chickpea are mentioned here. First, cultivars with high genetic potential combining resistance to major biotic and abiotic stresses have not been developed. Second, despite phenomenal increases in irrigation, cultivars for irrigated agriculture are not available. Third, suitable cultivars to replace fallow in the rice-based and wheat-based farming systems have not been bred. These are the major challenges confronting chickpea breeders. The authors summarize in this chapter past research in the area of genetics, cytogenetics and breeding of chickpea and hope that this information will prove useful in formulating and implementing genetic enhancement programmes of chickpea.

Origin and Distribution

De Candolle (1882) vaguely identified the region between Greece and the Himalayas as the place where chickpea originated. Vavilov (1926) suggested the Southwest Asia and Mediterranean region as the primary centre of origin with Ethiopia as the secondary centre. Van der Maesen (1984) pointed out that Anatolia in Turkey was the area where chickpea was believed to have originated.

Seeds of chickpea are generally found as damaged specimens in the archaeological excavations, especially in the carbonized state (van der Maesen, 1972). Helbaek (1959) earlier reported Palestinian finds of the 4th millennium BC as probably the earliest. Chickpea was probably grown indiscriminately along with grass pea, butter vetch, and other legumes. Ramanujam (1976), however, reported finds of chickpea radiocarbon dated at 5450 BC and evidence for its cultivation in the Mediterranean basin during 3000–4000 BC. The earliest record of chickpea in India dates back to 2000 BC, based on the excavations at Atranji Kher in Uttar Pradesh (Chowdhary *et al.*, 1970), but records have been obtained at much later dates (AD 200–150 BC) from the south. Northern and southern Indian names for chickpea suggest an independent introduction in north and south. Ramanujam (1976) suggested that chickpeas were received overland in northern India and overseas in the south.

From its original home in Western Asia, chickpea seems to have diverged in two directions. In the western parts of its distribution, where chickpea is grown under equable spring-summer environments, cultivars were developed having large, owl-head-shaped, light-coloured seeds with little wrinkling of seed coat. Plants were tall and leaflets and flowers were large. In the eastern and southern parts of its distribution, where the crop is grown in the cool dry season, cultivars with quick maturity, thin stems, and small leaflets and flowers were developed. They are also characterized by small, wrinkled, and dark-coloured seeds. It has been suggested that the western ecotype has been modified more by human selection than the eastern ecotype, which has received less attention (Rohewal *et al.*, 1966; Salimath *et al.*, 1984).

Recently, the crop has been introduced in other areas. From the eastern Peninsula, the Portuguese and Spanish travellers in the 16th century carried it to South and Central America and Mexico. The expatriate Indian population carried it to East and South Africa and West Indies. During the 18th century, large-seeded kabuli cultivars were introduced into India from the Mediterranean region.

The wild progenitor of cultivated chickpea was not identified with any degree of certainty in the beginning. Renfrew (1973) suggested that *C. judaicum* was the most closely related wild form of chickpea, but van der Maesen (1972) opined that *C. pinnatifidum, C. judaicum,* and *C. bijugum* were the close relatives of cultivated chickpea. However, morphologically and according to their seed protein profile, these were unrelated to the cultivated species. Two other wild species, *C. echinospermum* and *C. reticulatum*, found near southeast Turkey closely resembled the cultivated species in these aspects. A close similarity in the water-soluble seed protein of *C. arietinum* and *C. reticulatum* was demonstrated electrophoretically by Ladizinsky and Adler (1975). The seed protein profiles examined of cultivars within *C. arietinum* are uniform, indicating a monophylatic origin of the cultigen. Breeding experiments involving hybridization of cultivated chickpea with two wild species—*C. reticulatum* and *C. echinospermum*—and cytological examination of the hybrids by Ladizinsky and Adler (1976a) further confirmed that *C. reticulatum* could be the wild progenitor of cultivated chickpea. The two wild species were found to differ from each other by a major reciprocal translocation and their hybrid was completely sterile. *Cicer echinospermum* also appeared to differ from *C. arietinum* by the same major reciprocal translocation resulting in hybrid sterility. But meiosis in the hybrids involving four different lines of *C. arietinum* with *C. reticulatum* was normal and the hybrids were fertile. Recently, Singh and Ocampo (1993) were successful in producing fertile hybrids between crosses involving *C. arietinum* × *C. echinospermum* and *C. echinospermum* × *C. reticulatum.* Partial sterility was apparent in these crosses in F_1 and F_2 generations. But they did not find any sterility in crosses between *C. arietinum* and *C. reticulatum.*

Taxonomy

The word 'Cicer' is a derivative of the Greek word 'Krios' referring to the well-known Roman family name 'Cicero'. 'Arietinum' is also derived from the Latin 'aries', meaning 'ram', referring to the shape of the seed, which resembles a ram's head. The taxonomic breakdown of the genus given by Ladizinsky has been generally accepted (Witcombe and Erskine, 1984).

Grain legumes are all members of the subfamily Papilionoidae. They belong to three tribes: *Hedysareae* (e.g., groundnut), *Phaseolae* (e.g., pigeonpea, soybean and others), and *Vicieae* (*Vicia, Cicer, Lens, Pisum,* and *Lathyrus*). Recently, genus *Cicer* has been taken out from tribe *Vicieae* and put separately as monogenetic tribe, the *Cicereae*. Genus *Cicer* has 49 species of which 9 are annual. Chickpea. is the sole cultigen of any economic significance.

Biosystematic relationship between the cultigen and its annual wild relatives has been studied experimentally by Ladizinsky and Adler (1976b). Their study included *C. arietinum, C. echinospermum, C. judaicum, C. pinnatifidum, C. bijugum and C. cuneatum.* The chromosome number of all these was found to be 2n = 2x = 16. Based on their study, they classified the species into three crossability groups. *Cicer arietinum, C. reticulatum,* and *C. echinospermum* formed group I; *C. judaicum, C. pinnatifidum,* and *C. bijugum* formed group II, and *C. cuneatum* alone formed group III. Ladizinsky and Alder (1976b) attributed the difficulty in achieving successful hybrids between cultigen and *C. echinospermum* to cryptic structural hybridity in F_1 besides major reciprocal translocation. However, Singh and Ocampo (1993) have produced partially fertile F_1 hybrids. In terms of Harlan and de Wet's (1971) gene pools, the primary gene pool (GP_1) comprises *C. arietinum* (GP_1A, the domesticated component) and *C. reticulatum* (GP_1B, the wild component); the secondary gene pool (GP_2) apparently consists of *C. echinospermum*; and the remaining species tested can be assigned to the tertiary gene pool (GP_3).

Morphology

Chickpea is an annual, widely cultivated for the seeds. It has a deep taproot system (1 to 2 m). A great deal of variation exists in plant type, ranging from prostrate to erect growth habit. Generally, all plant parts except the corolla are glandular.

Stems may be straight or flexuous and ribbed. Plant height ranges from 20 to 80 cm. Leaves are imparipinnate with 11–13 or 15–17 leaflets. Rachis is 25–60 mm long, ending in a top leaflet. Arrangement of leaflets may be opposite or alternate. They are sub-sessile, obovate-oblong to elliptic, 10–15 mm long, and 4–12 mm wide. The base of leaflets is narrowly or broadly cuneate or rounded, top rounded or acuminate, margin serrate. Lower surface of the leaflets is more ribbed than the upper surface. Stipules may be ovate or obliquely triangular.

One or rarely two flowers are found in axillary racemes, peduncles are 13–17 mm long, ending in a small arista; bracts are small and triangular; pedicel is 6–13 mm long, straight when flowering and recurved when bearing fruits; calyx is glabrous at the base; size of calyx tube is 3–4 mm with lanceolate teeth. Corolla is veined, pinkish, purplish or red (fading into blue) or white, rarely greenish-white or blue. Diadelfous stamens (9 + 1) have 6–8 mm long filaments; anthers are basi-dorsifixed. Ovary is ovate, 2–3 mm long and 1–1.5 mm wide, generally with 2, sometimes 4, ovules; style is 3–4 mm long.

Pods are rhomboid-ellipsoid with glandular pubescence, 14–25 mm long and 8–15 mm wide. Seeds are ovate, globular or angular, beaked, 7–10 mm long and 5–8 mm wide; seed coat may be white, creamy, yellowish with an orange tinge, most often brown in various shades, black, or dull green. Seeds may be smooth or wrinkled. They are also characterized by a deep, greyish hilum with a conspicuously coloured margin.

Cytology

Studies on different cytological aspects of chickpea have been reviewed in detail by Bahl (1987). The chromosomes of *Cicer* are small, which makes the chromosome count and karyotype analysis difficult.

The somatic chromosome number of *C. arietinum* was earlier reported to be 14 (Dombrovsky-Sludsky, 1927; Rao, 1929; Dixit, 1932a; Singh, 1964; Furnkraz, 1968) as well as 16 (Milovidov, 1932; Dixit, 1932b; Iyengar, 1939; Ramanujam and Joshi, 1941; Meenakshi and Subramaniam, 1966; Phadnis *et al.*, 1968). With later reports (Sohoo *et al.*, 1970; Mercy *et al.*, 1974; Ladizinsky and Adler, 1976a, b; Ahmad and Godward, 1980; Sharma and Gupta, 1982a; Lavania and Lavania, 1983), the chromosome number of 2n = 16 has been accepted.

Based on several cytological studies, the characteristic features of chickpea cytology can be briefly summarized (Gupta and Bahl, 1983) as follows:

1) A pair of very long chromosomes, distinctly satellited and sub-metacentric.

2) Six pairs of metacentric to sub-metacentric chromosomes.

3) A pair of very short metacentric chromosomes.

Based on karyotype analysis of three lines of chickpea, Ahmad and Godward (1980) numbered chromosomes from 1 to 8 in order of decreasing size of the chromosomes. the size difference from pair one to pair eight being in the ratio of 3:1. Ahmad and Hymowitz (1983) recorded the total chromosome length at pachytene stage as 335.33 μm and also found that the chromosome size ranged from 30.53 to 58.05 μm.

Though there is generally homogeneity in respect of overall morphology of chromosome complements, small chromosome differences do exist, as evident from studies by Ladizinsky and Adler (1976a), and others.

Ramanujam and Joshi (1941) were the first to report artificial production of tetraploids, while Sen and Jana (1956) reported a spontaneous autotetraploid. Artificial induction of polyploids has also been reported by Akhtar (1954). Pundir *et al.* (1983) reported that seed treatment with 0.25 per cent colchicine for 4 hours was most effective in producing autotetraploids. As expected, the induced tetraploidy was accompanied by larger leaves, flowers, stomata, pollen grains, and seeds. Sharma and Gupta (1982b) found 0.1 per cent colchicine treatment for 4 hours to be the most suitable for inducing tetraploidy. They also studied the chromosome behaviour of autotetraploids. C_2 autotetraploids showed predominantly bivalent pairing and normal disjunction at anaphase I. Phadnis and Narkhede (1972) also reported the increasing trend towards bivalency in the autotetraploids. Morrison and Rajathy (1960) have, however, reported conflicting results with quadrivalent frequency remaining constant in autotetraploid generations.

Genetics

Qualitative Characters

Research on the genetics of chickpea was initiated in 1911 at the Imperial Agricultural Research Institute at Pusa, Bihar. However, only since the 1930s did reports appear on inheritance of a large number of simply inherited traits. Excellent reviews on the inheritance of simply inherited traits can be found from Argikar (1970), van der Maesen (1972) and Muehlbauer and Singh (1987).

Most of the Mendelian traits are of little consequence to breeders. Therefore, inheritance of a few important traits is reviewed here.

Several variations in leaf shape and type have been described. Ekbote (1937) described two recessive mutants, one with tiny leaf (*Tlv*) and the other with simple leaf (*Slv*). The mutant allele 'slu' was, however, unstable. A stable leaf mutant (*Slv*) was later reported by Singh (1962). Pal and Ramanujam (1944) and Ramanujam and Singh (1945) reported recessive mutant with narrow leaflets (*nlv*). Argikar (1952a, 1958) reported a recessive gene for alternate leaflet arrangement (*alv*) as compared to normal opposite leaflet arrangement. Argikar (1952b) also described a gigas condition controlled by a recessive gene (*glv*). From his studies, Argikar (1958, 1959) concluded that the genotype of normal leaf types in chickpea was *slv, slv, Tlv, Tlv, Nlv, Nlv, Glv, Glv, Alv, Alv*. Many other variations in leaf shape and arrangement have been reported that are not considered here.

Plant habit is a character of importance to breeders. It may vary from prostrate to erect. Erect growth is dominant over prostrate growth habit (Patil, 1959). Argikar and D'Cruz (1963) assigned gene symbol *'hg'* to the prostrate growth habit. Another closely related trait is the branching habit. Basically, two types exist, one with basal branching and the other with branching at a higher node. The later is also referred to as umbrella type. Basal branching was dominant over umbrella type (Ayyar and Balasubramanian, 1937).

Argikar and D'Cruz (1963) reported that foliage colour was controlled by two genes, *lg,* and *b.* Both *lg* and *b* are responsible for normal green foliage colour. The inheritance and the effects of the genes are confusing. For more details, the reader may refer to Muehlbauer and Singh (1987).

Khan and Akhtar (1934) were the first to report that the flower colour in chickpea is controlled by two genes, *B* and *P*. This *B* is different from the one described by Argikar and D'Cruz (1963) for foliage colour. Ayyar and Balasubramanian (1934) added a third factor *C* to the inheritance of flower colour. They reported that *C* is complementary to *B*, while *P* is supplementary to *B*. Thus, the homozygous dominant condition of all the three genes results in pink corolla; blue corolla results when *P* is recessive and white when *'B'* or *'C'* is recessive.

Several genes have been reported for seed coat colour (Singh and Ekbote, 1936; Balasubramanian, 1937, 1950; Brar and Athwal, 1970) and their interactions naturally result in a wide variety of phenotypes. Genes T^1, T^2, T^3 and T^4 are known to darken the testa colour. Besides, flower colour genes *B/b* and *P/p*, which have pleiotropic effect, are also known to affect seed coat colour.

Some of the pod characteristics have also been reported to be simply inherited. Double podding gene (S) as against the one flower or pod on each peduncle was reported by Singh (1965), Yadav *et al.* (1978), and other workers. Chickpea pod is generally found in the normal pendant position. But a recessive variaton, in which pods lie on the leaves, to the normal pendant position has been reported by Patil and Choudhary (1957) and Patil (1967) with gene symbol pd fr. Giant pod mutant with very large pod size, reported by Kumar *et al.* (1989), was found to be governed by two recessive genes. They did not assign any gene symbol.

Genetics of non-nodulation was first reported by Davis *et al.* (1986). Rupela and Sudarshan (1986) also found a non-nondulating mutant from chickpea genotype ICC 435, which was named ICC 435M. Singh *et al.* (1986) reported that non-nodulation in ICC 435M was genetically recessive and different from the non-nodulating gene in PM 233 (mutant reported by Davis *et al.*, 1986).

Quantitative Characters

A fair amount of information has been generated on the genetics of quantitative characters. Information on genetics of quantitative characters is reviewed under (1) nature of gene action, (2) character association, (3) heritability, and (4) heterosis.

Table 2
Type of gene action for various characters in chickpea (the number under each type of gene action indicates the source)

Character	Additive	Non-additive	Additive + Non- additive
Plant height	10, 14, 19, 21, 25, 28, 30, 41, 47	17, 22, 45	9, 26, 34, 39
Branches/plant	15, 28	20, 41, 42	
Pr. br./plant	1, 2, 3, 4, 11, 12, 25, 29, 31	24, 43, 47	10, 16, 19, 28, 30
Sec. br./plant	3, 11, 12, 29, 31	10, 17, 22, 43, 47	40
Plant spread	16, 22, 47	10	40
Days to flower	5, 7, 10, 14, 21, 28, 30, 40, 41	13, 15, 22	8, 9, 17, 19, 35, 37, 39, 47, 48
Days to maturity	10, 21, 25, 45	16, 37, 47	28, 34, 39, 40, 48
Pods/plant	16, 22, 29, 34, 46, 47	1, 8, 10, 11, 12, 13, 17, 25, 37, 41, 43, 45	4, 26, 28, 31, 32, 40
Seeds/pod	13, 16, 17, 19, 22, 28, 30, 36, 42, 47	8, 12, 21, 43, 45	9, 31, 38
100-seed weight	7, 11, 21, 28, 29, 32, 39, 43, 47, 48	13, 14, 17, 22	8, 9, 19, 24, 31, 38, 41
Seed yield/plant	1, 4, 8, 10, 11, 14, 21, 22, 23, 25	12, 13, 15, 16, 17, 19, 33, 41, 45, 47	18, 26, 28, 31, 32, 40
Bio. yield/plant	1	43	
Harvest index	15, 21, 23, 40	33, 43	
Protein content		44	6, 9

1 Zafar and Khan, 1968
2 Sandha and Athwal, 1971
3 Zafar and Abdullah, 1971
4 Dhaliwal and Gill, 1973
5 Chaudhary, 1974
6 Gupta *et al.*, 1974
7 Gupta and Ramanujam, 1974a
8 Gupta and Ramanujam, 1974b
9 Sandhu *et al.*, 1974
10 Asawa and Tiwari, 1976
11 Sandhu *et al.*, 1977
12 Singh *et al.*, 1977
13 Chaudhary *et al.*, 1978
14 Gowda and Bahl, 1978
15 Pande *et al.*, 1979
16 Bhatt and Singh, 1980a
17 Mandal,. 1980
18 Pande *et al.*, 1980
19 Rang, 1980
20 Salimath, 1980
21 Singh and Mehra, 1980
22 Ugale, 1980
23 Deshmukh and Bhapkar, 1981
24 Katiyar, 1981
25 Kunadia and Singh, 1981
26 Rawat *et al.*, 1981
27 Vaidya and Pandya, 1981
28 Singh, 1982
29 Singh *et al.*, 1982
30 Katiyar and Solanki, 1983
31 Malhotra *et al.*, 1983
32 Singh and Paroda, 1983
33 Singh and Sindhu, 1983
34 Malhotra and Singh, 1984
35 Salimath and Bahl, 1985a
36 Tiwari and Pandey, 1986
37 Patil *et al.*, 1987
38 Sandhu and Bhullar, 1987
39 Singh *et al.*, 1987
40 Vijaykumar, 1987
41 Yadavendra and Kumar, 1987
42 Arora, 1988
43 Salimath and Bahl, 1988
44 Salimath *et al.*, 1988
45 Salimath and Bahl, 1989
46 Kharrat *et al.*, 1991
47 Annigeri, 1992
48 ICRISAT, 1993

Nature of gene action. Most of the studies on gene action have been made using diallel and line × tester analysis, while a few reports are also available that have used generation mean analysis. The conclusions drawn from these studies have limited scope as the parental material used cannot be regarded as the true representative of the variation available in chickpea; the findings generally are applicable to the material from which the information was derived.

The type of gene action reported by various workers in the past is summarized in Table 2. The number of reports suggesting predominance of additive gene action is the same as the number indicating non-additive gene action; few reports are in favour of both additive and non-additive gene action for seed yield/plant. More reports are in favour of non-additive gene action for pods/plant and for additive gene action in case of 100 seed weight and seeds/pod. For number of primary branches, days to flower, and days to maturity, both additive and non-additive gene actions are reported. More reports indicate additive gene action for plant height than non-additive as well as both additive and non-additive gene action. There is one report each indicating additive and non-additive gene action for biological yield. For harvest index, as many as four reports suggest additive gene action, while two indicate non-additive gene action.

Character association. Study of relationship of seed yield with component traits is most important for the breeder to decide upon selection strategies. A brief review of the nature of associations in chickpea is presented in Table 3. The review is not exhaustive, but it does provide an insight into the kind of relationship that the component traits have with the seed yield in chickpea. The most striking aspect, as evident from the table, is about the positive correlation of pods/plant with seed yield. Almost every earlier report on correlation of pods/plant with seed yield in chickpea has confirmed the positive association between them. However, the same is not true with respect to 100-seed weight, which is another important yield-determining character. Some reports indicate positive correlation, while some suggest negative correlation of seed weight with seed yield. Most of the reports confirm the positive association of seeds/pod, number of secondary branches, plant height, harvest index, and total dry matter with seed yield. Some reports indicate positive correlation of days to flower and days to maturity with seed yield, while others reveal negative association. Pods per plant and seed weight are the two most important traits determining yield in chickpea. However, they seem to be correlated negatively with each other (Bahl and Jain, 1977; Mandal, 1977; Singh and Paroda, 1986).

Path analysis by various workers also highlights pods/plant as the most important trait, which has strong direct influence on seed yield (Sandhu and

Table 3
Association of seed yield with its component traits in chickpea

Seed yield with	Reports indicating	
	Positive association	Negative association
Pods/plant	Bahl *et al.*, 1976; Bahl and Jain, 1977; Setty *et al.*,, 1977; Jatsara *et al.*, 1978; Singh *et al.*, 1978; Ram *et al.*, 1980; Katiyar *et al.*, 1981; Tomar *et al.*, 1982; Islam *et al.*, 1984; Salimath and Bahl, 1986; Sindhu and Prasad, 1987; Malik *et al.*, 1988; Reddy and Rao, 1988; Sharma and Maloo, 1988; Khan *et al.*, 1989; Sandhu *et al.*, 1989; Mani and Bahl, 1990; Annigeri, 1992	
100-seed weight	Mandal and Bahl, 1980; Salimath and Bahl, 1986; Sindhu and Prasad, 1987; Malik *et al.*, 1988; Singh *et al.*, 1990; Annigeri, 1992.	Khosh-Khui and Niknejad, 1972; Dabholkar, 1973; Singh *et al.*, 1973; Chand *et al.*, 1975; Khan and Choudhary, 1975; Rastogi and Singh, 1977; Raju *et al.*, 1978; Singh and Paroda, 1986.
Sec. branches	Bahl and Jain, 1977; Jatsara *et al.*, 1978; Mandal and Bahl, 1980; Katiyar *et al.*, 1981; Tomar *et al.*, 1982; Sandhu *et al.*, 1989; Mani and Bahl, 1990.	Singh *et al.*, 1978, Sindhu and Prasad, 1987.
Plant height	Khosh-Khui and Niknejad, 1972; Khan and Choudhary, 1975.	Phadnis *et al.*, 1970; Annigeri, 1992.
Seeds/pod	Phadnis *et al.*, 1970; Chand *et al.*, 1975; Setty *et al.*, 1977; K.P. Singh *et al.*, 1977; Raju *et al.*, 1978.	Annigeri, 1992.
Harvest index	Bahl and Jain, 1977; Gupta and Lal, 1981; Katiyar and Singh, 1987; Mishra *et al.*, 1988.	Mandal and Bahl, 1980.
Total dry matter	Mishra, 1972; Bahl and Jain, 1977.	
Days to flower	Mandal, 1977; Raju *et al.*, 1978; Adhikari and Pandey, 1982.	
Days to mature	Katiyar *et al.*,1977.	Setty *et al.*, 1977.
Biomas yield	Singh *et al.*, 1990a.	
Harvest yield	Singh *et al.*, 1990a.	

Singh, 1970: Gowda, 1972; Bahl *et al.*, 1976; Jatsara *et al.*, 1978; Salimath and Bahl, 1986; Sharma and Maloo, 1988; Sandhu *et al.*, 1989; Anilkumar, 1991). Among the other traits, number of branches has been found to be an important yield-contributing trait in the path analysis (Sandhu and Singh, 1972; Singh *et al.*, 1973; Salimath and Bahl, 1986; Sandhu *et al.*, 1989). Under a Mediterranean environment, biomass yield followed by harvest index was found to be significantly associated with speed yield (Singh *et al.*, 1990a).

Heritability. Most of the reports on heritability in chickpea are based on broad-sense heritability estimates. Plant height is reported to be moderate (Khosh-Khui and Niknejad, 1972; Mandal, 1977; Raju *et al.*, 1978; Pandey and Tiwari, 1983) to highly heritable (Chandra, 1968; Gupta and Lal, 1981). High heritability has been reported for branches per plant (Chandra, 1968; Mishra, 1972). Most of the reports indicate that days to flowering is a highly heritable character (Chandra, 1968; Raju *et al.*, 1978; Mandal and Bahl, 1980). Contrary to this, days to maturity has been reported to be moderately heritable by Rastogi and Singh (1977), while Pandey and Tiwari (1983) indicated it to be a highly heritable trait. Most of the studies have indicated high heritability for pod number (Sandhu and Singh, 1970; Malhotra and Singh, 1973; Patil and Phadnis, 1977; Raju *et al.*, 1978). Seed weight is another character which shows high heritability (Chandra, 1968; Niknejad *et al.*, 1971; Patil and Phadnis, 1977; Raju *et al.*, 1978; Mandal and Bahl, 1980). Even the narrow-sense heritability for seed weight was very high as reported by Salimath and Patil (1990). Seed yield/plant was reported as moderately heritable by Khosh-Khui and Niknejad (1973), Mandal and Bahl (1983), and Anilkumar (1991). But some have reported it to be highly heritable (Sandhu and Singh, 1970; Rastogi and Singh, 1977; Mishra *et al.*, 1988; Sandhu *et al.*, 1989).

Heterosis. Ever since the first report (Pal, 1945) on heterosis in chickpea appeared, a large number of workers have reported heterosis for yield and other agronomic characters (Argikar, 1970; Singh *et al.*, 1973; Gupta and Ramanujam, 1974a; Gowda and Bahl, 1976; Mandal and Bahl, 1979; Singh and Mehra, 1980; Singh and Ramanujam, 1981; Mandal and Bahl, 1984; Salimath and Bahl, 1985a; Vijaykumar, 1987; Bahl and Kumar, 1989; Dangaria, 1991; Bejiga and Singh, 1991; Annigeri, 1992). An overall review on the extent of heterosis for different characters has been summarized by Bahl (1988) and is given in Table 4. Most of these studies are based on single plant observations taken in space-planted nurseries.

As evident from Table 4, a very high degree of heterosis has been observed for seed yieid/plant followed by pods/plant as compared to other traits. For seed yield/plant, the magnitude of heterosis was higher

Table 4
Range and magnitude of mild-parent and better parent heterosis (percentage) in chickpea (Bahl, 1988)

Character	Range of heterosis over	
	Mild-parent	Better parent
Branches/plant	–56.5 to +96.0	–61.3 to +78.4
Pr. br./plant	–68.9 to +70.0	–70.0 to +99.8
Sec. br./plant	–50.3 to +59.6	–58.0 to +56.3
Plant height	–30.7 to +48.2	–39.7 to +36.2
Days to flower	–31.8 to +17.1	–36.8 to +24.0
Pods/plant	–82.9 to +192.0	–87.3 to +132.8
100-seed weight	–56.9 to +79.8	–63.8 to +38.0
Seeds/pod	–32.7 to +53.1	–46.8 to +57.2
Seeds yield/plant	–86.6 to +251.9	–89.8 to +199.8

in desi × kabuli crosses than in desi × desi or kabuli × kabuli crosses (Kunadia and Singh, 1980; Mandal and Bahl, 1984).

Genetics of Resistance to Biotic and Abiotic Stresses

The important biotic stresses in case of chickpea include those caused by diseases like wilt, dry root rot, and ascochyta blight and insect-pests, particularly pod borer. The genetics of resistance to these biotic stresses as well as that of cold tolerance is briefly reviewed in the following paragraphs.

Wilt. Ayyar and Iyer (1936) were the first to report that resistance to wilt [*Fusarium oxysporum* Schlecht. emend Snyd. and Hans f.sp. *ciceri* (Padwick) Snyd. & Hans.)] is governed by a single recessive gene. This was later confirmed by Pathak *et al.* (1975), Haware *et al.* (1980), Tewari *et al.* (1981), Kumar and Haware (1982), and Sandhu *et al.* (1983). Other studies indicated the evidence for two recessive genes for the control of wilt (Lopez Garcia, 1974; Upadhyaya *et al.*, 1983a, b). The recessive gene for wilt resistance in late wilter K 850 is independent of the recessive gene identified in another late wilter, C 104 (Singh *et al.*, 1987). Smithson *et al.* (1983) suggested that three independent genes are responsible for wilt resistance. Most of these studies on wilt resistance have been made against race 1. Some recent reports are now available regarding resistance to race 2 (ICRISAT, 1993). The fact that at least four races have been reported from India (Haware and Nene, 1982) suggests the complicated nature of inheritance of wilt if two or more races are considered. Studies in this regard appear to have not been attempted so far.

Dry root rot. Resistance to dry root rot [*Rhizoctonia bataticola* (Taub.)]

is governed by a single dominant gene as reported by Anand Rao and Haware (1987).

Ascochyta blight. Inheritance of resistance to ascochyta blight [*Ascochyta rabiei* (Pass.) Labr.] has been reviewed by Singh and Reddy (1991). Resistance is governed either by one or more independent dominant or recessive genes as reported by Tiwari and Pandey (1986), Singh and Reddy (1989), and many other workers. Singh and Reddy (1989) have found different dominant genes responsible for resistance to be allelic. Inheritance of resistance has also been reported to be quantitative (ICRISAT, 1994) with a significant vertical component.

Pod borer. Not many reports are available on the inheritance of resistance to pod borer (*Helicoverpa armigera*). However, this trait appears to be under the control of additive genes (Gowda *et al.*, 1985).

Cold tolerance. The study on inheritance of cold tolerance reported by Malhotra and Singh (1990) suggested that it was governed by both additive and non-additive gene effects, with a preponderance of the former. Cold tolerance was dominant over susceptibility and was controlled by at least five sets of genes. Its heritability in the narrow sense was 87.9 per cent. In another study, Malhotra and Singh (1991) confirmed the previous finding of the importance of both additive and non-additive gene effects. They found the presence of genic interactions in addition to additive and dominance gene effects through the generation mean analysis. Among the interactions, additive × additive and dominance × dominance with duplicate epistasis were present. Cold tolerance was dominant over susceptibility to cold. Selection for cold tolerance would be more effective if dominance and epistatic effects were reduced after a few generations of selfing.

Genetic Improvement

India is the largest producer of the crop, but its yield is low (0.7 t/ha) as compared to that of Australia, Egypt, Israel, and Italy, where it is more than 1 t/ha. The land races of chickpea in the Indian subcontinent have been selected to grow under marginal conditions of low soil fertility and moisture stress. Swaminathan and Jain (1973) suggested that under these conditions, which are not very different from the wild state, adaptability of the species to such environments becomes more important. Chickpea genotypes have adapted themselves to these conditions by acquiring bushy, spreading, and indeterminate growth habit and photothermosensitivity (Bahl *et al.*, 1979). The reasons for low productivity have been summarized by

Bahl (1981) and Singh (1987). The major concern of breeding is, therefore, to increase the genetic potential for yield.

History of Chickpea Breeding

Breeding work on chickpea improvement appears to have been initiated by government institutions in India in the 1920s. According to Bahl (1981), the early improvement can be considered in three distinct phases. During the first phase from the early part of the century to the late 1950s, improvement work was mainly based on simple selection from local land races, and to a limited extent by artificial hybridization. This work resulted in the identification of several improved cultivars: NP 17, NP 25, NP 58 from IARI; Pb 1 (kabuli), Pb 7, S-26 and G 24 from Punjab; T1, T2, T3, T4 (kabuli) and K5 from Uttar Pradesh; RS 10 from Rajasthan; Dohad Yellow from Gujarat, Adt. T-V No. 10, EB 28, A-1-8, D8, Gwalior 2, and Ujjain 2 from Madhya Pradesh; Chaffa, Gulab, N70, N29, N30, and N31 from Maharashtra; Kadale 2 and 3 from Karnataka and Sel 75 and Sel 98 from West Bengal.

The second phase covers the period from mid-1960s, when the All India Co-ordinated Pulse Improvement Project was established, till 1977–78. This period was characterized by the increased exchange and flow of material from one region to another, unlike in the first phase. Several high yielding cultivars were released during this phase. Some prominent cultivars among these are: C 235, G 130, G 543, and L 550 (kabuli) from Punjab; C 214, H 208, H 355 and C 104 (kabuli) from Haryana; Radhey and K 850 from Uttar Pradesh; RS 11 from Rajasthan; JG 62, JG 221 and JG 74 from Madhya Pradesh; BDN 9-3 from Maharashtra, Annigeri from Karnataka; and B 100 and B 115 from West Bengal. The period from 1978-79 onwards may be considered the third phase, during which research efforts have increased enormously. The cultivars bred during this phase were superior to the ones developed during the second phase and widely adapted. Some of the prominent cultivars released are Pusa 209, Pusa 212, Pusa 240, Pusa 244, Pusa 256, Pusa 261, Pusa 267 (kabuli), Pusa 329, and Pusa 372 from IARI, ICCV 4, ICCV 10, and ICCV-2 (kabuli) from the International Crops Research Institute for Semi-Arid Tropics (ICRISAT), and PGB 1, Gaurav, GNG 146, RSG 44, Avrodhi, Udai, JG 315, Vishwas, and Vijay from several agricultural universities.

Despite the progress made in the cultivar development, the average productivity has remained low in India. There are production gaps between yields obtained from different management practices as evident from the yields obtained at research stations, those obtained at farmers' fields with improved and local cultivars, and those recorded at state level. Bahl (1981) and Bahl and Salimath (1992) have analysed these production gaps and observed that the average productivity can be increased appreciably if seeds of the improved varieties are popularised and improved technology is transferred to the farmers. However, much ground is still to be covered in order to raise the productivity of this crop to a level that is competitive and attractive to farmers.

Breeding Objectives

In chickpea, the major breeding objective is to achieve high yield and stability. High yield potential can be achieved by improving biomass as well as its favourable partitioning (Singh *et al.,* 1990a). Another way of raising yield potential is by breeding cultivars responsive to inputs like fertilizer and irrigation. This approach is being given more attention in recent years. Extending adaptation of chickpea in space and time (Bahl *et al.,* 1990; Singh, 1990) is also expected to increase chickpea production.

The objective of stability can be realized by breeding for resistance to various biotic and abiotic stresses. Incorporation of resistance or tolerance to different biotic and abiotic stresses would help in realization of the true genetic potential of the genotypes for high yield as well as in stabilization of yield levels.

Breeding Approaches, Strategies and Methods

Chickpea is a self-pollinated crop and thus the breeding methods applicable for any autogamous crop are employed. Basically, genetic variation is created on which selection is practised. Chickpea breeders use well-known methods of augmenting variation, which include introduction, hybridization, and mutations.

Germplasm utilization. Progress in any crop improvement programme depends directly on the magnitude of variation available in the germplasm collections of the crop species. In India, the National Bureau of Plant Genetic Resources (NBPGR) maintains about 9,400 chickpea germplasm accessions. The global and regional responsibilities have been entrusted to ICRISAT and ICARDA for desi and kabuli germplasm, respectively. ICRISAT now holds about 15,939 accessions representing 42 countries, while ICARDA maintains about 6,804 from 34 countries (Mangesha *et al.,* 1990). The passport data for the entire collection have been maintained. The collection has been evaluated for 25 morpho-agronomic traits at ICRISAT (Pundir *et al.,* 1988b) and 29 traits at ICARDA (Singh *et al.,* 1983, 1991). The international institutes also maintain accessions of wild annual *Cicer* spp. The species include *C. bijugum* K.H. Rech, *C. chorassanicum* (Bge) M. pop., *C. cuneatum* Hochst. ex Rich, *C. echinospermum*

Table 5
Range of variation for some important traits in chickpea germplasm (Pundir *et al.*, 1988b)

Character	Range
Days to 50% flowering	33–107
Flower duration (days)	13–75
Plant height (cm)	14–96
Plant canopy width (cm)	13–124
Days to maturity	84–169
No. of basal primary branches	0.3–15.7
No. of basal secondary branches	0–13.7
No. of pods/plant	3–238
No. of seeds/pod	1–3.2
100-seed weight (g)	3.8–59.1
Seed yield (kg/ha)	70–5130
Seed protein content (%)	12.1–29.6

P.H. Davis, *C. judaicum* Boiss, *C. pinnatifidum* Jaub. & Sp., *C. reticulatum* Ladiz., and *C. yamashitae* Kitamura.

Evaluation of data reveals some diverse and useful genetic characteristics. The range of variation available for some traits in the ICRISAT chickpea germplasm is presented in Table 5.

Further germplasm diversification is possible by more exploration in the chickpea-growing areas or nations that are not well represented in the present germplasm assembly. It can also be done by careful observation and evaluation of spontaneous mutants reported from time to time, such as lobed vexillum (Rao and Pundir, 1983), polycarpy and twin-podded types (Pundir *et al.*, 1988a), thick stem, open and short bushy mutant (Dahiya *et al.*, 1984a), giant pod mutant (Kumar *et al.*, 1989), and cold-tolerant mutant (Haq and Singh, 1984).

Selection of parents for hybridization. In any crop breeding programme, a popular cultivar of the region or area is taken as a reference point over which improvement is sought either in productivity or in any specific trait. Therefore, generally, one of the parents chosen in hybridization programme is a locally adapted genotype. Published evaluation reports of the germplasm assembly, annual reports from national centres, NBPGR, and international institutes, and other periodicals serve as valuable sources to identify genotypes as parents for the crossing programme based on specific objectives. Biometrical methods such as multivariate analysis, lines × tester, or diallel mating designs may also be used to identify potential parents expected to result in productive segregating populations.

Floral characteristics and crossing technique. Chickpea is an autogamous crop with extremely low level of natural outcrossing ranging from 0 to 1.92 per cent (Khosh-Khui and Niknejad, 1971; Gowda, 1981; Malhotra and Singh, 1986) over a range of environments. The flowers open in the forenoon and close in the afternoon. Every flower opens for two to three successive days (Ayyar and Balasubramanian, 1935). Eshel (1965) has divided the development processes of bud and flower of chickpea into five stages as described below:

1) Closed bud: Stigma immature, anther filaments elongation insufficient

2) Hooded bud: Stigma receptive, filaments sufficiently elongated but still below stigma

3) Half-open flower: Stigma and anthers at the same height, pollination takes place

4) Fully open flower: Anthers shrivelled, fertilization takes place

5) Fading flower: Post-fertilization stage

Pollen grains germinate in about 30 min. after pollination and pollen tubes reach the base of the ovary in about 4 hours after pollination (Malti and Shivanna, 1983). Fertilization is reported to take place 24 hours after pollination. Functional specialization of stigmatic cells has been reported by Turano Margaret *et al.* (1983). Through microscopic and histochemical studies, they found that only certain papillate cells, which produce a localized secretion that reacts positively to an esterase test, have a role in the retention and hydration of pollen grains.

Artificial hybridization techniques have been reviewed by Argikar (1970), van der Maesen (1972), Smartt (1976), and Auckland and van der Maesen (1980).

For emasculation, one should select a flower bud due to open the following day. The selected flower bud is held at the base between the thumb and the forefinger. The front sepal is stripped off with the help of forceps and the keel is pushed downwards. The exposed anthers are then gently removed. After that, a coloured cotton thread or a tag is tied around the stem below the emasculated bud. Pollination is generally done the following day. Half-open flowers are the best source of mature pollen. From the intended male parent, such buds are collected in a petri dish or a glass tube. At the time of pollination, the standard and wings of the already selected buds are removed and the upper portion of the keel is lightly pressed between the thumb and the forefinger to shed enough pollen on the stigma of the emasculated flower. After the pollination is complete, the tag tied below the pollinated flower is labelled for the cross attempted.

It is a tedious process with generally a low rate of success ranging from 15 to 50 per cent depending upon season and skill of workers. However, the success rate can be enhanced in the following ways: (1) mechanical injury should be avoided while the flower bud is handled at the time of emasculation or pollination; (2) large flower buds should be selected for emasculation; (3) success is reported to be greater on the lateral buds than on the terminal ones (Sindhu *et al.*, 1981); (4) in northern India, better success is obtained in crossing after the first pod is formed (Bahl and Gowda, 1975); (5) the timing of pollination and fertilization also decides the success rate. Early in the season, when the temperatures are low, emasculation in the afternoon followed by pollination on the following morning (0800–1300 hours) gives better results in northern India. However, as the season advances, and the day temperature rises, morning emasculation followed by immediate pollination or pollination in the same afternoon is more effective. Khosh-Khui and Niknejad (1971) reported 97.5 per cent hybrid seed set when pollination is done just after emasculation. Singh and Auckland (1975) and Pundir and Reddy (1984) also reported better success with simultaneous emasculation and pollination in southern India. In Ethiopia, emasculation in the morning (0800–1000 hours) followed by pollination in afternoon (1200–1400 hours) gave the highest hybrid seed set (Bejiga and Tessema, 1981).

Handling of crosses and segregating populations. The general practice in autogamous crops is to reject poor crosses at F_1 level and carry forward only the promising hybrids. Singh and Auckland (1975) have suggested that in case of chickpea the F_2 populations may be classified as 'very promising', 'promising', and 'difficult to ascertain' on the basis of their overall performance. The pedigree method is used for 'very promising', modified bulk method for 'promising', and bulk method for 'difficult to ascertain' types of populations.

In fact, it may be ideal to identify potential crosses on the basis of information on the level of F_1 heterosis and inbreeding depression in F_2. The crosses exhibiting high positive heterosis for seed yield or other important traits preferably over a commercial check coupled with the absence of any significant inbreeding depression are likely to throw superior transgressive segregants whose performance may be fixed by a simple selection scheme.

Chickpea being a self-pollinated crop, pedigree, bulk, back-cross, or various modifications of these methods may be used for cultivar development. Details of these methods can be found in standard plant breeding texts (Allard, 1960; Briggs and Knowles, 1967; Fehr, 1987). However, for different situations some specific selection methods could prove more useful. In case of breeding for resistance to biotic stresses, pedigree or back-cross and sometimes bulk-pedigree methods are better suited. For abiotic stresses, generally bulk-pedigree or modified bulk methods are useful. In case of interspecific hybridization and desi × kabuli introgression, back-cross method is more useful.

The usefulness of early generation testing is often debated. It was found advantageous by some (Dahiya *et al.*, 1984b; Salimath and Bahl, 1985b), while Rahman and Bahl (1986) reported it to be useful for seeds/pod and seed weight but not for pods/plant and seed yield. Some studies have compared the efficiency of different selection methods in chickpea. Lal *et al.* (1973) emphasized the importance of pedigree method in breeding for yield, while Byth *et al.* (1980) suggested that it be replaced by bulk method, since they found that pedigree method was not well suited in breeding for higher yield.

To improve seed yield in chickpea, selection for component traits like seed size or seeds/pod may be better than selection on the basis of yield *per se* (Bisen *et al.*, 1985; Salimath and Bahl, 1985b). To find out efficiency of direct and indirect selection for yield in chickpea, Kumar and Bahl (1992) compared yield performance of F_6 lines, derived by exercising selection pressure for four successive generations (F_2–F_5), for yield and its components and concluded that indirect selection for yield via pod number and seed weight was more efficient than direct selection for yield.

Some studies have assessed the relative efficiency of different forms of bulk method (selected bulk, random bulk) and single seed descent (SSD) method. Selected bulk method, which is a combination of pedigree and bulk method, was found to be more efficient than random bulk and SSD method by Rahman and Bahl (1985) and Bahl and Vinod (1991). These authors further concluded that SSD method should be preferred to random bulk since it is less influenced by sampling and natural selection. More investigations are needed to generate information on selection methods suitable for different situations. Singh *et al.* (1994) found the bulk-pedigree method to be simple and effective in breeding cultivars resistant to cold and ascochyta blight.

Plant Type Considerations

Improved plant type has played a very important role in recent years in raising the yield plateau in cereals and in certain legumes. However, little thinking has been given to this aspect in chickpea. Chickpea, which continues to be grown under marginal conditions, has a morphophysiological structure suited for survival rather than for high productivity. It has coadaptive gene complexes conserved by genetic linkages (Swaminathan and Jain, 1973; Jain, 1975; Bahl *et al.*, 1979). Therefore, the breeder must change the physiological make-up by restructuring the plant type, with features such as compact, photothermoinsensitive and determinate types, and thus bring about wider adaptability and high productivity.

The information from correlations (Bahl and Jain, 1977; Salimath and Bahl, 1986) and path coefficient analysis (Bahl *et al.*, 1976) helps to find out the structural yield components of a chickpea plant that should be appropriately incorporated into an improved plant type. Bahl (1986, 1988) proposed an ideal plant type in chickpea with the following attributes (Fig. 2), based on the relevant information available from (a) studies on intraspecific variation (Bahl, 1980; Salimath *et al.,* 1984; Mani, 1987), (b) desirable features of desi and kabuli types and tall cultivars from the USSR (Bahl, 1981), and (c) morphophysiological architecture of tall and bushy types (Bahl, 1983; Bahl *et al.,* 1984; Yadav and Bahl, 1990).

1) Medium tall or tall with upright and compact habit

2) Optimum number of primary and secondary branches with minimum of tertiary and late order branches

3) Photothermoinsensitivity

4) Determinate growth habit

5) Well-developed nodules with functional nodules during reproductive phase of the plant

Such a plant type is expected to intercept more sunlight, accommodate larger plant population per unit area, and enhance biological yield and harvest index. Recombination breeding involving traditional desi types, Mediterranean kabuli types and Russian tall types would help to develop genotypes with compact habit (Fig. 2).

The perception of the plant type may vary to some extent depending on the target area, since chickpea is grown in diverse agroecological zones. For example, in drought-prone areas, the genotype should have rapid emergence, spreading growth habit, and early maturity, while for irrigated conditions late-maturing types may be desirable to take advantage of the long growing season (Singh, 1987). Similarly, for mechanized harvest, tall and erect types are desired.

Saxena and Johansen (1990) described the constraints of the physical environment of the major growing areas of chickpea in South Asia and then proposed the functional traits to overcome these constraints, while describing the required plant ideotype. For example, in drought-stressed environments, denser and longer root system and smaller leaf size are considered advantageous. Similarly, Sedgley *et al.* (1990) have described the chickpea ideotype for winter sowing in Mediterranean environments, which would be erect with limited branching habit, early in flowering, and tolerant to suboptimal preflowering temperatures.

Breeding for Disease Resistance

As many as 47 diseases are reported on chickpea (Nene and Reddy, 1987). The few serious ones are fusarium wilt, ascochyta blight, dry root rot, botrytis gray mould (*Botrytis cinerea* pers., ex. Fr.), and stunt (bean leaf roll virus). Soil-borne diseases like wilt root rot are serious in lower latitudes (0–20°) where chickpea is grown under short, warm, and dry season,

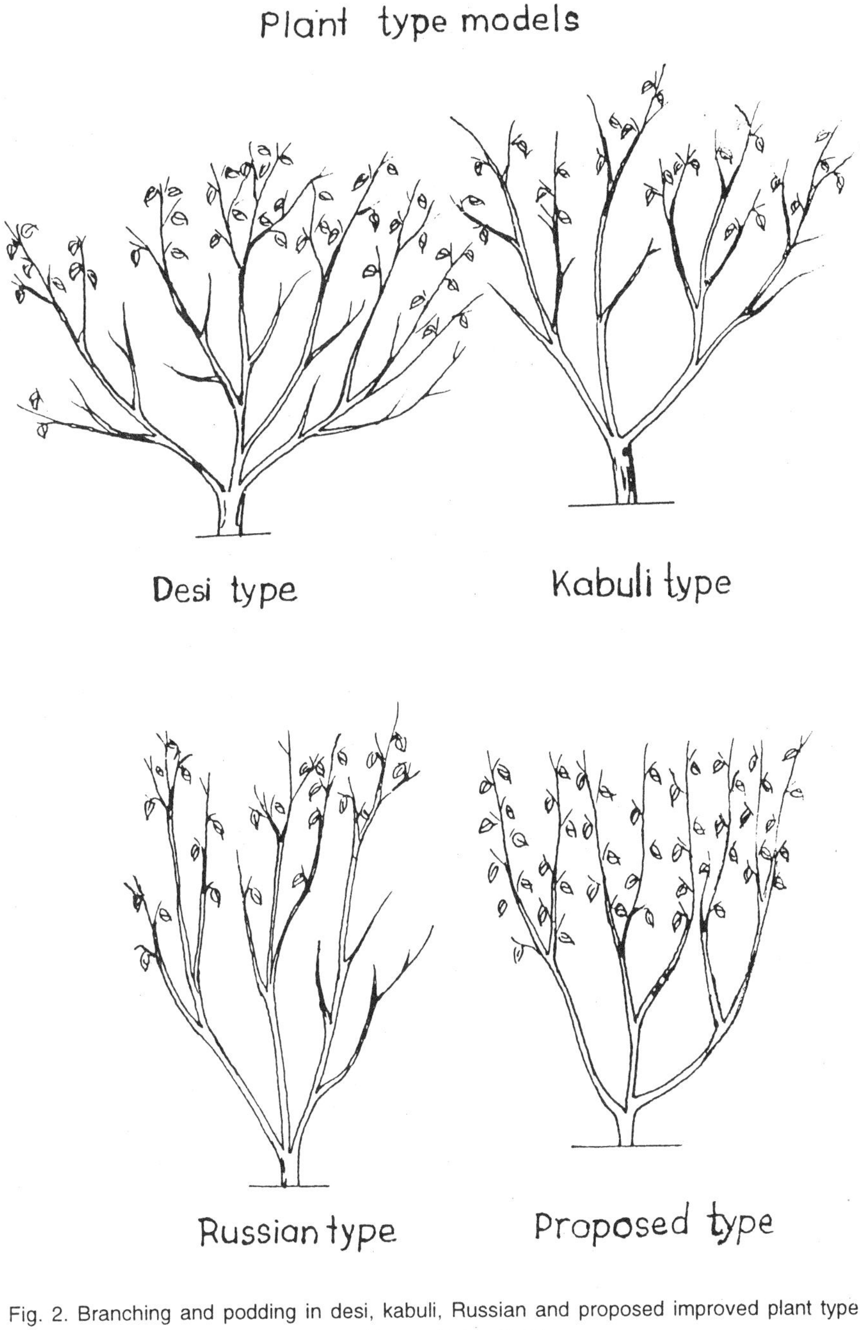

Fig. 2. Branching and podding in desi, kabuli, Russian and proposed improved plant type in chickpea (Bahl, 1988).

whereas foliar diseases, particularly ascochyta blight, assume serious proportions in higher latitudes (20–40°) where chickpea is grown under long, cool, and wet seasons. Only wilt and ascochyta blight are globally important (Singh, 1990). Chemical measures to control these diseases are mostly impractical and uneconomical. Therefore, host-plant resistance seems to be the best strategy to control diseases. Progress made in disease resistance breeding has been summarized by Nene and Reedy (1987) and Singh and Reddy (1991). A brief account of breeding for resistance against the two major diseases is given below.

Fusarium wilt. The disease has been reported from India, Pakistan, Myanmar, Mexico, Peru, the United States, and other countries (Nene, 1978). It is a typical vascular disease that causes xylem browning and blackening. Expression of symptoms is more rapid at high temperature (> 30°C). Root systems do not show any visible symptoms. Leaves droop but retain their green colour. The disease occurs in patches, which grow in size if the same land is planted with chickpea year after year.

Existence of races of the pathogen complicates breeding for resistance to this disease. As many as seven physiological races (four from India and three from Spain) have been reported. Of these, race 1 may survive for 7 years in the soil, while races 2 and 3 for 5 years, and race 4 for 6 years (ICRISAT, 1994). With two recessive genes and possibly a partially dominant gene conferring resistance, breeding for wilt resistance should not cause insurmountable problems, particularly with the availability of well-developed wilt-sick plots, broad-based, stable and resistant sources (e.g., ICC-10803, -11550, -11551, -11322 and -11323), and standard screening techniques (Nene *et al.,* 1981). Lines GL 87078 and GL 87079, possessing resistance to all the four races of wilt reported to be prevalent in India, are now available (Singh *et al.,* 1994).

The existence of races along with other soil-borne diseases creates problems for resistance breeding. Therefore, an effective resistance breeding programme should take care of these factors. The objective of resistance breeding may be achieved by pyramiding of genes resistant for different races. Bahl *et al.* (1990) have suggested breeding of near isogenic lines (NILs) resistant to different races of the wilt pathogen by following back-cross approach with the aim of ultimately blending the appropriate NILs according to the races of pathogen prevalent in a particular region. The ultimate goal will be to generate a continuous flow of NILs that carry different resistance genes.

Effective field screening and laboratory procedures at ICRISAT have helped to identify sources of resistance against wilt. As a result, it has been possible to incorporate resistance into agronomic backgrounds In India, several high yielding and wilt-resistant cultivars, such as Pusa 212, Pusa

244, WR 315, Avrodhi, ICCV-2, and ICCV-10, have been developed and released. In Mexico, large-seeded, wilt-resistant varieties, such as Surutato-77, Sonora-80, and Santo Domingo have been bred and released for commercial cultivation. In Tunisia, Amdoun1 has been released as a large-seeded kabuli chickpea.

Ascochyta blight. Ascohyta blight has been reported from 26 countries (Nene, 1984). Unlike wilt, which occurs regularly, blight generally appears in a cycle of about 5 years. When it occurs, it persists for 2–3 years. Temperatures of 15–25°C, cloudy and rainy days, and winds particularly after flowering are favourable for the development and spread of the disease. That is how the disease develops whenever winter-sown chickpeas in northwestern India and Pakistan and spring-sown crop in the Mediterranean region receive rains during the crop season (Reddy *et al.*, 1990). The symptoms appear on all the above-ground parts, including leaves and pods, and in extreme cases on seeds also. Circular spots typically appear on leaves and pods, elongated lesions on stems, which are generally girdled, and deep cankerous lesions on seeds.

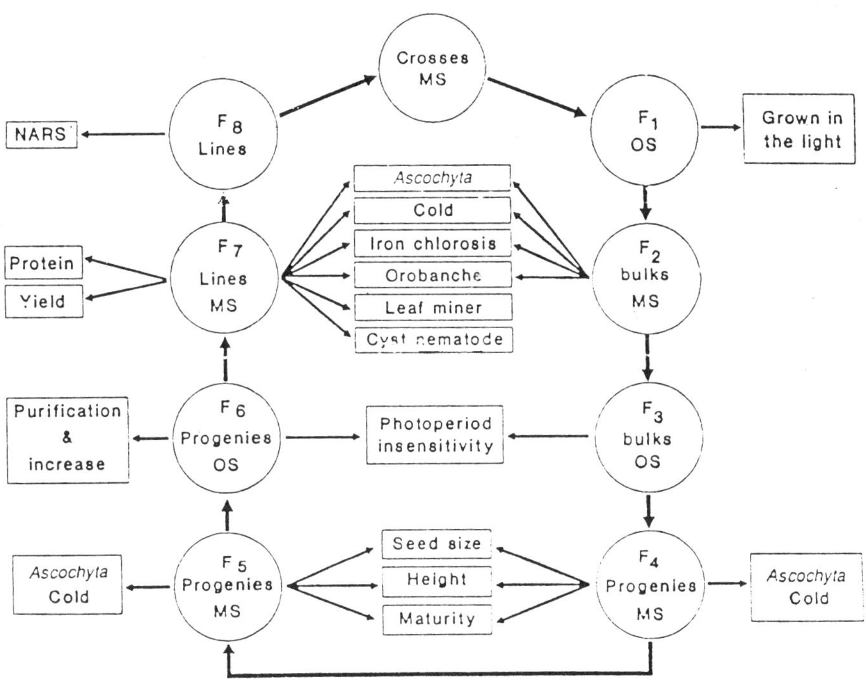

Fig. 3. Bulk-pedigree method for breeding chickpeas resistant to *Ascochyta* blight and other stresses. MS, Main season; OS, off-season. Source : Singh and Reddy (1991)

Without resistance to ascochyta blight, chickpea cannot be grown during winter in the Mediterranean region. Therefore, breeding for resistance to this disease has received major attention at ICARDA. Studies at ICARDA have helped in the identification of 6 races in Syria and 13 races in the Mediterranean region (Reddy and Kabbabeh, 1985; Reddy *et al.,* 1992). Also five non-location-specific resistant sources—ILC 200, ILC 6482, ICC 4475, ICC 6328, and ICC 12004—were identified. Resistance sources of diverse origin were crossed to develop stable sources of resistance. Some of the lines coming out from this programme, which was initiated in 1989–90, showed a rating of 2 on a 1–9 scale where 1 = free from damage and 9 = all plants killed. It is believed that some of these lines possess higher and more stable resistance than the resistance sources available at ICARDA (ICRISAT, 1994). Pyramiding genes for resistance, breeding for partial, (i.e., horizontal) resistance, and transfer of resistant genes from related wild species are some of the strategies suggested for breeding resistance to this disease (Singh and Reddy, 1991). Bulk pedigree method is generally followed at ICARDA in breeding for blight and other stresses. A generalized scheme is shown in Fig. 3. Following this method, over 2000 blight-resistant lines have been bred and 18 countries have released 51 cultivars from the material furnished from ICARDA (ICARDA, 1994).

It has been conclusively established that races of ascochyta blight reported from India and Pakistan are more aggressive than those present in the Mediterranean region (Singh *et al.,* 1984). Therefore, resistance breeding against this disease is more difficult in this region, though lines possessing resistance in the vegetative stage (not in the podding stage) to isolates of *A. rabiei* present in India have been reported (Singh *et al.,* 1988).

Breeding for Pest Resistance

Pod borer. Although about 54 insect pests have been reported on chickpea (Reed *et al.,* 1987), only pod borer (*Heliothis* spp., *Helicoverpa* spp.) assumes global importance. Techniques have been developed for screening against pod borer (Lateef and Reed, 1981, 1985), but the screening needs to be repeated for at least three years for reliable results.

So far, no race differences have been reported for pod borer. Lateef and Sachan (1990) have listed several sources that have shown good level of resistance across the agroecological zones of India. Other sources that have recently been reported to be resistant (ICRISAT, 1994) are ICCL 87315, 87316, and 87317, which showed less than 2 per cent damage. Some are also high yielders (ICRISAT, 1990). Most of the *Helicoverpa*-resistant lines are highly susceptible to fusarium wilt and ascochyta blight. Effort are being made to combine genes for resistance to pod borer and fusarium wilt. In ICCL 86111, for example, resistance to both these stresses has been combined (Singh *et al.*, 1990b). Singh *et al.* (1994) have described the procedure to breed for multiple stress resistance in chickpea.

Leaf miner. Leaf miner is an important pest on chickpea in Mediterranean region. Reed *et al.* (1987) and Weigand and Tahhan (1990) have described the research done there on leaf miner resistance. The yield loss due to leaf miner ranges from 7 to 36 per cent with an average of 18 per cent and the damage to chickpea is confined to countries with a Mediterranean climate. Three resistant lines, ILC 3800, ILC 5901, and ILC 7738, have been identified (ICARDA, 1994). The screening technique still needs some improvement. Only a limited progress has been made in breeding for leaf miner resistance.

Breeding for Drought Tolerance

Drought is the most important abiotic stress of chickpea. Saxena *et al.* (1993) reported yield loss due to drought of the order of 30 to 60 per cent in India. However, Singh (1993) found even up to 100 per cent yield loss due to drought. Methods have been developed to screen chickpeas tolerant to drought using line-source sprinkler system (Saxena *et al.*, 1993) in tropics and delayed sowing (Singh *et al.*, 1994) during spring in the Mediterranean region. A desi line, ICC 4958 (Saxena *et al.*, 1993), and a kabuli line, FLIP 87-59C (Singh *et al.*, 1994), have been identified as sources of drought tolerance. Efforts have been made to breed cultivars with tolerance to drought, but to date the major achievement is lacking. This area should receive major attention in the future.

Breeding for Cold Tolerance

Cold tolerance is extremely important for winter sowing of chickpea in the Mediterranean region. Singh *et al.* (1989, 1994) have developed an easy and reliable screening technique to evaluate chickpea germplasm. They have evaluated more than 10,000 germplasm lines and have identified several cold-tolerant lines, including ILC 8262, ILC 8617, and FLIP 87-82C. They have used the cold-tolerant sources in breeding and have developed several cold tolerant breeding lines. These breeding lines tolerate cold up to –10°C and 60 days in a year. The methods of breeding cold-tolerant chickpeas is simple. The high-yielding chickpea lines are crossed with the cold-tolerant lines. The material is grown in bulk from F_2 to F_4 generations and sown one month earlier than the normal sowing time to enable the plants to grow to late vegetative stage before the onset of severe winter. Plants affected by cold are uprooted. The F_5 and F_6 generations are grown as progeny rows. Progenies affected by cold are discarded. ICARDA has found this method satisfactory.

Exploitation of Wild Species

The genus *Cicer* includes 49 species, 9 of which are annuals. These annual wild species possess several genes for characters that would be useful in chickpea improvement (Table 6). In fact, wild species are the only sources of resistance so far found for seed beetle and cyst nematode and they also possess a higher level of resistance than the cultivated species for fusarium wilt, leaf miner, and cold.

Table 6
Characters in wild *Cicer* species useful for chickpea improvement (Singh and Verma, 1994 and others)

Species	Characters
C. bijugum	Resistant to ascochyta blight, wilt, leaf miner, seed beetle, cyst nematode, and cold
C. chorassanicum	Resistant to ascochyta blight, and leaf miner
C. cuneatum	Resistant to seed beetle, fusarium wilt, and leaf miner, 3 seeds/pod
C. echinospermum	Resistant to leaf miner, seed beetle, cold, and wilt
C. judaicum	Resistant to fusarium wilt, gray mould, ascochyta blight, leaf miner, seed beetle, and cold
C. pinnatifidum	Resistant to wilt, cyst nematode, seed beetle, and cold
C. reticulatum	Promising against cold, seed beetle, cyst nematode, and wilt
C. yamashitae	Resistant to leaf miner
C. montbretii	Resistant to ascochyta blight, 3–7 seeds/pod

Among the wild species, *C. reticulatum* is easily crossable with cultivated species and there are many reports indicating the successful transfer of desirable characters from *C. reticulatum* to the cultivated species. Singh *et al.* (1984) obtained early and cold-tolerant lines from crosses between *C. arietinum* and *C. reticulatum.*

Some reports indicate successful hybridization with other wild species also, particularly with *C. echinospermum* (ICARDA, 1990: Shiela *et al.*, 1992; Singh and Ocampo, 1993; ICRISAT, 1994). Isozyme studies in the F_2 progenies of these crosses suggested closer relationship of *C. arietinum* and *C. echinospermum* (Gaur and Slinkard, 1990a, b). Pundir and van der Maesen (1983) obtained one hybrid seed from 304 pollinations between *C. arietinum* and *C. cuneatum.* Verma *et al.* (1990, 1991) have reported successful hybridization of *C. arietinum* with *C. reticulatum, C. pinnatifidum, C. judaicum,* and *C. bijugum.*

The derivatives of crosses of the cultivated species with *C. echinospermum* and *C. reticulatum* provide evidence for the presence of 'yield genes' in wild *Cicer* species that can be transferred to the cultigen for increasing seed yield (ICRISAT, 1994). F_1 heterosis in those crosses was up to 250 per cent and the F_4 and F_5 derivatives outyielded the best control cultivar by a margin of 50–60 per cent.

Use of mentor pollen, hormone treatment, and embryo rescue are now being tried to attempt crosses with the wild species. Some success has been obtained in interspecific hybridization by treatment with hormone solution comprising GA 8 mg/ml, KN 1 mg/ml, and NAA 0.5 mg/ml to the base of pollinated pistils, which increased pod retention. Five hybrid seeds were obtained from *C. arietinum* × *C. pinnatifidum* after hormone application (ICRISAT, 1993).

Segregants from crosses between *C. arietinum* and the wild species may suffer from pod dehiscence, reticulate and hard seed coat, unattractive and unacceptable seed colour, prostrate growth habit, non-uniform maturity, and/or unpalatable seed. Selection pressure should be applied from early segregating generations to eliminate undesirable genes coming from wild species. Nevertheless, successful hybridization between cultivated and wild species of chickpea has opened up exciting possibilities of effective transfer of genetic material across the species barriers.

New Breeding Strategies and Approaches

Desi kabuli introgression. The utility of incorporation of the alien germplasm into the breeding populations by germplasm introgression has often been suggested as a procedure for increasing genetic variability. In fact, some examples now suggest that even quantitative characters can be transferred by introgression. For instance, earliness of six-row barleys and the tillering capacity of two-row barleys have been successfully combined. Likewise, the septoria resistance and cold tolerance of winter wheats and the rust resistance and bread-making qualities of spring wheats have been combined.

Within the cultivated chickpea, kabuli and desi are considered two distinct groups of cultivars (Bahl, 1980; Hawtin and Singh, 1980; Salimath *et al.,* 1984), although evidence in favour of subspecific differentiation is meagre. Mani (1987) assessed divergence pattern in 200 entries, representing 100 each of desi and kabuli types from six major chickpea-growing regions, over the 10 characters by non-hierarchial Euclidean cluster analysis, apart from graphical representation by canonical and principal component scores. The 200 entries, grouped into 15 clusters, showed a tendency to form separate clusters for desi and kabuli types. The study confirms the presence of specific genotypes with differential expression in respect of specific agronomic traits forming the two subgroups within chickpea.

Kabuli type generally has enhanced expression for certain useful quantitative traits, while desi type excels in respect of a different set of traits. For instance, kabulis are characterized by higher number of primary branches, high seed weight, erect habit, and cold tolerance, while desi germplasm is a good source of genes for seeds/pod, secondary branches, pods per plant, and drought resistance. Therefore, crosses between them can help in transferring desirable characters from one subgroup to the other.

To investigate the possibility of kabuli and desi introgression, Ugale and Bahl (1983) studied the pattern of transgressive segregation in three single crosses, each involving a desi and kabuli parent. Transgressants for yield and its components were studied in selfed generations (F_2s, and F_3s) and progenies of backcrosses (BC_1F_2 and BC_2F_2) using both desi and kabuli as recurrent parents. Relative proportions of transgressive segregants for different agronomic characters were higher in progenies of back-crosses than in selfing series. This work established two things: the effective introgression of desi-kabuli germplasm and the relative efficiency of back-cross approach over selfing generations in recovering the higher frequency of transgressants.

Several high yielding and widely adapted varieties, such as L-104, L-144, L-550, and Pusa 267 among kabuli types and Pusa 244, Pusa 256, and Pusa 261 among desis, have emanated from crosses in which desi and kabuli germplasm have been recombined in single, two-way, or double crosses (Bahl, 1988). Desi-kabuli introgression holds promise for chickpea improvement, which must be exploited through consistent and concerted research efforts in future (Bahl *et al.*, 1990; Bahl and Salimath, 1992).

Adaptation to late-sown conditions in Northern India. With increase in irrigation facilities and availability of input responsive, high yielding and more remunerative cultivars of crops like wheat, chickpea started receiving less attention from farmers. In fact, chickpea production has declined over the years (Jodha and Subba Rao, 1987). However, where limited irrigation is available for *rabi* crops particularly as a second crop after maize, cotton, rice or millet, chickpea can find a place. But the yields of traditional chickpea cultivars are reduced when sown late. Therefore, chickpea breeders particularly in northern India have to make efforts to develop cultivars suitable for late-sown situations (Bahl and Salimath, 1992).

The fact that the growth habit has played an important role in domestication and evolution of pulse crops (Smartt, 1976) gave a clue to look for the differential response of different plant types for their adaptability to late-sown conditions. Yadav (1986) studied 10 bushy types of Indian origin and 9 tall types emanating from crosses having a tall genotype from the former USSR in their parentage, for possible genetic differentiation through multivariate analysis based on the pooled data of two years with two environments in each year. The 19 entries grouped into 5 clusters formed separate clusters for tall and bushy types. Based on clear demarcation between the two types, it was concluded that either tall types diverged from bushy types during the course of evolution or the two types possess differentiated characters as a result of area-specific adaptations. Bahl *et al.* (1991a) emphasized the role of area-specific adaptation based on their study of 329 chickpea lines from six chickpea-growing regions of the world.

Bahl *et al.* (1984) and Yadav and Bahl (1990) evaluated bushy and tall types under normal- and late-sown conditions and found tall types to be better adapted for late-sown conditions. Probably tall types have evolved to withstand the adverse effects of late germination and thus provide opportunities for nonconventional cropping patterns. Pusa-261, a tall variety bred at IARI, is suitable for late-sown conditions in northern India (Bahl *et al.*, 1991b). Yielding ability of such tall types may be further improved by reducing the internodal distance to increase the number of pod-bearing nodes (Khanna-Chopra and Sinha, 1987) by recombination breeding.

Adaptation to early-sown conditions in peninsular India. The growing conditions of chickpea in peninsular India are quite different from those in the north. In the south, chickpea is generally grown as a rainfed crop either on rainy season fallows or after a short-duration millet or pulse crop under receding soil moisture conditions. The general absence of winter rains coupled with receding soil moisture conditions lead to drought stress. In addition, the rising temperatures from February onwards result in forced maturity of the crop. The growing duration is thus short (90–110 days). These conditions involving a short-growing season and limited moisture are responsible for low productivity. These problems can be partly overcome by advancing the sowing time by a month, i.e., mid-September instead of October, to provide slightly longer growing duration and better germination and crop establishment due to better soil moisture conditions in September. Yield advantage of 35 to 120 per cent by early sowing has been reported over normal sowings (ICRISAT, 1986). Screening of germplasm and breeding lines for adptation to early sowing showed that several genotypes of medium maturity, notably P 1329, P 1067-1, P-18, and P-4089-1, consistently outyielded the control cultivar Annigeri at ICRISAT when sown in September (Singh *et al.*, 1983). Constraints associated with early sowing, such as high temperature and excessive rains, hinder germination and cause new disease problems, like collar rot and *Colletotricum* blight in the seedling stage. More research is required to overcome these problems to identify varieties suitable for early planting. The possibility of advancing the sowing date and thus increasing yield also exists in Ethiopia and other countries, provided cultivars resistant to ascochyta blight are developed.

Winter sowing in the Mediterranean region. Faba bean (*Vicia faba* L.), pea (*Pisum sativum* L.), lentil (*Lens culinaris* Medik.), and chickpea are the important cool season food legumes in the Mediterranean region. Chickpea is normally grown as a spring-sown crop, whereas other main food legumes are grown as winter-sown crops. In case of winter-sown food legumes, germination and crop establishment occur before the onset of cold weather, while vegetative growth takes place with rising temperatures in the spring. However, in spring-sown chickpea, growth and development takes place under conditions of increasing temperature, day length, and aridity (Hawtin and Singh, 1984).

Work on the possible yield advantage by winter planting of chickpea was started at ICARDA during the 1970s and has been summarized by Singh (1987). Researches carried out at ICARDA have clearly shown that chickpea can be successfully grown as a winter crop in the Mediterranean region, provided blight-resistant and cold-tolerant varieties are made available (Hawtin and Singh, 1984). Yield advantage (50–100 per cent) of winter-sown over spring-sown chickpea has been demonstrated in all countries of the region (Singh, 1990). However, because of agrobiological and socioeconomic constraints the adoption rate of winter-sown chickpea is rather slow.

Biotechnology in Chickpea Improvement

A detailed map of the *Cicer* genome is a prerequisite for the different kinds of approaches now being evolved to study plant genetics as well as for the genetic improvement of chickpea. Such a map is expected to be based on joint segregation of morphological, isozyme, and RFLP markers. Isozyme polymorphisms have been identified in interspecific crosses between *C. arietinum* and *C. reticulatum* (Gaur and Slinkard, 1989; Muehlbauer *et al.*, 1990). Gaur and Slinkard (1991) identified 7 linkage groups involving 26 isozymes and 2 morphological trait loci. Weeden *et al.* (1994) have reported the possibility of application of RADP analysis in chickpea. The chickpea map is now being extended using RFLP and RADP markers.

Pea, lentil, and chickpea have common linkage groups consisting of homologous genes, and hence linkages found in one genus can be used to predict similar linkages in related genera in the development of linkage maps (Kazan *et al.*, 1993). *Pisum* is an extensively mapped genus and provides clues for developing gene maps of *Cicer* L. A detailed gene map could then be used in tagging genes of interest with molecular markers to facilitate their transfer to acceptable backgrounds and also in the identification of quantitative trait loci.

Kahl *et al.* (1994) have reviewed the potentiality of DNA fingerprinting and DNA amplification fingerprinting in chickpea breeding and the control of ascochyta blight.

A lot of research efforts are still needed to produce transgenic chickpea plants. Srinivasan *et al.* (1988) and Islam and Riazuddin (1992, 1994) have reported susceptibility of chickpea to *Agrobacterium tumefaciens*, which is now an established vector for genetic engineering of plants. But further progress in achieving an efficient regeneration system and optimization of parameters for transformation is needed for the production of transgenic chickpea plants.

Techniques such as embryo rescue and ovule culture also need to be perfected which could become useful in transferring desirable genes from wild *Cicer* species that are not easily crossable with the cultivated species.

Future Research Needs

Since the early 1960s, major efforts have been made in case of cereals and industrial crops, in both developed and developing countries to increase the availability of inputs such as fertilizer, pesticides, fungicides, herbicides, irrigation, and farm machinery to mechanize operations from seeding to harvest. Plant breeders working on these crops throughout the world have made efforts to develop cultivars responsive to the application of fertilizer and irrigation, resulting in phenomenal increase in seed yield. However, chickpea production and yield have remained almost static because researchers have continued to breed cultivars for dry-land conditions with little or no inputs. Over the years, farmers have assigned better lands to cereals and cash crops and chickpea has been relegated to marginal lands. If chickpea yield and production have to increase then we suggest the following measures to be adopted.

1. *Development of cultivars for high-input conditions.* There has been substantial increase lately in irrigated areas. Chickpea cultivars suited to irrigated conditions have not been bred. When the cultivars developed for other conditions are tried under irrigated conditions, the crop shows excessive growth and lodges. Quite often the irrigated chickpea is infested by new diseases and pests. Also, the chickpea crop does not generally respond to fertilizer applications. Singh and Singh (1994) have described the methods to tackle the problems of irrigation and fertilizer application and develop cultivars responsive to these inputs.

2. *Development of cultivars for paddy fallows.* It is estimated that out of 43 million ha covered by the rice crop in Bangladesh, India, Myanmar, and Pakistan between 20°N and 30°N latitude, about 30 million ha can be cropped during the *rabi* season. Currently only 10 million ha are cultivated. If suitable cultivars are bred, part of the 20 million ha land could easily be brought under chickpea cultivation. Details of problems associated with the breeding of chickpea for fallow replacement have been discussed by Singh and Singh (1994).

3. *Breeding chickpea with multiple-strees resistance.* Present-day cultivars are resistant to none of the stresses or resistant to only one or two stresses, whereas the crop suffers from many stresses in most areas.

Therefore, to stabilize chickpea yield, attempts should be made to breed cultivars resistant to all important stresses of a given area.

4. *Introduction of chickpea to non-traditional areas.* Just as winter-sown chickpea was introduced in the Mediterranean region, chickpea can be introduced in many new areas in India, for example, as replacement of rice fallow. Also, chickpea can be double-cropped with maize or cotton or pearl millet where one or two supplemental irrigations are available. Again, breeders have to develop cultivars suited to such conditions. Efforts have so far met with little success. More research input is needed to introduce chickpea in two-crop rotation in the farming systems.

5. *Development of cultivars with resistance to ascochyta blight.* Susceptibility of chickpea cultivars to ascochyta blight has reduced the area and production of chickpea in northwestern India. Research inputs in this direction have been minimal as compared to the importance of the disease. More research efforts are needed to revive chickpea cultivation on a large scale in this important area of the country.

6. *Exploitation of alien germplasm.* Bahl and Salimath (1992) have emphasized the role of alien germplasm as well as tertiary gene pool to broaden the genetic base of chickpea. The introgression of exotic germplasm from both cultivated and wild species can help to transfer useful gene complexes to the breeding populations. This line of work, which has received meagre attention in the past, offers tremendous potential, considering the novel biotechnological tools now available for affecting free flow of germplasm across the crossability barriers.

REFERENCES

Adhikari, G., and Pandey, M.P. 1982. Component analysis of seed yield in chickpea. *Crop Improv.* 9: 69–74.

Ahmad, F., and Hymowitz, T. 1983. The fine structure of chickpea (*Cicer arietinum* L.) chromosomes as revealed by pachytene analysis. *Theor. Appl. Genet.* 86(5): 637–641.

Ahmad, S., and Godward, M.B.E. 1980. Cytological studies on the cultivars of *Cicer arietinum* L. *Pakistan Caryologia* 33: 55–68.

Akhtar, M. 1954. A study on the introduction of polyploidy in gram. *Proc. Pakistan Sci. Conf.* 6: 124–125.

Allard, R.W. 1960. *Principles of Plant Breeding.* John Wiley and Sons, Inc., New York, 485 pp.

Anand Rao, P.K., and Haware, M.P 1987. Inheritance of dry root rot (*Rhizoctonia bataticola*) resistance in chickpea (*Cicer arietinum* L.). *Plant Breeding* 98: 349–352.

Anilkumar, T.V. 1991. Genetic variability for photosynthetic and yield related traits in chickpea (*Cicer arietinum* L.). M.Sc. (Agri.) thesis, UAS, Dharwad.

Annigeri, B.S. 1992. Genetics of yield and maturity characters in chickpea (*Cicer arietinum* L.). M.Sc. (Agri.) thesis UAS, Dharwad.

Argikar, G.P. 1952a. Occurrence and inheritance of 'filicoid' from in *Cicer arietinum* L. *Curr. Sci.* 21: 76.

Argikar, G.P. 1952b. Segregation for cotyledon colour in *C. arietinum* L. *Poona Agric. College Mag.* 42: 102–108.
Argikar, G.P. 1958. Inheritance of tiny-leaved and alternifolia forms in *Cicer arietinum* L. *Curr. Sci.* 27: 183–184.
Argikar, G.P. 1959. Another case of interaction of factors in *Cicer arietinum* L. *Curr. Sci.* 28: 34–35.
Argikar, G.P. 1970. Gram. In *Pulse Crops of India*. P. Kachroo, ed. ICAR, New Delhi, pp. 54–135.
Argikar, G.P. and D'Cruz, R. 1963. Genetic studies in gram. *J. Indian Bot. Soc.* 42: 401–405.
Arora, P.P. 1988. Line x tester analysis for combining ability in Bengal gram. *Him. J. Agric. Res.* 14: 6–12.
Asawa, B.M., and Tiwari, A.S. 1976. Analysis of genetic architecture in segregating population of gram (*Cicer arietinum* L.). *Z. Pflanzenzuchtg*, 77: 251–256.
Auckland, A.K., and van der Maesen, L.J.G. 1980. Chickpea. In *Hybridization of Crop Plants*. W.R. Fehr and H.H. Hadely eds. American Society of Agronomy, Wisconsin, pp. 249–259.
Ayyar, V.R., and Balasubramanian, R. 1934. The inheritance of certain colour characters in gram (*Cicer arietinum* L.). *Proc. Indian Acad. Sci.* 4: 1–26.
Ayyar, V.R. and Balasubramanian, R. 1935. Anthesis and pollination in Bengal gram (*Cicer arietinum* L.). *Madras Agric. J.* 23: 170–178.
Ayyar, V.R., and Balasubramanian, R. 1937. Inheritance of branching habit in gram (*Cicer arietinum* L.) *Madras Agric. J.* 25: 105–106.
Ayyar, V.R., and Iyer, R.B. 1936. A preliminary note on the mode of inheritance of reaction to wilt in *Cicer arietinum* L. *Proc. Indian Acad. Sci.* 3: 438–443.
Bahl, P.N. 1980. Kabuli-desi introgression and genesis of new plant type in chickpea. In Proc. Workshop on Chickpea Improvement, ICRISAT, 28 Feb. to 2 Mar. 1979, Hyderabad, India, pp. 75–80.
Bahl, P.N. 1981. Yield potential of improved cultivars and plant type concept in chickpea. In Proc. FAI Group Discussion on Increasing Pulse and Oilseed Production in India, New Delhi, pp. 32–40.
Bahl, P.N. 1983. Performance of tall chickpea genotypes under normal and late-sown conditions. *Int. Chickpea Newsletter* 8: 8–9.
Bahl, P.N. 1986. Breaking yield barriers in chickpea (*Cicer arietinum* L.). *J. IARI Post Grad. School* 9: 1–6.
Bahl, P.N. 1987. Cytology of chickpea. In *The Chickpea*. M.C. Saxena and H.B. Singh eds. CAB International, Wellingford, Oxon, U.K., pp. 83–98.
Bahl, P.N. 1988. Chickpea. In *Pulse Crops*. B. Baldev, S. Ramanujam and H.K. Jain, eds. Oxford and IBH Publishing Co., New Delhi, pp. 95–131.
Bahl, P.N., and Baldev, B. 1981. Production technology for chickpea. *Indian Fmg.*, 31: 67–71.
Bahl, P.N., and Gowda, C.L.L. 1975. Pod setting in crosses of Bengal gram. *Indian J. Genet.* 35:13–16.
Bahl, P.N., and Jain, H.K. 1977. Association among aronomic characters and plant ideotype in chickpea (*Cicer arietinum* L.). *Z. Pflanzenzuchtg.* 79: 154–159.
Bahl, P.N., and Kumar, J. 1989. Evaluation and utilization of high yelding hybrids of chickpea. *Indian J. Genet.* 49: 53–58.
Bahl, P.N., and Salimath, P.M. 1992. Chickpea improvement—Present status and future strategies. In *New Frontiers in Pulses Research and Development*. Proc. Nat. Symp. 10–12 Nov. 1989. J.N. Sachan, ed. Directorate of Pulses Research, Kanpur, India. pp. 33–43.
Bahl, P.N., and Vinod, S. 1991. Evaluation of methods of generation advance in chickpea (*Cicer arietinum* L.). Proc. Symp. Grain Legumes. Indian Society of Genetics and Plant Breeding, New Delhi. pp. 351–358.
Bahl, P.N., Mehra, R.B., and Raju, D.B. 1976. Path analysis and its implications for chickpea breeding. *Z. Pflanzenzuchtg.* 77: 67–71.

Bahl, P.N., Singh, S.P., Hayat Ram, Raju, D.B. and Jain, H.K. 1979. Breeding for improved plant architecture and high protein yields. In FAO/IAEA Int. Symp. Seed Improvement in Cereals and Grain Legumes, Vienna, Austria, pp. 297—307.

Bahl, P.N., Singh, K.P., and Singh, Daljeet. 1984. Evaluation of tall chickpea genotypes for normal and late sowings. *Indian J. Agric. Sci.* 54: 110–113.

Bahl, P.N., Salimath, P.M., Malik, B.A., Dahiya, B.S., Deshmukh, R.B., and Rangaswamy, P. 1990. New approaches in chickpea breeding and prospects in the nineties. In *Chickpea in the Nineties*: Proc. Second Int. Workshop on Chickpea Improvement, 4–8 Dec. 1989, ICRISAT Centre, Patancheru, India, pp. 225–232.

Bahl, P.N., Kumar, J., and Raju, D.B. 1991a. Genetic variation and adaptations in chickpea. *Plant Breeding* 106: 164–167.

Bahl, P.N., Raju, D.B., Kumar, J., and Salimath, P.M. 1991b. Pusa 261-A blight resistant gram variety for late planting. *Indian Fmg. 40 (12): 15.*

Balasubramanian, R. 1937. Inheritance of characters in gram. Foliage colour and rough seed coat. *Madras Agric. J.* 25: 207–208.

Balasubramanian, R. 1950. The association of seed coat colour in gram. *Madras Agric. J.* 37: 379–384.

Bejiga, G., and Singh, K.B. 1991. Estimation of heterosis in *Kabuli* chickpea. *Acta Agronomica* 40: 145–150.

Bejiga, G., and Tessema, T. 1981. Identification on the optimum time of emasculation and pollination to increase percentage of hybrid seed set in chickpea. *Eth. J. Agric. Sci.* 3: 129–134.

Bhatt, D.D., and Singh, D.P. 1980. Combining ability in chickpea. *Indian J. Genet.* 40: 456–460.

Bisen, M.S., Singh, S.P., and Rao, S.K. 1985. Effectiveness of selection methods in chickpea *Cicer arietinum* L. under different environments. *Theor. Appl. Genet.* 70 (6): 661–666.

Brar, H.S., and Athwal, D.S. 1970. Identification of genes controlling seed colour in *Cicer arietinum* L. *Indian J. Genet.* 30: 690–703.

Briggs, F.N., and Knowles, P.F. 1967. *Introduction to Plant Breeding.* Reinhold Publishing Corporation, New York, Amsterdam, London, 426 pp.

Byth, D.E., Green, J.M., and Hawtin, G.C. 1980. ICRISAT/ICARDA chickpea breeding strategies. In Proc. Int. Workshop on Chickpea Improvement, ICRISAT, 28 Feb. to 2 Mar. 1979, Patancheru, India, pp. 11–27.

Chand, H., Srivastava, L.S., and Trehan, K.B. 1975. Estimate of genetic parameters and path coefficient analysis in gram (*Cicer arietinum* L.). *Madras Agric. J.* 62: 178–181.

Chandra, S. 1968. Variability in gram. *Indian J. Genet.* 28: 206–210.

Chaudhary, D.S. 1974. Studies on heterosis and combining ability in Bengal gram (*Cicer arietinum* L.). M.Sc. thesis, Haryana Agric. Univ., Hissar, India.

Chaudhary, D.S., Gupta, V.P., and Chandra, S. 1978. Selection based on heterosis, combining ability and early generation testing among crosses of chickpea (*Cicer arietinum* L.). *Legume Res.* 1:77–86.

Chowdhary, K.S., Saraswat, K.S., Hasan, S.N., and Gaur, R.G. 1970. 4000–3500 years old barley, rye and pulses from Atranjikhera. *Sci. and Cult.,* 37: 531–533.

Dabholkar, A.R. 1973. Yield components in *Cicer arietinum* L. *J.N.K.V.V. Res. J.* 11: 408.

Dahiya, B.S., Lather, V.S., Solanki, I.S., and Kumar, R. 1984a. Useful spontaneous mutants in chickpea. *Int. Chickpea Newsletter* 11: 408.

Dahiya, B.S., Waldia, R.S., Kaushik, L.S., and Solanki, I.S. 1984b. Early generation yield testing versus visual selection in chickpea (*Cicer arietinum* L.). *Theror. Appl. Genet.* 68 (6): 525–529.

Dangaria, C.J. 1991. Genetics of nitrogen fixation parameters, yield and yield attributes in chickpea (*Cicer arietinum* L.). Ph.D. thesis, Univ. Agri. Sci., Dharwad, India.

Davis, T.M., Foster, K.W., and Philips, D.A. 1986. Inheritance and expression of three genes controlling root nodule formation in chickpea. *Crop. Sci.* 26: 719–723.

De Candolle, A. 1882 L' Origine des plantes cultivees. Paris, p. 258.

Deshmukh, R.B., and Bhapkar, D.G. 1981. Heterosis and combining ability for harvest index and agronomic characters in chickpea. *Legume Res.* 4:19–22.

Dhaliwal H.S., and Gill, A.S. 1973. Studies on heterosis, combining ability and inheritance of yield and yield components in a diallel cross of Bengal gram (*Cicer arietinum* L.). *Theor. Appl. Genet.* 43: 381–386.

Dixit, P.D. 1932a. Studies in Indian pulses. A note on the cytology of kabuli and desi gram types. *Indian J. Agric. Sci.* 2–4: 385–390.

Dixit, P.D. 1932b. Studies in Indian pulses. A case of gigantism in gram (*Cicer arietinum*). *Indian J. Agric Sci.* 2–4: 391–408.

Dombrovsky-Sludsky, I., 1927. La cinese somatique de *Cicer arietinum zh. Russk. Bot. Obsheh.* 12: 163–172.

Ekbote, R.B. 1937. Mutations in gram (*Cicer arietinum* L.). *Curr. Sci.* 5: 648–649.

Eshel, Y. 1965. Studies on flowering, seed setting and ripening of chickpea (*Cicer arietinum* L.). Ph.D. thesis, Hebrew Univ., Jerusalem.

F.A.O., 1992. FAO Production Year Book. Rome, Vol. 46.

Fehr, R.W. 1987. Principles of Cultivar Development, Vol. 1, *Theory and Technique.* Macmillan Publishing Co., New York, Collier Macmillan Publishers, London.

Furnkraz, D. 1968. Contribution to the systematics and cytology of the genus *Cicer* l. The European species Oester. *Bot. Zeitschr.* 115: 400–410.

Gaur, P.M., and Slinkard, A.E. 1989. New isozyme markers for chickpea. *Int. Chickpea Newsletter* 23: 5–8.

Gaur, P.M., and Slinkard, A.E. 1990a. Inheritance and linkage of isozyme coding genes in chickpea. *J. Hered.* 81: 455–461.

Gaur, P.M., and Slinkard, A.E. 1990b. Genetic control and linkage relations of additional isozyme markers in chickpea. *Theor. Appl. Genet.* 80: 648–656.

Gaur, P.M., and Slinkard, A.E. 1991. An isozyme map of chickpea. *Int. Chickpea Newsletter* 24: 25–28.

Gowda, C.L.L. 1972. Path coefficient analysis for yield and yield components in Bengal gram (*Cicer arietinum* L.). M.Sc. thesis, G.B. Pant Univ. Agric. and Tech., Pantnagar UP, India.

Gowda, C.L.L. 1981. Natural outcrossing in chickpea. *Int. Chickpea Newsletter* 5: 6.

Gowda, C.L.L., and Bahl, P.N. 1976. Heterosis in chickpea. *Indian J. Genet.* 36: 265–268.

Gowds, C.L.L., and Bahl, P.N. 1978. Combining ability in chickpea. *Indian J. Genet.* 38: 245–251

Gowda, C.L.L., Lateef, S.S., Smithson, J.B., and Reed, W. 1985. Breeding for resistance to *Heliothis armigera* in chickpea. In Nat. Semi. Breeding Crop Plants for Resistance to Pests and Diseases, 25–27 May 1983, Coimbatore, Tamil Nadu, India, Tamil Nadu Agric. Univ., pp. 36–39.

Gupta, P.K., and Bahl, J.R. 1983. Cytogenetics and origin of some pulse crops. In *Cytogenetics of Crop Plants.* M.S. Swaminathan, P.K. Gupta and U. Sinha, eds. Macmillan India Ltd., Delhi, Bombay, Calcutta, Madras, pp. 405–440.

Gupta, V.P., and Lal, S. 1981. Development allometry and plant type in chickpea. *Int. Chickpea Newsletter* 4: 8–9.

Gupta, V.P., and Ramanujam, S. 1974a. Cumulative gene effects and genetic parameters of heterosis in F_1 and F_2 generations of chickpeas (*Cicer arietinum* L.). *Genetika, Belgrade* 6: 263–275.

Gupta, V.P., and Ramanujam, S. 1974b. Genetic achitecture of yield and its components in chickpea. *Indian J. Genet.*, 34A: 793–799.

Gupta, V.P., Ramanujam, S., and Kaul, A.K. 1974. Genetic analysis in respect of protein content and protein quality in chickpea (*Cicer arietinum* L.). *Genetika, Belgrade* 6: 231–245.

Haq, M.A., and Singh, K.B. 1994. Induction of cold tolerance in *kabuli* chickpea (*Cicer arietinum* L.) through induced mutations. *Mutation Breeding Newsletter,* IAEA, Vienna, Austria 41: 6–7.

Haware, M.P., and Nene, Y.L. 1982. Races of *Fusarium oxysporum* L. sp. *Ciceri. Plant Disease* 66: 809–810.

Haware, M.P., Kumar, J., and Reddy, M.V. 1990. Disease resistance in kabuli-desi chickpea introgression. In Proc. Int. Workshop on Chickpea Improvement, ICRISAT, Hyderabad, pp. 67–69.

Hawtin, G.C., and Singh, K.B. 1980. Kabuli-desi introgression: Problems and prospects. In Proc. Int. Workshop on Chickpea Improvement, ICRISAT, Hyderabad, pp. 51–60.

Hawtin, G.C., and Singh, K.B. 1984. Prospects and potential of winter sowing of chickpea in the mediterranean region. In *Ascochyta Blight and Winter Sowing of Chickpeas.* M.C. Saxena and K.B. Singh, eds. Martinus Nijhoff/Dr. W. Junk Publishers, the Hague, the Netherlands, pp. 7–16.

Helbaek, H. 1959. Domestication of food plants in the old world. *Science* 130-3372: 365–372.

ICARDA, 1990. Food legumes Improvement Programme. in Annual Report for 1989. ICARDA, Aleppo, Syria, pp. 24–25.

ICARDA, 1994. Legume Programme Annual Report for 1993. ICARDA, Aleppo, Syria.

ICRISAT, 1986. Chickpea Breeding Report 32. Report of work, June 1985 to May 1986. Legumes Programme, ICRISAT, Patancheru, AP, India.

ICRISAT, 1990. ICRISAT Annual Report 1989. Patancheru, AP, India.

ICRISAT, 1993. Annual Report 1992. Patancheru, AP, India.

ICRISAT, 1994. Annual Report 1993. Patancheru, AP, India.

Isalam, A.K.M.R., and Riazuddin, S. 1992. Transformation of chickpea tissues by *Agrobacterium tumefaciens* in *ex vitro* condition. *Int. Chickpea Newsletter* 26: 11–12.

Islam, A.K.M.R., and Riazuddin, S. 1994. Influence of genotype and age of seedling on chickpea response to *Agrobacterium tumefaciens. Int. Chickpea and Pigeonpea Newsletter.* 1: 23–25.

Islam, M.Q., Begum, K., and Kaul, A.K. 1984. Phenotypic variability and correlation studies in kabuli chickpea. *Bangladesh J. Agric. Res.* 9(1): 33–37.

Iyengar, N.K. 1939. Cytological investigations on genus *Cicer. Ann. Bot. New Series,* 3–10: 271–305.

Jain, H.K. 1975. Breeding for yield and other attributes in grain legumes. *Indian J. Genet.* 35: 169–187.

Jatsara, D.S., Ram, C., Chandra, S., and Singh, A. 1978. Correlations and path analysis in segregating populations of chickpea (*Cicer arietinum* L.). *Indian J. Agric. Res.* 12: 219–222.

Jodha, N.S., and Subba Rao, K.V. 1987. Chickpea: World importance and distribution. In *The Chickpea.* (M.C.) Saxena, and K.B. Singh, eds. CAB International, Wallingford, Oxon, UK.

Kahl, G., Kaemmer, D., Weising, K., Kost, S., Weigand, F., and Saxena, M.C. 1994. The potential of gene technology and genome analysis for cool season food legume crops: theory and practice. *Euphytica* 73: 177–189.

Katiyar, R.P. 1981. Inheritance of seed weight in chickpea. *Pulse Crops Newsletter* 1: 5–6.

Katiyar, R.P., and Singh, H.G. 1987. Inter-relationship of physiological attributes with seed yield and its components in chickpea. *Farm Sci. J.* 2: 125–130.

Katiyar, R.P., and Solanki, R.K. 1983. Combining ability and gene action for yield and its components in chickpea (*Cicer arietinum* L.). 1991 abstract in Proc., XV Int. Cong. of Genet., New Delhi, 12–21 Dec. 1983. Oxford and IBH Publishing Co., New Delhi, India.

Katiyar, R.P., Prasad, J., Singh, A.B., and Ram, L. 1977. Association analysis of grain yield and its components in segregating populations of chickpea. *Indian J. Agric. Sci.* 47: 325–327.

Katiyar, R.P., Sood, O.P., and Kalia, N.R. 1981. Selection criteria in chickpea. *Int. Chickpea Newsletter* 4: 5–6.

Kazan, K., Muehlbauer, F.J., Weeden, N.F., and Ladizinsky, G. 1993. Inheritance and linkage relationships of morphological and isozyme loci in chickpea (*Cicer arietinum* L.). *Theor. Appl. Genet.* 86(4): 417–426.

Khan, A.R., and Akhtar, A.R. 1934. The inheritance of petal colour in gram. *Agriculture and Livestock in India* 4: 127–155.

Khan, M.A., and Choudhary, H.A. 1975. Inter-relationships between yield and other plant characters in gram (*Cicer arietinum* L.). *J. Agric. Res.* 13: 589–592.

Khan, I.A., Malik, B.A., and Basheer, M. 1989. Studies on floral biology and its association with yield in chickpea. *Sarhad J. Agric.* 5(4): 377–380.

Khanna-Chopra, R., and Sinha, S.K. 1987. Chickpea: Physiological aspects of growth and yield. In *The Chickpea.* M.C. Saxena, and K.B. Singh, eds. CAB International, Wallingford, Oxon, U.K., pp. 163–189.

Kharrat, M., Gill, J., and Cubero, J.I. 1991. Genetics of grain yield components in chickpea (*Cicer arietinum* L.). *J. Genet. and Breed.* 45: 87–92.

Khosh-Khui, M., and Niknejad, M. 1971. Hybridisation methods in chickpea. *Israel J. Agric. Res.* 21: 81–82.

Khosh-Khui, M., and Niknejad, M. 1972. Plant height and width inheritance and their correlation with some of the yield components in chickpea (*Cicer arietinum* L.). *J. Agric. Sci.* 78: 37–38.

Khosh-Khui, M., and Niknejad, M. 1973. Heritability and number of genes responsible for yield per plant in reciprocal crosses of chickpea (*Cicer arietinum* L.). *Curr. Sci.* 41: 296–297.

Kumar, J., and Bahl, P.N. 1992. Direct and indirect selection for yield in chickpea. *Euphytica* 60: 197–199.

Kumar, J., and Haware, M.P. 1982. Inheritance of resistance to *Fusarium* wilt in chickpea. *Phytopathology* 72: 1035–1036.

Kumar, J., Bahl, P.N., and Raju, D.B. 1989. Giant pod mutant and its inheritance in chickpea. *Int. Chickpea Newsletter* 21: 6–7.

Kunadia, B.A., and Singh, D.P. 1980. Heterosis and inbreeding depression in *Kabuli* gram (*Cicer arietinum* L.). *Int. Chickpea Newsletter 3: 3–4.*

Kunadia, B.A., and Singh, D.P. 1981. Combining ability in *Kabuli* gram (*Cicer arietinum* L.). *Int. Chickpea Newsletter* 4: 6–7.

Ladizinsky, G., and Adler, A. 1975. The origin of chickpea as indicated by seed protein electrophoresis. *Israel J. Bot.*, 24: 183–189.

Ladizinsky, G., and Adler, A. 1976a. The origin of chickpea (*Cicer arietinum* L.). *Euphytica,* 25: 211–217.

Ladizinsky, G., and Adler, A. 1976b. Genetic relationship among the annual species of *Cicer* L. *Theor Appl. Genet.* 48: 197–203.

Lal, S., Singh, S.N., Lal, S.B., and Tripathi, B.D.H. 1973. Improvement in gram through pedigree method of breeding. In Proc. Tenth Workshop on *Rabi* Pulses AICRPP (ICAR), Bose Institute, 10–14 Sept., Calcutta, ICAR, New Delhi, India, pp. 70–80.

Lateef, S.S., and Reed, W. 1981. Development of a methodology for open-field screening for insect pest resistance in pigeonpea. In Proc. Int. Workshop on Pigeonpeas, 15–19 Dec. 1980, ICRISAT Centre, India, Vol. 2. Patancheru, AP, India, pp. 315–322.

Lateef, S.S., and Reed, W. 1985. A suggested system of rating pigeonpea and chickpea entries for field resistance to *Heliothis armigera.* Nat. Sem. Breeding Crop Plants for Resistance to Pests and Diseases, 25–27 May 1983, Coimbatore, India, Tamil Nadu Agric. Univ. pp. 127–131.

Lateef, S.S., and Sachan, J.N. 1990. Host-plant resistance to *Hlicoverpa armigera* in different agroecological contexts. In *Chickpea in the Nineties.* Proc. Second *Int.* Workshop Chickpea Improv. 4–8 Dec. 1989, ICRISAT Centre, India, pp. 181–190

Lavania, V.C., and Lavania, S. 1983. Karyotype studies in Indian pulses. *Genetica Agraria.* 37: 299–308.

Lopez Garcia, H. 1974. Inheritance of the character resistance to wilt (*Fusarium* sp.) in chickpea (*Cicer arietinum* L.) under field conditions. *Agricultural Technica en Mexico* 3: 286–289.

Malhotra, R.S., and Singh, K.B. 1973. Genetic variability and genotype environment interaction in Bengal gram. *Indian J. Agric. Sci.* 43: 914–917.

Malhotra, R.S., and Singh, K.B. 1984. Detection of epistasis and estimation of additive and dominance variation in chickpea (*Cicer arietinum* L.). *Z. Pflanzenzuchtg.* 7: 186–193.

Malhotra, R.S., and Singh, K.B. 1986. Natural cross pollination in chickpea. *Int. Chickpea Newsletter* 14: 3–5.

Malhotra, R.S., and Singh, K.B. 1990. The inheritance of cold-tolerance in chickpea. *Genetics and Breeding* 44: 227–230.

Malhotra, R.S., and Singh, K.B. 1991. Gene action for cold-tolerance in chickpea. *Theor. Appl. Genet.* 82: 598–601.

Malhotra, R.S., Singh, K.B., and Lal, B. 1983. Combining ability for yield and its components in chickpea (*Cicer arietinum* L.). *Indian J. Genet.* 43: 149–151.

Malik, B.A., Kahn, I.A., and Malik, M.R. 1988. Genetic variability and correlations among metric traits in chickpea. *Pakistan J. Agric. Res.* 9(3): 352–354.

Malti and Shivanna, K.R. 1983. Pollen pistil interaction in chickpea. *Int. Chickpea Newsletter* 9: 10–11.

Mandal, A.K. 1977. Studies on variability, interrelationship, path coefficient analysis and heterosis in chickpea (*Cicer arietinum* L.). M.Sc. thesis, IARI, New Delhi.

Mandal, A.K. 1980. Heterosis and combining ability analysis in chickpea (*Cicer arietinum* L.). Ph.D. thesis, IARI, New Delhi.

Mandal. A.K., and Bahl, P.N. 1979. Heterosis in different parental combinations in chickpea (*Cicer arietinum* L.). *Trop. Grain Legume Bull.* 15: 14–17.

Mandal, A.K., and Bahl. P.N. 1980. Estimates of genetic variability and correlations in chickpea. *Ann. Agric. Res.* 1: 130–140.

Mandal, A.K., and Bahl, P.N. 1983. Estimates of genetic variability and correlations of harvest index in chickpea. *Int. Chickpea Newsletter* 8: 11–12.

Mandal, A.K., and Bahl, P.N. 1984. Heterosis in diverse crosses of chickpea. *Indian J. Genet.* 44: 173–176.

Mangesha, M.H., Holly, L., Pundir, R.P.S., and Thomas, T.A. 1990. Chickpea genetic resources–Present and future in chickpea in the nineties. Proc. Second Int. Workshop on Chickpea Improvement, 4–8 Dec. 1989, ICRISAT Centre, Patancheru, AP, India, pp. 51–56.

Mani, M. 1987. Intraspecific variation and combining ability in chickpea (*Cicer arietinum* L.). Ph.D. thesis, IARI, New Delhi, India, 133 pp.

Mani, M., and Bahl, P.N. 1990. Components of productivity in contrasting types of chickpea. *J. Genet.* 44: 163–167.

Meenakshi, G., and Subramaniam, M.K. 1966. Triploidy in *Cicer arietinum* L. *Caryologia* 19: 163–166.

Mercy, S.T., Kakar, S.N., and Chowdury, J.B. 1974. Cytological studies in three species of the genus *Cicer. Carylogia* 39: 379–382.

Milovidov, P.F. 1932. A case of mixoploidy in *Cicer arietinum* L. *Preslia* 11: 62–66.

Mishra, R., Rao, S.K., and Koutu, G.K. 1988. Genetic variability, correlation studies and their implication in selection of high yielding genotypes in chickpea. *Indian J. Agric. Res.* 22(1) 51–57.

Mishra, S.S. 1972. Relationship between yield attributes and yield in gram (*Cicer arietinum* L.). *Madras Agric. J.* 59: 186–191.

Morrison, J.W., and Rajathy, P. 1960. Frequency of quadrivalents in autotetraploid plants. *Nature* 187: 528–530.

Muehlbauer, F.J., and Singh, K.B. 1987. Genetics of chickpea. In *The Chickpea*. M.C. Saxena and K.B. Singh, eds. CAB International, Wallingford, Oxon, UK, pp. 99–125.

Mauehlbauer, F.J., Simon,. C.J., Spacth, S.C., and Haddad, N.L. 1990. Genetic improvement of chickpea: Key factors to be considered for a breakthrough in productivity. In *Chickpea in the Nineties*: Proc. Second Int. Workshop Chickpea Improvement, 4–8 Dec. 1989, ICRISAT Centre, India, Patancheru, AP, pp. 209–216.

Nene, Y.L. 1978. A world list of pigeonpea (*Cajanus cajan* [L.] Millsp.) and chickpea (*Cicer arietinum* L.) pathogens. ICRISAT, Hyderabad, India, Pulse Pathology Progress Rep. 3: 1–15.

Nene, Y.L. 1984. A review of ascochyta blight of chickpea (*Cicer arietinum* L.) In *Ascochyta Blight and Winter Sowings of Chickpea*. M.C. Saxena, and K.B. Singh, eds. Martinus Nijhoff/Dr. W. Junk Publishers, the Hague, the Netherlands, pp. 17–34.

Nene, Y.L., and Reddy, M.V. 1987. Chickpea diseases and their control. In *The Chickpea*. M.C. Saxena and K.B. Singh, eds. CAB International, Wallingford, Oxon, UK, pp. 232–270.

Nene, Y.L., Haware, M.P., and Reddy, M.V. 1981. Chickpea diseases: resistance screening techinques. Int. Bull. No. 10. ICRISAT, Patancheru, India.

Niknejad, M., Khosh-Khui, M., and Ghorashy, S.R. 1971. Inheritance of seed size in chickpea (*Cicer arietinum* L.). *Crop Sci.* 1: 768–769.

Pal, B.P. 1945. Studies in hybrid vigour—notes on manifestation of hybrid vigour in gram, sesamum, chillies and maize. *Indian J. Genet.* 5: 106–121.

Pal, B.P., and Ramanujam, S. 1944. Plant breeding and genetics at the Imperial Agricultural Research Institute, New Delhi. *Indian J. Genet.* 4: 43–53.

Pande, K., Pandya, B.P., and Jain, K.C. 1979. Diallel analysis of yield and yield components in Bengal gram. *Indian J. Agric. Res.* 13: 187–194.

Pande, K., Jain, K.C., and Pandya, B.P. 1980. Epistasis, á basis in the genetic analysis of yield in chickpea. *Egyptian J. Cytol.* 9: 297–305.

Pandey, R.L., and Tiwari, A.S. 1983. Heritability and genetic gain in chickpea. *Int. Chickpea Newsletter* 9: 5–6.

Pathak, M.M., Singh, K.P., and Lal, S.B. 1975. Inheritance of resistance to wilt (*Fusarium oxysporum* f. sp. *ciceri*) in gram. *Indian J. Farm Sci.* 3: 10–11.

Patil, J.A. 1959. Inheritance study in gram. *Curr. Sci.* 28: 508.

Patil, J.A. 1967. Inheritance of pod deposition and blue flower colour in gram (*Cicer arietinum* L.). *Poona Agric. College Mag.* 57: 85–86.

Patil, J.A., and Choudhary, B.B. 1957. A pod mutant in gram (*Cicer arietinum* L.). *Curr. Sci.* 26: 250.

Patil, V.N., and Phadnis, B.A. 1977. Genotypic variability and its implication in selection of gram (*Cicer arietinum* L.). *J. Maharashtra Agric. Univ.* 2: 121–123.

Patil, V.D., Nerkar, Y.S., and Dayama, D.K. 1987. Inheritance of quantitative characters in chickpea. *Indian J. Agric. Sci.* 57: 78–81.

Phadnis, B.A., and Narkhede, M.N. 1972. Chromosome behaviour in colchicine induced autotetraploids of *Cicer arietinum* L. *Cytologia* 37: 415–421.

Phadnis, B.A., Narkhade, M.N., and Thombre, M.V. 1968. Further studies in polysomaty in root tips of *Cicer arietinum* L. *Proc. Indian Acad. Sci.* B. 68: 279–283.

Phadnis, B.A., Ekbote, A.P., and Ainchwar, S.S. 1970. Path coefficient analysis in gram (*Cicer arietinum* L.). *Indian J. Agric. Sci.* 40: 1013–1016.

Pundir, R.P.S., and Reddy, K.N. 1984. Comparison of cross-pollination methods in chickpea. *Int. Chickpea Newsletter* 11: 9–11.

Pundir, R.P.S., and van der Maesen, L.J.G. 1983. Interspecific hybridization in *Cicer*. *Int. Chickpea Newsletter* 8: 4–5.

Pundir, R.P.S., Rao, N.K., and van der Maesen, L.J.G. 1983. Induced autotetraploidy in chickpea (*Cicer arietinum* L.). *Theor. Appl. Genet.* 65: 119–122.

Pundir, R.P.S., Mangesha, M.H., and Reddy, K.N. 1988a. Occurrence and genetics of natural mutant of chickpea having twin flower peduncles and polycarpy. *J. Hered.* 79: 479–481.

Pundir, R.P.S., Reddy, K.N., and Mangesha, M.H. 1988b. ICRISAT chickpea germplasm catalogue. Evaluation and analysis. ICRISAT, Patancheru, AP, India, 94 pp.

Rahman, M.A., and Bahl, P.N. 1985. Comparison of single seed descent, mass selection and random bulk methods in chickpea. *Indian J. Genet.* 45: 186–193.

Rahman, M.A., and Bahl, P.N. 1986. Evaluation of early generation testing in chickpea. *Z. Pflanzenchutg.* 97: 82–85.

Raju, D.B., Mehra, R.B., and Bahl, P.N. 1978. Genetic variability and correlations in chickpea. *Trop. Grain Legume Bull.* 14: 35–39.

Ram, C., Choudhary, M.A., Chandra, S., and Jatsara, D.S. 1980. Association in segregating populations of chickpea. *Indian J. Genet.* 40: 117–121.

Ramanujam, S. 1976. Chickpea (*Cicer arietinum* L.). In: Evolution of crop plants. N.W. Simmounds, ed. Longman, London, New York, pp. 157–159.

Ramanujam, S., and Joshi, A.B. 1941. Colchicine induced polyploidy in crop plants. *Indian J. Agric. Sci.,* 11: 835–849.

Ramanujam, S., and Singh, H.B. 1945. Narrow leaf—another leaf mutation in gram. *Indian J. Genet.* 5: 46–50.

Rang, A. 1980. Genetic analysis of grain yield, its components and quality in chickpea (*Cicer arietinum* L.). Ph.D. thesis, PAU, Ludhiana, India.

Rao, N.K., and Pundir, R.P.S. 1983. Inheritance and linkage relationships of a new lobed vexillum mutant in chickpea. *J. Hered.* 74: 300.

Rao, N.S. 1929. Further contributions to the cytology of some crop plants of South India. *J. Indian Bot. Soc.* 18: 222–229.

Rastogi, K.B., and Singh, L. 1977. Heritability, phenotypic and genotypic correlation coefficients of some of the characters in the F_2 populations of gram. *Crop. Improv.* 4: 191–197.

Rawat, M.C., Tiwari, A.S., and Pandey, R.L. 1981. Combining ability analysis in Bengal gram (*Cicer arietinum* L.). *J. Res. Assam Agric. Univ.* 2: 151–155.

Reddy M.Y., and Kabbabeh, S. 1985. Pathogen variability in *Ascochyta rabiei* in Syria and Lebanon. *Phytopath. Medit.* 24: 265–266.

Reddy, M.V., Nene, Y.L., Singh, Gurdip and Bashir, M. 1990. Strategies for management of foliar diseases of chickpea. In *Chickpea in the Nineties.* Proc. Second Int. Workshop on Chickpea Improvement, 4–8 Dec. 1989, ICRISAT Centre, Patancheru, AP, India, pp. 117–127.

Reddy, M.V., Singh, K.B., and Malhotra, R.S. 1992. Multilocation evaluation of chickpea germplasm and breeding lines for resistance of Ascochyta blight. *Phytopath. Medit.* 31: 59–66.

Reddy, R.K., and Rao, S.K. 1988. Analysis of yield factors in segregating populations in chickpea. *Legume Res.* 11(1): 27–31.

Reed, W., Cardona, C., Sithanantham, S., and Lateef, S.S. 1987. Chickpea insect pests and their control. In *The Chickpea.* M.C. Saxena and K.B. Singh, eds. Commonwealth Agricultural Bureaux International, Wallingford, Oxon, UK, pp. 283–318.

Renfrew, J.M. 1973. *Palaeoethnobotany.* Methuen, London.

Rohewal, S.S., Ramanujam, S., and Mehra, K.L. 1966. Plant type in Bengal gram. *Indian J. Genet.,* 26: 255–261.

Rupela, O.P., and Sudarshan, M.R. 1986. Identification of non-nodulating spontaneous mutant in chickpea. *Int. Chickpea Newsletter* 15: 13–14.

Salimath, P.M. 1980. Genetic diversity, combining ability and selection response in chickpea (*Cicer arietinum* L.). Ph.D. thesis, IARI, New Delhi.

Salimath, P.M., and Bahl, P.N. 1985a. Heterosis and combining ability for earliness in chickpea (*Cicer arietinum* L.). *Indian J. Genet.* 45: 97–100.

Salimath, P.M., and Bahl, P.N. 1985b. Early generation selection in chickpea . III. Predicted and realised gains. *Exp. Genet.* 1(2): 59–62.

Salimath, P.M., and Bahl, P.N. 1986. Association analysis and plant ideotype in chickpea (*Cicer arietinum* L.). *Exp. Genet.* 2(1–2): 41–46.

Salimath, P.M., and Bahl, P.N. 1988. Genetic analysis of seed yield and its attributes in chickpea (*Cicer arietinum* L.). *Mysore J. Agric. Sci.* 22: 31–38.

Salimath, P.M., and Bahl, P.N. 1989. Combining ability studies in crosses involving tall and dwarf types in chickpea (*Cicer arietinum* L.). *Indian J. Genet.* 48: 29–34.

Salimath, P.M., and Patil, S.S. 1990. Genetic study in F_3 and F_4 generations of chickpea. *Indian J. Genet.* 50(4): 378–381.

Salimath, P.M., Bahl, P.N., and Mehra, R.B. 1984. Genetic diversity in chickpea (*Cicer arietinum* L.). *Z. Pflanzenzuchtg.* 92: 52–60.

Salimath, P.M., Bahl. P.N., and Kumar, J. 1988. Genetic analysis of seed protein content in chickpea (*Cicer arietinum* L.). *Legumes Res.* 11(2): 31–38.

Sandha, G.S., and Athwal, D.S. 1971. Inheritance of branching in gram (*Cicer arietinum* L.). *J. Res. Punjab Agric. Univ.*, Ludhiana 8: 143–152.

Sandhu, J.S., Singh, K.P., and Slinkard, A.E. 1983. Inheritance of resistance to *Fusarium* wilt in chickpea. *J. Hered.* 74: 68.

Sandhu, S.S., Keim, W.F., Hodges, H.F., and Hyquist, W.E. 1974. Inheritance of protein and sulphur content in seeds of chickpea. *Crop Sci.* 14: 649–652.

Sandhu, T.S., and Bhullar, B.S. 1987. Combining ability analysis for grain yield and other quantitative traits in chickpea. *Crop Improv.* 14: 95–97.

Sandhu, T.S., and Singh, N.B. 1970. Genetic variability, correlation and segression studies in gram (*Cicer arietinum* L.). *J. Res. Punjab Agric. Univ.* 7: 423–427.

Sandhu, T.S., and Singh, N.B. 1972. Correlation, path analysis and discriminant function selection in *Cicer arietinum* L. *J. Res. Punjab. Agric. Univ.* 9: 417–421.

Sandhu, T.S., Brar, H.S., and Arora, B.S. 1977. Combining ability from a diallel cross of chickpea *Cicer arietinum* L. *Crop. Improv.* 4: 11–17.

Sandhu, T.S., Gumber, R., Bhullar, B.S., and Bhatia, R.S. 1989. Genetic analysis of grain protein, grain yield and its components in chickpea. *J. Res. Punjab Agric. Univ.* 26(1): 86–88.

Saxena, M.C. 1988. Food legumes in the Mediterranean type of environment. In *Nitrogen Fixation by Legumes in Mediterranean Agriculture*: Proc. workshop, 14–17 April 1986, Aleppo, Syria D.P. Back and L.A. Meteron, eds. Martinus Nijhoff, Dordrecht, the Netherlands, pp. 11–23.

Saxena, M.C. 1990. Problems and potential of chickpea production in the nineties. In *Chickpea in the Nineties.* Proc. Second Int. Workshop on Chickpea Improvement, 4–8 Dec. 1989, ICRISAT Centre, Patancheru, AP, India, pp. 13-25.

Saxena, N.P., and Johansen, C. 1990. Chickpea ideotypes for genetic enhancement of yield and yield stability in South Asia. In *Chickpea in the Nineties:* Proc. Second Int. Workshop on Chickpea Improvement, 4–8 Dec. 1989, ICRISAT, Patancheru, AP, India, pp. 81-85.

Saxena, N.P., Krishnamurthy, L. and Johansen, C. 1993. Registration of a drought resistant chickpea germplasm. *Crop Sci.* 6: 1424.

Sedgely, R.H., Siddique, K.H.M., and Walton, G.H. 1990. Chickpea idotypes for mediterranean environments. In *Chickpea in the Nineties:* Proc. Second Int. Workshop on Chickpea Improvement, 4–8 Dec. 1989, ICRISAT, Patancheru, AP, India.

Sen, N.K., and Jana, M.V. 1956. A spontaneous autotetraploid gram (*Cicer arietinum* L.). *Curr. Sci. 25: 231–232.*

Setty, A.N., Patil, M.S., and Hiremath, K.G. 1977. Genetic variability and correlation studies in *Cicer arietinum* L. *Mysore J. Agric. Sci.* 11: 131–134.

Sharma, P.C., and Gupta, P.K. 1982a. Karyotypes in some pulses crops. *Nucleus* 25: 181–185.

Sharma, P.C., and Gupta, P.K. 1982b. Chromosome associations in autotetraploid chickpea. *Int. Chickpea Newsletter* 6: 3–4.

Sharma, P.P., and Maloo, S.R. 1988. Variability parameters in chickpea (*Cicer arietinum* L.)., *Madras Agric. J.* 75: 330–333.

Sheila, V.K., Moss, J.P., Gowda, C.L.L., and Van Rheenen, H.A. 1992. Interspecific hybridization between *Cicer arietinum* L. and wild cicer species. *Int. Chickpea Newsletter* 27: 11–13.

Sindhu, J.S., and Prasad, S. 1987. Causation and association analysis in chickpea. Research and Development Report, CSA Univ. Agric. and Technol. Kanpur 4(2): 136–140.

Sindhu, J.S., Lal, S.B., and Singh, R.P. 1981. Studies on the factors determining crossing success in chickpea (*Cicer arietinum* L.). *Pulse Crops Newsletter* 1: 21–22.

Singh, B. 1962. Inheritance studies on a new spontaneously originated simple leaf mutant in *Cicer arietinum* L. *Sci. Cult.* 28: 138–139.

Singh, B. 1964. Report. Government Research Farm, Kanpur, Agricultural Research, ICAR, 4: 143–144.

Singh, B. 1965. Genetics of double podded mutant in gram (*Cicer arietinum* L.). *Sci. Cult.* 31: 145–146.

Singh, B.D., Jaiswal, H.K., Singh, R.M., and Singh, A.K. 1984. Isolation of early flowering recombinants from the interspecific cross between *Cicer arietinum* and *Cicer reticulatum. Int. Chickpea Newsletter* 11: 14–16.

Singh, G., Singh, G., and Kaur, L. 1988. Chickpea response to various races of *Ascochyta rabiei. Int. Chickpea Newsletter* 19: 10–13.

Singh, H., and Ekbote, R.B. 1936. The inheritance of seed characters in gram (*Cicer arietinum* L.). *Indian J. Agric. Sci.* 6: 1087–1104.

Singh, H., Brar, J.S., and Dahiya, B.S. 1977. Combining ability by full-sib and half-sib analysis in *Cicer arietinum* L. *Mysore J. Agric. Sci.* 11: 275–280.

Singh, H., Kumar, J., Smithson, J.B., and Haware, M.P. 1987. Complementation between genes for resistance to race 1 of *Fusarium*. In *Genetics and Plant Pathogenesis*. R.R. Day and G.J. Jellies, Blackwell Scientific Publications, Oxford, UK, pp. 339–342.

Singh, K.B. 1987. Chickpea breeding. In *The Chickpea*. M.C. Saxena and K.B. Singh, eds. CAB International, Wallingford, Oxon, UK, pp. 127–162.

Singh, K.B. 1990. Winter chickpea: Problems and potential in the Mediterranean region. In *Present Status and Future Prospects of Chickpea Crop Production and Improvement in the Mediterranean Countries*. M.C. Saxena, J.I. Cuboro, and J. Wery, eds. Options Mediterranéennes, Serie A: Seminaires Mediterranéens, Numero 9, CIHEAM, Zaragoza, Spain, pp. 25–34.

Singh, K.B. 1993. Problems and prospects of stress resistance breeding in chickpea. In *Breeding for Stress Tolerance in Cool-Season Food Legumes*. K.B. Singh and M.C. Saxena, eds. John Wiley and Sons, Chichester, UK, pp. 17–35.

Singh, K.B., and Auckland, A.K. 1975. Chickpea breeding at ICRISAT. In Proc. Int. Workshop Grain Legumes, ICRISAT, Hyderabad, pp. 3–17.

Singh, K.B., and Ocampo, B. 1993. Interspecific hybridization in annual *Cicer* species. *J. Genet. Breed.* 47: 199–204.

Singh, K.B., and Reddy, M.V. 1989. Genetics of resistance to blight in four chickpea lines. *Crop Sci.* 29: 657–659.

Singh, K.B., and Reddy, M.V. 1991. Advances in disease resistance breeding in chickpea. *Adv. Agron.* 45: 191–222.

Singh, K.B., and Singh, O. 1994. Prospects of creating higher yield potential in chickpea. Paper presented at the Int. Symp. Pulses Research, 2–6 April 1994, New Delhi, India.

Singh, K.B., and Verma, M.M. 1994. Wild *Cicer* species: Problems and prospects. Paper presented at the Int. Symp. Pulses Research, 2–6 April 1994, New Delhi, India.

Singh, K.B., Malhotra, R.S., and Singh Herbans, 1978. Correlation and path coefficient analysis in chickpea. *Indian J. Agric. Res.* 12: 44–46.

Singh, K.B., Malhotra, R.S., and Respana, B.L. 1982. Inheritance studies for yield and its components in chickpea. *Genet. Agr.* 36: 231–245.

Singh, K.B., Malhotra, R.S., and Witcombe, J.R. 1983. Kabuli chickpea germplasm catalogue. International Centre for Agricultural Research in the Dry Areas, Aleppo, Syria, 284 pp.

Singh, K.B., Reddy, M.V., and Nene, Y.L. 1984. International testing of chickpeas for resistance to ascochyta blight. *Plant Dis.* 68: 782–784.

Singh, K.B., Malhotra, R.S., and Respana, B.L. 1987. Inheritance of yield and its components in chickpea. *Indian J. Hered.* 19: 30–38.

Singh, K.B., Malhotra, R.S., and Saxena, M.C. 1989. Chickpea evaluation for cold tolerance under field conditions. *Crop Sci.* 29: 282–285.

Singh, K.B., Bejiga, G., and Malhotra, R.S. 1990a. Association of some characters with seed yield in chickpea. *Euphytica* 49: 83–88.

Singh, K.B., Kumar J., Haware, M.P., and Lateef, S.S. 1990b. Disease and pest resistance breeding: which way to go in the nineties? In *Chickpea in the Nineties*: Proc. Second Int. Workshop on Chickpea Improvement, 4–8 Dec. 1989. ICRISAT Centre, Patancheru, AP, India, pp. 223–238.

Singh, K.B., Helly, L., and Bejiga, G. 1991. A catalogue of *Kabuli* chickpea germplasm: An evaluation report of winter sown *Kabuli* chickpea, land races, breeding lines and wild *Cicer* species. International Centre for Agricultural Research in the Dry Areas (ICARDA), Aleppo, Syria, 398 pp.

Singh, K.B., Malhotra, R.S., Halila, M.H., Knights, E.J., and Verma, M.M. 1994. Current status and future strategy in breeding chickpea for resistance to biotic and abiotic stresses. *Euphytica* 73: 137–149.

Singh, K.P., and Sindhu, J.S. 1983. Gene action for harvest index and grain yield in chickpea. 1131 (abstract). In Proc. XV Int. Cong. Genetic., New Delhi, 12–21 Dec. 1983. Oxford and IBH Publishing Co., New Delhi, India.

Singh, K.P., Singh, V.P., and Chaudhary, D.B. 1977. Path coefficient analysis in chickpea (*Cicer arietinum* L.). *Z. Pflanzenzuchtg.* 79: 219–223.

Singh, L. 1980. All India Co-ordinated Pulses Research Project. Chickpea Improvement. In Proc. Int. Workshop on Chickpea Improvement, ICRISAT, Hyderabad, India, pp. 191–196.

Singh, L., Tomar, G.C., and Mishra, P.K. 1973. Variability, interrelationships and path coefficients for some quantitative characters in Bengal gram (*Cicer arietinum* L.). *SABRAO Newsletter* 5: 23–28.

Singh, Onkar. 1982. Genetic analysis of irradiated and non-irradiated diallel population in chickpea (*Cicer arietinum* L.). Ph.D. thesis, HAU, Hissar.

Singh, Onkar, and Paroda, R.S. 1983. Genetic analysis of irradiated and non-irradiated diallel populations in chickpea. 1134 (abstract). In Proc. XV Int. Con. Genet., New Delhi, 12–21 Dec. 1983. Oxford and IBH Publishing Co., New Delhi, India.

Singh, O., and Paroda, R.S. 1986. Association analysis of grain yield and its components in chickpea following hybridization and a combination of hybridization and mutagenesis. *Indian J. Agric. Sci.* 56(2): 139–141.

Singh, O., Smithson, T.B., and Gowda, L.L.L. 1983. P1329 a superior chickpea cultivar adapted to early sowing in peninsular India. *Int. Chickpea Newsletter* 9: 8–9.

Singh, Onkar, Rupela, O.P., and van Rheenen, H.A. 1986. Genetics of non-nodulation in chickpea. *Int. Chickpea Newsletter* 15: 5–6.

Singh, R.P. 1990. Status of chickpea in the world. *Int. Chickpea Newsletter* 22: 10–16.

Singh, S.P., and Mehra, R.B. 1980. Genetic analysis of yield and yield components in Bengal gram. *Indian J. Genet.* 40: 482–489.

Singh, S.P., and Ramanujam, S. 1981. Gene action and heterosis in Bengal gram. *Indian J. Genet.* 41: 150–153.

Smithson, J.B., Kumar, J., and Singh, Harijit. 1983. Inheritance of resistance to *Fusarium* wilt in chickpea. *Int. Chickpea Newsletter* 9: 21–22.

Smartt, J. 1976. Comparative evolution of pulse crops. *Euphytica* 25: 139–143.

Sohoo, M.S., Athwal, D.S., and Chandra, S. 1970. Colchicine induced polyploidy in chickpea (*Cicer arietinum* L.). *Theor. Appl. Genet.* 40: 163–168.

Srinivasan, M., Gupta, N., and Chopra, V.L. 1988. *Agrobacterium* mediated transformation of chickpea. *Int. Chickpea Newsletter* 19: 2–3.

Swaminathan, M.S., and Jain, H.K. 1973. Food legumes in Indian agriculture. In *Nutritional Improvement of Food Legumes by Breeding*. Max Milner, ed. PAG, UN, New York, pp. 69–82.

Tewari, A.S., Pandey, R.L., Mishra, P.K., and Kotashane, S.R. 1981. Studies on wilt inheritance in gram (*Cicer arietinum* L.). *J.N.K.V.V. Res. J.* 15: 133–134.

Tiwari, S.K., and Pandey, M.P. 1986. Combining ability analysis for yield and its components in chickpea. *Egyptian J. Genet. Cytol.* 15: 69.

Tomar, G.S., Mishra, Y., and Rao, S.K. 1982. Path analysis and its implications in selection of high yielding chickpea (*Cicer arietinum* L.). *Indian J. Plant Physiol.* 25: 127–132.

Turano Margaret, J., Baird Lisa, M., and Webseter Barbana, D. 1983. Characteristics of the stigma of chickpea. *Crop Sci.* 23: 1033–1036.

Ugale, S.D. 1980. Incorporation of germplasm from kabuli to desi and vice versa in chickpea (*Cicer arietinum* L.). Ph.D. thesis, IARI, New Delhi.

Ugale, S.D., and Bahl. P.N. 1983. Incorporation of germplasm from kabuli to desi and vice versa in chickpea. Abstract. In Proc. XV Int. Cong. Genetics, 12–21 Dec. 1983, New Delhi, India. Oxford and IBH Publishing Co., New Delhi, 646 pp.

Upadhyaya, H.D., Haware, M.P., Kumar, J., and Smithson, J.B. 1983a. Resistance to wilt in chickpea. I. Inheritance of late wilting in response to Race 1. *Euphytica* 32: 747–752.

Upadhyaya, H.D., Haware, M.P., Kumar, J., and Smithson, J.B. 1983b. Resistance to wilt in chickpea. II Further evidence for two genes for resistance to Race 1. *Euphytica* 32: 749–755.

Vaidya, P.K., and Pandya, B.P. 1981. Combining ability in kabuli gram (*Cicer arietinum* L.). *Genet. Agr.* 35: 177–186.

van der Maesen, L.J.G. 1972. *Cicer,* L., A monograph of the genus, with special reference to the chickpea (*Cicer arietinum* L.), its ecology and cultivation. Veenman and Zoner, Wageningen.

Vavilov, N.I. 1926. Studies on the origin of cultivated plants. Leningrad, 129: 238.

Verma, M.M., Sandhu, J.S., Brar, H.S, and Brar, J.S. 1990. Crossability studies in different species of *Cicer* (L.). *Crop Improv.* 17: 179–181.

Verma, M.M., Sandhu, J.S., and Ravi, K. 1991. Wide hybridization in chickpea. In Proc. Golden Jubilee symp. Genetic Research and Education: Current Trends and the Next Fifty Years, abstracts, Vol. III. Indian Society of Genetics and Plant Breeding. Indian Agricultural Research Institute, New Delhi, pp. 686–687.

Vijaykumar, C.H.M. 1987. Variability, genotypic response to *Rhizobium* treatment and gene action for yield attributes and some quality characters in chickpea (*Cicer arietinum* L.). M.Sc. (Agri.) thesis, UAS, Dharwad.

Weeden, N.F., Timmerman, G.M., and Lu, J. 1994. Identifying and mapping genes of economic significance. *Euphytica* 73: 191–198.

Weigand, S., and Tahhan, O. 1990. Chickpea insect pests in the Mediterranean zones and new approaches to their management. In *Chickpea in the Nineties*. Proc. Second Int. Workshop on Chickpea Improv., 4–8 Dec. 1989, ICRISAT Centre, India, pp. 169–175.

Witcombe, J.R., and Esskine, W. (eds.) 1984. Genetic Resources and Their Exploitation—Chickpeas, Faba Beans and Lentils. Martinus Nijhoff/Dr. W. Junk Publishers, the Hague, 256 pp.

Yadav, L.N., Mahadik, C.N., and Dixit, S.S. 1978. Inheritance of double podded character and petal colour in gram (*Cicer arietinum* L.). *Sci. Cult.* 4: 547.

Yadav, S.S. 1986. Morphogenetic architecture of plant type in chickpea (*Cicer arietinum* L.). Ph.D. thesis, IARI, New Delhi, India, 193 pp.

Yadav, S.S., and Bahl, P.N. 1990. Morphophysiological architecture of structural and yield components in bushy and tall chickpea (*Cicer arietinum* L.). In. Proc. Int. Cong. Plant Physiology, Vol. 2, S.K. Sinha, P.V. Sane, S.C. Bhargava and P.K. Agarwal eds. 15–20 Feb. 1988, New Delhi, India, Society for Plant Physiology and Biochemistry, pp. 299–306

Yadavendra, J.P., and Kumar, S. 1987. Combining ability in chickpea. *Indian J. Genet.* 47: 67–70.

Zafar, A.M., and Abdullah, M. 1971. Diallel analysis for some economic characters in gram. *J. Agric. Res., Pakistan* 9: 14–24.

Zafar, A.M., and Khan, M.A. 1968. Genetic studies on gram hybrids and their parents. *J. Agric. Res., Pakistan* 6: 33–41.

2. PIGEONPEA

K.B. Saxena
Senior Scientist (Breeding)

Laxman Singh
Principal Scientist (Breeding)
Genetic Enhancement Division, International Crops Research Institute for the Semi-Arid Tropics (ICRISAT), Asia Centre, Patancheru, P.O., A.P. 502324, India

Introduction

Pigeonpea grows well in subtropical and tropical environments extending between 30°S and 30°N latitudes. In India, it is mainly cultivated as rainfed dryland crop in about 3.7 m ha. India, Kenya, Uganda, Malawi, Tanzania, Myanmar, Puerto Rico, the Dominican Republic, and Venezuela are the major pigeonpea-producing countries. Within India, which accounts for more than 90 per cent of the world pigeonpea production and area, Maharashtra is the largest pigeonpea-producing state, with about 1016,100 ha annually under this crop. Uttar Pradesh, where long-duration cultivars are generally grown on 521,600 ha, has the highest annual production (Table 1). In the states of Maharashtra, Andhra Pradesh, and Karnataka, which occupy about 50 per cent of the total area, pigeonpea is grown under inadequate rainfall conditions (around 660 mm) (Reddy and Virmani, 1981). In Andhra Pradesh, medium-duration cultivars are grown under intercropping on lighter soils and in rainfed areas resulting in an average yield of about 368 kg/ha. In Uttar Pradesh, pigeonpea is cultivated in the Indo-Gangetic alluvium and deep Vertisols with average yield of over 1072 kg/ha. From 1950 to 1978, area under pigeonpea in India ranged from 2.2 to 2.6 m ha, after which it has shown increasing trends. The present pigeonpea area reflects a 70 per cent increase over 1950 acreage. This significant increase in pigeonpea area may partly be attributed to the introduction of short-duration genotypes that extended pigeonpea cultivation in non-traditional areas such as Punjab, Rajasthan, Haryana, western Uttar Pradesh, and Tamil Nadu. In Punjab alone, the area under pigeonpea has increased from 1900 ha in 1972 to 12,900 ha in 1992 with an average yield of 992 kg/ha. Similarly, in Haryana the area under pigeonpea has increased from 2200 ha in 1976 to 51,800 ha in 1992. Due to high and stable returns, the pigeonpea crop has almost completely replaced the high-input cotton crop in Bharuch, Vadodara, Sabarkantha, and Surat districts of Gujarat state.

Table 1
Crop area, production and productivity in pigeonpea growing states of India, 1991-92

State	Area ('000 ha)	Production ('000 t.)	Productivity (kg/ha)
Maharashtra	1016	363	357
Karnataka	531	176	331
Madhya Pradesh	452	346	765
Uttar Pradesh	522	560	1072
Gujarat	397	240	604
Andhra Pradesh	338	124	368
Tamil Nadu	131	95	726
Orissa	168	111	662
Bihar	66	91	1379
Haryana	52	53	1023
Rajasthan	22	6	248
Punjab	13	13	992
Others	21	12	571
All India	3727	2190	588

(*Source:* Agricultural Situation in India, July, 1993)

Wide variation exists in flowering and maturity periods of pigeonpea. Based on days to 50 per cent flowering, Green *et al.* (1979) classified the pigeonpea types into 10 maturity groups (Table 2). Since specific adaptation with respect to maturity is very important in this crop, the maturity classification helps in organizing breeding activities. This classification may not hold true universally as the phenology of pigeonpea is highly influenced by photoperiod, temperature, and their interaction. Therefore, local breeders need to study the flowering behaviour of the lines before selecting parents for crosses and developing breeding strategies.

Table 2
Maturity classification of pigeonpea at Patancheru (17 N)

Maturity group	Days to 50% flower	Reference variety
0	< 60	Pant A 3
I	61–70	Prabhat
II	71–80	UPAS 120
III	81–90	T 21
IV	91–100	ICP 6
V	101–110	BDN 1
VI	111–130	C 11
VII	131–140	HY 3C
VIII	141–160	ICP 7065
IX	> 160	NP (WR) 15

Source: Green *et al.*, 1979.

Origin and Distribution

The place of origin of pigeonpea is disputed and has been the subject of speculation. The presence of pigeonpea seeds in the Egyptian tombs of the XII Dynasty indicated that it was cultivated there about 2400 BC. Van Rheede (1686), Linnaeus (1737), Burman (1737), and Jacquin (1772) mentioned pigeonpea as a native of the East Indies. Rumphius (1747) suggested that pigeonpea was native to Malaya. The Swiss botanist de Candolle (1813), while establishing the genus, mentioned 'India Orientalis' as its home and tropical America as another area of cultivation. Later in 1884, he suggested tropical African origin (Zanzibar to Guinea). Some other workers (Purseglove, 1968; Zeven and Zhukovsky, 1975) also opined that pigeonpea is a native of Africa, from where it spread to India and other parts of the globe.

Vavilov (1939) considered India a primary centre of origin of the cultivated pigeonpea. Oza (1972) reviewed the subject and concluded that India was the original home of pigeonpea before it spread to the new world. According to De (1974) the pigeonpea originated in peninsular India and was carried to Africa before 2000 BC. Reviewing the literature on this aspect, Royes (1976) considered that the dispute about the origin of pigeonpea has settled in favour of India. Brucher (1977), however, advocated the African origin of pigeonpea because of the presence of a single endemic West African species, *Cajanus kerstingii* Harms. In a recent review, van der Maesen (1980) concluded that 'floristic, linguistic, and cytological evidence points to an Indian origin of the pigeonpea, from where it was most probably distributed to Africa at least two millennia B.C.' He also feels that Africa is a definite secondary centre of origin, since some pigeonpea characters found in African germplasm are not present in the materials from the Indian subcontinent.

Reddy (1973) and De (1974) postulated that the genus *Cajanus* probably originated from an advanced *Atylosia* species through selection of single-gene mutation. Hains (1920) described the morphology of a wild relative of pigeonpea (*A. cajanifolia*) that resembles cultivated pigeonpea in all respects except for the presence of large strophiole (a vegetative outgrowth on the hilum end of the seed) on its seeds. The prominent strophiole has also been found on several accessions of pigeonpea (van der Maesen, 1980). Recently, *A. cajanifolia* has been collected by the Germplasm Botanist of the International Crops Research Institute for the Semi-Arid Tropics (ICRISAT) from the forests of central India, where the locals call it *Ban-Arhar*, meaning 'wild pigeonpea'. *Atylosia cajanifolia* can be pollinated from pigeonpea pollen transferred onto its stigma by insects to produce natural hybrids. Morphological and cytological similarities of *A. cajanifolia* and *C. cajan* led van der Maesen (1980) to postulate that *A. cajanifolia* is presumably the link between the genera *Atylosia* and *Cajanus.*

Taxonomy

Tuvarai and *tuvari*, the oldest names of pigeonpea in the Dravidian language, are quoted in the text *Gathasaptasati* published between 300 and 400 AD (De, 1974). According to van der Maesen (1976), the first scientific nomenclature without very clear illustrations of the crop was by Bauhin and Cherla during 1650–1651. They called pigeonpea *Arbor trifolia indica*, while van Rheede (1686) named it *thora paerou*, meaning 'common *dhal*' in Malayalam language.

According to Sloane (1725), 'pigeonpea', the present-day popular English name of this crop, was first used by Plukenet (1692) in Barbados, where it was grown in barren soils mainly for feeding pigeons. The name *Cajanus* is derived from the Malaysian name Katjang. Linnaeus (1753) gave the first binomial nomenclature to pigeonpea as *Cytisus cajan.* Since then, various taxonomists have modified its scientific name from time to time. Thothathri and Jain (1981) and van der Maesen (1986) have written excellent reviews on this aspct. At present, the following taxonomical classification is accepted universally.

Tribe Phaseoleae
Subtribe Cajaninae
Familay Leguminosae
Subfamily Pipilionaceae
Genus *Cajanus*
Species *cajan*

Bionomials such as *Cytisus cajan* L., *Cytisus pseudo-cajan* Jacq., *Cajanus cajan* Huth., *Cajanus cajan* Druce., *Cajanus inodorum* Medic, *Cajanus pseudo-cajan* Schi. & Gull., *Cajanus bicolor* DC., *Cajanus flavus* DC, *Cajanus indicus* Spreng, and *Cajanus luteus* Bello, found in the old literature, can be considered synonyms of the presently accepted binomial *Cajanus cajan* (L.) Millsp.

There are 13 closely related genera within the subtribe Cajaninae and the genus *Cajanus* has always been considered to be closely related to the wild species of the genus *Atylosia.* In fact, morphologically these two genera appear to be quite distinct; *Cajanus* is cultivated, while *Atylosia* species are found in uncultivated, wild, and weedy forms. According to van der Maesen (1986) a species found in Senegal, Togo, and Ghana by Hutchinson and Dalziel (1958) and described as *Cajanus kerstingii* is a misnomer and should have been classified to the genus *Atylosia.* Baker (1876) separated *Cajanus* and *Atylosia* on the basis of the prominent strophiole present on *Atylosia* seeds. The genus *Atylosia* W. and A. forms a secondary gene pool consisting of 34 species.

Van der Maesen (1981) reviewed the relationship between *Cajanus* and *Atylosia* and judged *Atylosia* to be congeneric with *Cajanus*; therefore, they can be merged. The taxonomy of the genera *Cajanus* and *Atylosia* has been revised recently by van der Maesen (1986), and on the basis of morphological, cytological, and chemo-taxonomical data he merged genus *Atylosia* into *Cajanus*. According to him, instead of 13 there are now 11 genera in subtribe *Cajaninae* and the genus *Cajanus* embraces 32 species endemic to Indian, African, and Australian continents.

Cytology

The Genus Cajanus

Roy (1933) was the first to report chromosome number ($n = 11$) in gametophytic tissues of pigeonpea. Krishnaswami and Ayyangar (1935) reported that the basic chromosome number of the Phaseoleae tribe is 11. Naithani (1941) studied mitosis in pigeonpea and reported the somatic chromosome number as $2n = 22$. He also reported the presence of prochromosome at the resting phase. He described pigeonpea somatic chromosomes as very small ranging from 1.35 to 2.7 μm in length with no morphological difference between varieties. Normal pairing at metaphase I was observed by Kumar *et al.* (1945), Bhattacharjee (1956), and Reddy and De (1983). Kumar *et al.* (1958) observed one nucleolus with chromosomes attached in meiotic cells. Deodikar and Thakar (1956) described the length of each chromosome and position of primary and secondary constrictions and reported that the number of satellited chromosomes in pigeonpea varied from two to zero, depending on the variety. Sinha and Kumar (1979) found a considerable variation with respect to arm ratio, total chromatin length, and ratio of shortest to longest chromosome. They studied 13 pigeonpea varieties and classified them into 7 groups on the basis of karyotype similarity. Reddy (1981a, b, c) prepared karyotypes of pigeonpea using pachytene analysis and identified chromosomes on the basis of relative length, arm length, position of centromere, nucleolar association, and amount and distribution of hetero- and euchromatin. Akinola *et al.* (1972) reported $2n = 22$ in 95 diverse pigeonpea accessions collected from various parts of the world.

The Genus Atylosia

Of the 30 *Atylosia* species available, about half have been studied cytologically. I.S. Dundas (personal communication) summarized these studies and concluded that all the *Atylosia* species studied had the same chromosome number as pigeonpea. Kumar *et al.* (1958) reported the presence of one nucleolus in *A. lineata*. Somatic karyotypes of *A. lineata* (Deodikar and Thakar, 1956), *A. sericea* (Deodikar and Thakar, 1956), and *A. scarabaeoides* (Roy and De, 1965) and pachytene karyotype of *A. lineata* (Reddy, 1981a), *A. sericea* (Reddy, 1981b), and *A. scarbaeoides* (Reddy, 1981c) have been developed and the chromosomes of these species can be distinguished at pachytene on the basis of arm length, ratio, position of secondary constrictions, and amount and distribution of hetero- and euchromatin.

Polyploidy

Pathak and Yadav (1951) reported the spontaneous occurrence of less vigorous hexaploid (2n = 66) and tetraploid (2n = 44) plants under field conditions. Bhattacharjee (1956) and Kumar *et al.* (1945) induced polyploidy using colchicine and compared their cytology and morphology with those of diploids. In general, the tetraploids were poor in growth and vigour. Saxena *et al.* (1982) made detailed study of the effects of spontaneous chromosome doubling on some vegetative and reproductive characteristics of pigeonpea. They found that the tetraploids were shorter, with fewer nodes, thicker leaves and stems, and lower shoot dry weight per plant, than those of the diploids. The size of flowers, and of most floral parts, was substantially greater for the tetraploids than for the diploid progeny. The tetraploids had fewer but larger stomata, larger pollen with relatively higher pollen sterility, and poor pod set. They also observed that the position of the stigma in relation to the anthers was highly variable among the tetraploid flowers and in most cases the stigma was located above the ring of anthers.

A few cases of aneuploidy have also been reported in pigeonpea. D'Cruz and Jadhav (1972) reported the occurrence of a plant with 2n = 23. The chromosome numbers ranging from 2n = 8 to 2n = 28 were observed by Bajaj *et al.* (1980) in mixaploid anther-derived callus. It is believed that no identified stocks of aneuploids presently exist (Dundas, 1990).

Morphology

Botanically, pigeonpea is a short-lived perennial shrub with a life span of about 10–12 years (Remanandan, 1981). However, it is generally cultivated as an annual crop. A large variation exists for plant type and growth habit. Traditionally, indeterminate tall types and lately short determinate types are cultivated. Recently, Saxena *et al.* (1992a) reported the development of dwarf genotypes with annual growth habit. The angles of primary branches determine the spread of plant, which in nature varies from compact with acute branch angle to spreading types having obtuse branches. In dense sowings and in intercropping systems the branching is reduced considerably due to competition (Sheldrake, 1984). Plant height in pigeonpea varies from 30 to 400 cm and the aged perennial types may even grow taller. The expression of plant biomass, height, and branches of a genotype is greatly influenced by the environmental conditions. Plantings done during shorter days result in small plants with considerable reduction in height, branching, and biomass production.

The first couple of leaves of a young seedling are simple and opposite. The subsequent leaves are trifoliate, having laceolate to elliptic leaflets that are acute at both ends. The pinnate leaves are hairy on both surfaces and vary in colour with various shades of green with minute resin glands beneath. The leaves are arranged in a 2/5 phyllotaxis on the main axis and on branches. Petioles are ribbed and end in a leaflet. At the base of the petiole and leaflets pulvini are present. The activity of pulvini helps in adjusting the leaf/leaflets angles throughout the day. When the sunlight is intense and especially when the plants are under water stress, the leaflets in the exposed canopy tend to take up a position parallel to the incident light. At night the leaflets are folded vertically upward (Sheldrake, 1984). As the leaves approach maturity, an abscission zone develops at the base of the petiole and leaflets. Saxena and Sharma (1981) reported large variation for leaf weight, petiole weight, petiole length, leaf area, and specific leaf weight in 27 pigeonpea genotypes of diverse maturity groups.

The pigeonpea plant has a deep taproot system often growing over 2 m deep with numerous rootlets. The most extensive development of roots takes place in the upper 60 cm soil layer (Sheldrake and Narayanan, 1979a). A number of nodules are formed on the roots by rhizobia belonging to the cowpea group. Nodule formation in pigeonpea is initiated through the development of infection thread in the root hairs. Pigeonpea nodules have a terminal meristem and a green pigmented senescent zone develops from the base of the nodule (Thompson *et al.,* 1981). The bacteroid zone of pigeonpea nodules is generally pink due to leghemoglobin. The number of nodules on field-grown plants fluctuates throughout the growing season but shows a clear tendency to decline towards the end of the reproductive phase (Sheldrake and Narayanan, 1979a). There is variability among germplasm lines in nodulation and associated characters (Thompson *et al.*, 1981). Nodule formation and development are also affected by soil type, season, and duration of the cultivar.

Generally, the inflorescence is an axillary raceme varying in length from 4 to 12 cm. In the indeterminate cultivars flowering proceeds acropetally, while in the determinate types it is basipetal. On an average, four to six floral buds are found in bunches. In a few cultivars, flowering starts at nodes behind the apex and proceeds acropetally as well as basipetally (Sheldrake, 1984). In some genotypes a few flowers can also be seen on the main stem. Flowering in certain genotypes continues even after pod formation and maturity. The rachis is small and bracts are deciduous, frequently with glandular hairs. The pedicels are green and 1–2 cm long with dense pubescence and glandular hairs. Calyx is companulate. Corolla typically is papilionaceous. Standard petal is bi-auriculate. Wing petals are obovate and the keel is boat-shaped, dorsally split. The stamens are diadelphous, having free vexillary stamen. The longer and shorter filaments are alternately arranged. Anthers are ellipsoid and dorsifixed. Stigma is capitate, while the ovary is sessile. The majority of the flowers open between 8 a.m. and 3 p.m. At the base of pedicel, there is an abscission zone and most of the flowers drop off without setting pods (Howard *et al.,* 1919; Pandey and Singh, 1981) and only 10–20 per cent of the flowers develop into pods. Depending on the genotypes the flowering may be synchronous or unsynchronous. The phenotypic expression of the inflorescene can be modified markedly by photoperiod, temperature, sowing time, and plant spacing.

For about the first 21 days after anthesis the pod wall develops more

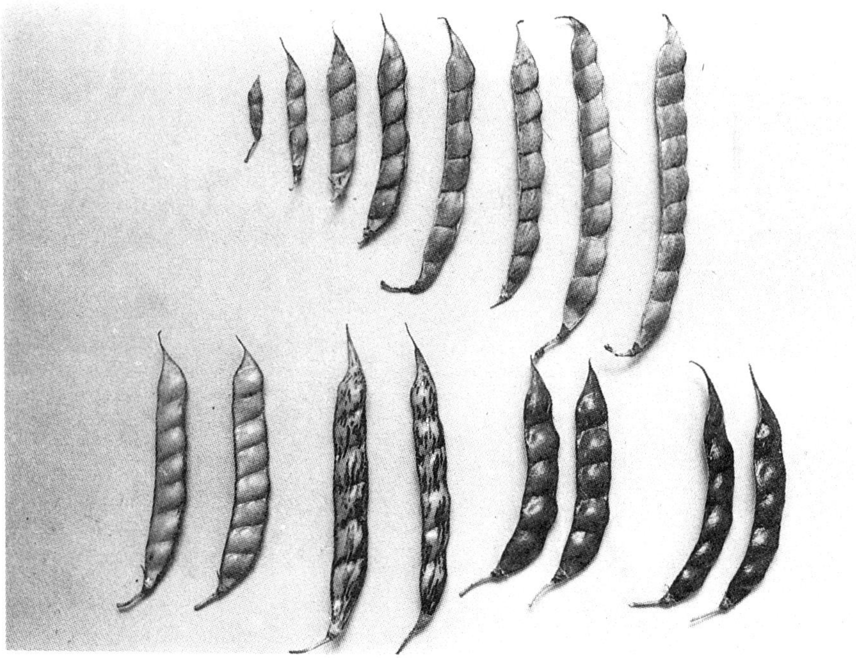

Fig. 1. Variation in pigeonpea pod colour and size (Photo: ICRISAT)

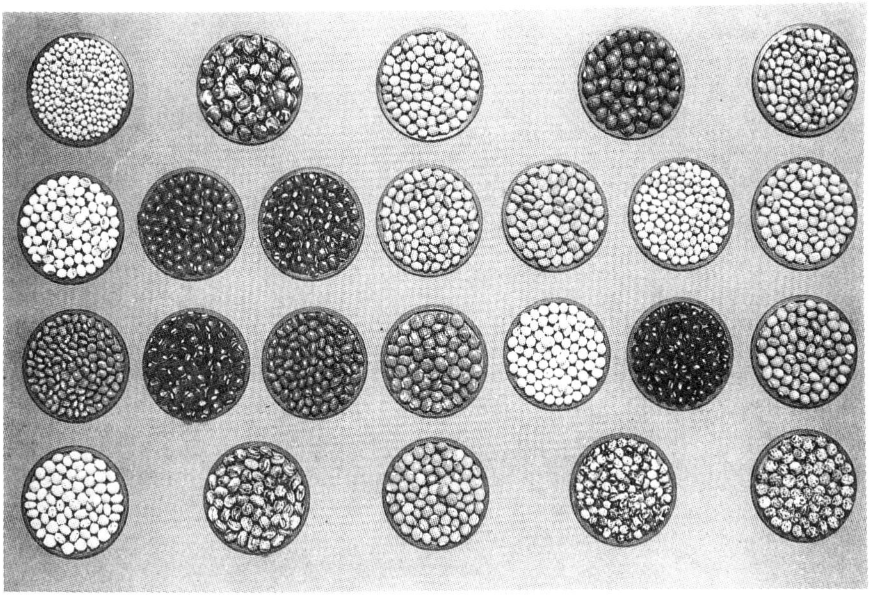

Fig. 2. Pigeonpea variation in seed size and colour (Photo: ICRISAT)

rapidly than the young seeds, but thereafter, it shows little further growth (ICRISAT, 1975). Fully developed indehiscent pods are flattened, slightly curved, or straight with diagonal constrictions between seeds and terminating in a slender beak. Pod colour varies from yellowish-green to dark green with purple blotches of varying intensities. Pods contain two to nine seeds (Fig. 1) and in large pods some ovule abortion may also be observed. The pod wall contains hairs and is well supplied with secretory ducts containing tannin-like substances (Sheldrake, 1984).

Pigeonpea seeds are non-endospermic and have two cotyledons. The seeds of pigeonpea vary greatly (Fig. 2) with respect to their size (2–24 g/100 seeds), shape (round, oval, or flatened), and colour (white, brown, red, purple, or black). The seeds have a smooth surface with a small white hilum. The cotyledons are light yellow. In general, there is no seed dormancy and the germination is hypogeal.

Reproductive Behaviour

Natural Cross-pollination

The flowers of pigeonpea are cleistogamous and generally favour self-pollination. However, unlike other legumes, a considerable degree of natural outcrossing has also heen reported in this crop. Occurrence of natural outcrossing poses problems in developing pure lines and in maintaining the purity of released cultivars and germplasm accessions. On the contrary, outcrossing in pigeonpea can be profitably used for developing high-yielding hybrids (Saxena *et al.*, 1986b) and for the improvement of breeding populations (Khan, 1973: Ariyanayagam *et al.*, 1991). As a result of frequent out-crossing the existing standard adapted pigeonpea cultivars have become heterogeneous for several important agronomic characters (Gupta *et al.*, 1981).

Natural outcrossing in pigeonpea takes place as a result of frequent insect visitation from one flower to another within and across the fields. Onim (1981) listed 24 species of insects capable of affecting cross-fertilization in pigeonpea. Of these, *Magachile* spp. and *Apis mellifera* are the main pollinating vectors (Williams, 1977; Onim, 1981).

Howard *et al.* (1919) were the first to report outcrossing in pigeonpea and since then several reports have appeared in the literature. In a recent review Saxena *et al.* (1990a) described a large variation (0–70%) in natural outcrossing in pigeonpea (Table 3). At a particular site a combination of factors determine the extent of outcrossing in pigeonpea. These include the number of insect pollinators present in relation to the number of flowers, the flowering habit of the varieties, the location of the field in relation to the insect habitat or barrier crops, and environmental factors such as temperature, humidity, and wind velocity and direction (Bhatia *et al.*, 1981).

Table 3
Per cent natural outcrossing in pigeonpea recorded in various countries

Country	% outcrossing	Reference
India	0–70	Howard *et al.* (1919)
		Deshmukh and Rekhi (1963)
		Bhatia *et al.* 1981)
		Saxena *et al.* *(1990a)*
Kenya	12–47	Onim (1981)
Hawaii	14–16	Wilsie and Takahashi (1934)
Puerto Rico	5–6	Abrams (1967)
Trinidad	26	Ariyanayagam (1976)
Australia	2–40	Byth *et al.* (1982)

Byth *et al.* (1982) reported < 2 per cent outcrossing in cv Royes and attributed it to the floral modification (wrapped flower) of the cultivar. This trait, however, is ineffective in different environments (Saxena *et al.*, 1987c). Another floral mutant (Fig. 3), isolated from an intergeneric cross between *Cajanus cajan* and *Atylosia lineata* (Saxena *et al.*, 1992b), has been

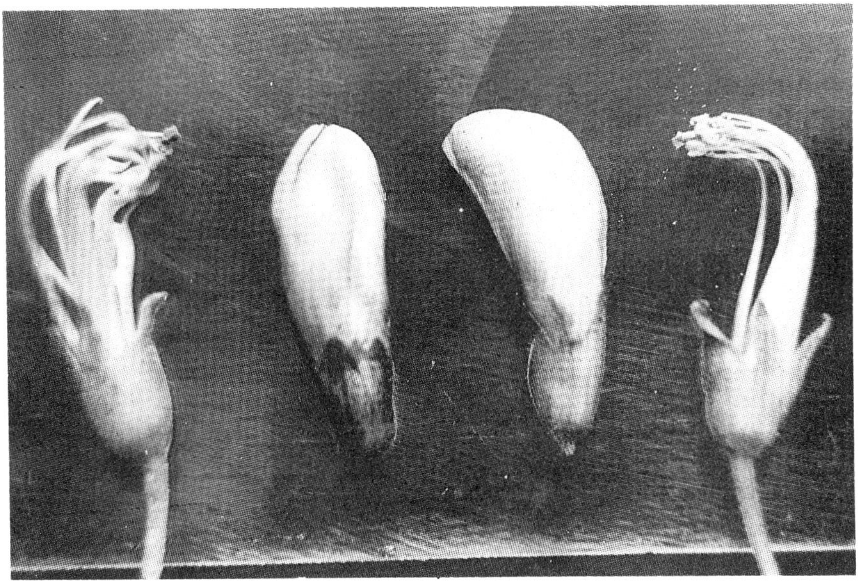

Fig. 3. Partly cleistogamous (A, B) and normal (C, D) flowers of pigeonpea (Photo: ICRISAT)

reported to restrict outcrossing to 1 or 2 per cent (Saxena *et al.*, 1990c; Saxena and Jeysekera, 1992c).

Photoperiod Responses

Pigeonpea genotypes are quantitative short-day plants (Sengupta, 1955) and vary widely in the length of their growth cycle ranging from 100 to 300 days. Because of its photosensitivity, planting date has a marked effect on the phenology of plant. Singh *et al.* (1971) in India, Spence and Williams (1972) in Trinidad, Abrams and Julia (1973) in Puerto Rico, and Akinola and Whiteman (1974) in Australia studied the effect of planting dates and observed that plantings in the months with shorter day lengths flowered in less time and had reduced plant height and branching than planting under longer days. Photoperiod-sensitive cultivars flower more rapidly when sown after the longest day (Whiteman *et al.*, 1985). Like soybean (Byth, 1968) early-maturing pigeonpea cultivars are less sensitive to changes in day length than the medium- and late-maturing genotypes (Green *et al.*, 1979). Wallis *et al.* (1981a) studied flowering responses of 37 early-maturing lines of pigeonpea under normal and extended photoperiods (16 hours) in Australia (27°S). They concluded that selection for earliness of flowering in field conditions at 27° to 29° latitudes does not necessarily identify material insensitive to day length. Turnbull *et al.* (1981) studied the effects of photoperiod and temperature on floral initiation and floral bud development in early-maturing pigeonpea lines under controlled environmental conditions. They found that the floral initiation occurred earliest under low temperature (24/16°C) in a 16-hour photoperiod. Reduction in day length to 8 hours and increase in temperature both resulted in delays in floral initiation. They also reported that the rate of development of the floral primordia increased with increase in both day length and temperature. From the controlled environment studies McPherson (1985) concluded that the minimum duration from sowing to flower bud initiation occurred between 20° and 24°C while the period from sowing to flowering was at a minimum between 20° and 28°C.

Saxena (1981) reported that under extended photoperiod (16 hours) three major genes, *Ps 3, Ps 2,* and *Ps 1,* control flowering in the sensitive parent MS4A, and hierarchically, *Ps 3* over-rides the expression of *Ps 2* and *Ps 2* over-rides *Ps 1.* Saxena *et al.* (1984) reported a pigeonpea mutant that failed to reach reproductive stage and remained vegetative for a full year.

Male Sterility

Reddy *et al.* (1977) identified five aberrant floral types in 3659 pigeonpea germplasm and 3557 sublines grown at ICRISAT Centre in 1974. The most interesting of these male-sterile types was characterized by translucent anthers caused by non-separation of tetrads associated with a persistent tapetum and intercellular walls of the two adjacent microsporangia. A single recessive allele (ms1) was responsible for this from of male sterility (Reddy *et al.,* 1978). Wallis *et al.* (1981b) reported another form of male sterility that had shrivelled, arrowhead-shaped, non-dehiscent, brown anthers. In this male-sterile source the sterility was found to be due to degeneration of the tapetum by vacuolation that occurred during the first division of meiosis (Dundas *et al.,* 1981). Saxena *et al.* (1983) reported that the brown-anthered male sterility was controlled by another recessive gene (ms2). A third male-sterile source in photoperiod-insensitive genetic background was reported by Dundas *et al.* (1982). They also reported that in this male-sterile genotype, degeneration of the microsporogenesis takes place at early prophase stage. Saxena *et al.* (1981a) reported the presence of a partial male-sterile system in pigeonpea that was caused by partial collapse of fully developed tetrads. In this type the amount of pollen grain produced within each flower was very small.

Pod Setting

In pigeonpea about 90 per cent of the flowers drop without setting pods (Pandey and Singh, 1981). Rawson and Constable (1981) reported that within a pigeonpea plant photosynthates are produced in excess of that required for pod setting. Analyses of growth have shown that all the assimilates accumulated during the reproductive phase are not partitioned into pods, rather a high proportion of it goes into vegetative parts (Sheldrake and Narayanan, 1979a). Sheldrake and Narayanan (1979b) also found no differences in the pods formed earlier or later with respect to either seeds/pod or seed weight. These studies have shown that in pigeonpea neither the source nor the sink is limiting. Sheldrake (1979) developed a working hydrodynamic model that illustrates some of the main factors that might be involved in the control of pod set in pigeonpea. Sheldrake (1984) opined that pest damage, inadequate pollination, competition from earlier-formed pods, and other sinks within the plant and intrinsic physiological mechanisms connected with the perennial nature of the plants are the factors responsible for poor pod set in pigeonpea. Saxena *et al.* (1988) reported a mutant with an open carpel in pigeonpea. A considerable level of floral abscission induced by low temperature has also been reported in pigeonpea (Saxena *et al.*, 1985).

Production Systems

Pigeonpea is remarkable in that its duration and growth habit are a continuum of short-duration, near annual types to long-duration perennial types. This nature of the species allows it to adapt to a wide range of cropping systems and soil and climatic variations in semi-arid tropics. The traditional production systems involve inter- or mixed-cropping of medium to long-duration pigeonpea types with a wide range of crops. Development of short-duration genotypes created a potential for commercial mechanized production of pigeonpea as a monocrop in multiple cropping systems. More prevalent and potential production systems are discussed hereafter.

Intercropping

Traditionally, in northern India, eastern, central and southern Africa, and the Caribbean, long-duration pigeonpea cultivars are grown as an intercrop or mixed crop with sorghum, pearl millet, or maize. Pigeonpea in this system occupies land for 8–9 months, whereas the companion crops are harvested within the first 3–4 months. This practice aims to exploit the land, moisture, and sunlight resources optimally in the areas where a second crop is not possible. There is a considerable variation not only in companion crop but also in the proportions of the companion crops and pigeonpea. Mehta (1970) and Singh (1980) have described some of these situations. Singh and Shrivastava (1976) made an extensive survey of the prevailing pigeonpea cultivation systems in Madhya Pradesh (India). They also defined at least nine distinct plant types from over 900 collections, differing in height, spread, branching, adaptation, and other characteristics. Intercropped long-duration pigeonpeas have value for vegetable purposes (in Africa, the Caribbean, and some parts of India) and for dry grain, besides being an important source of fuel wood and construction material. It is essential to consider its value in this regard when economic comparisons are made with other cropping systems.

Intercropping systems with medium-duration (6–7 months) pigeonpea are well adapted to dryland agriculture where the growing period is constrained by soil moisture. These are proven systems of maximizing stability of agricultural production in central and peninsular India (Rao and Willey, 1980). The environment is characterized by high variability in the incidence of rainfall. For subsistence farmers at least, the stability of production over seasons is a more important consideration than simply pursuing high yields. The characteristics, advantages, and disadvantages of this system have been discussed by Willey (1985). Traditionally, a cereal such as sorghum or millet has been the companion crop of medium-duration pigeonpea in this system. The farmers generally regard the cereal as main crop and adjust plant populations and cultural operations accordingly.

Medium-duration pigeonpea is also intercropped with other crops such as upland rice, cotton, castor, groundnut, cowpea, soybean, urdbean, and mungbean. The range of production systems involving pigeonpea as an intercrop has been summarized by Singh (1980). There seem to be a number of intercropping possibilities under this system in which the differing phenologies and water use characteristics of pigeonpea and the shorter-duration companion crops can effectively maximize and stabilize production in harsh environments. Many of the intercropping systems with medium-duration pigeonpea that have been evolved in India would be suitable for some of the semi-arid, monsoonal environments of Africa. Mechanization of such a system should be possible by adjusting spacings and proportions. As they are generally grown in arid environments, where sources of fuel and fodder are usually scarce, the medium-duration pigeonpeas are important sources of these products. Agronomic studies have shown that the equivalent of up to 40 kg/ha fertilizer N can be provided by pigeonpea for a following cereal crop (Whiteman *et al.*, 1985).

Short-duration pigeonpea (4–5 months) can also be intercropped with short-statured grain legumes such as mungbean, urdbean, cowpea, soybean, and groundnut. Saraf and Amarchand (1980) reported no reduction in grain yield of short-duration pigeonpea cultivar T.21 from 2.7 t/ha when intercropped with mungbean or urdbean, with the latter crops yielding 0.8–1.0 t/ha, which was *at par* with their sole crop yields.

Sole Cropping

The advent of short-duration genotypes has permitted the use of pigeonpea as a sole crop in double or multiple cropping systems, as distinct from its traditional use as a long season crop. One such system developed in northern India over the last decade is double cropping of short-duration pigeonpea varieties (UPAS 120, ICPL 151, Manak, AL 15, etc.) with wheat; the pigeonpea is grown during the monsoon period and is harvested by November, leaving enough time for the field preparation and sowing of winter wheat. In this system, pigeonpea grain yields exceeding 2 t/ha and wheat grain yields exceeding 5 t/ha may be achieved (Panwar and Yadav, 1981).

Monocrops of short-statured short-duration pigeonpea are amenable to mechanization. This particularly applies to Africa and the Caribbean where pigeonpea cultivation now depends almost entirely on manual operations. In Australia, mechanized cultivation of pigeonpea has been achieved for commercial production (Byth *et al.*, 1981). There is a scope of extending this system to drier environments by using pigeonpea genotypes of even shorter duration (< 100 days) that can escape drought stress. In such areas in Africa and Asia, low-yielding, short season catch crops are generally grown.

Multiple Harvest System

In environments with warm winters (minimum temperature > 15°C) it is possible to exploit the perennial characteristics of short-duration pigeonpea by obtaining ratoon harvests. Chauhan *et al.* (1984) reported up to 5.2 t/ha yield in such a production system in peninsular Indian environments. In this case three harvests were taken, the first in September, the second in November, and the third in March. The amount of seed harvested generally declined with subsequent harvests. This system gives much higher yields than the traditional medium-duration genotypes in this environment over a similar time period (Chauhan *et al.*, 1987). ICPL 87, a short-duration variety having good ratoonability, was released under the name 'Pragati' for peninsular India in 1986.

In Sri Lanka, a production system based on the ratooning of short-duration varieties is becoming very popular in the bimodal rainfall areas. In this system the crop is planted with the onset of rains in the month of October. Harvesting is done in March by ratooning the crop about 30 cm above the ground level. A good canopy develops during the second (short) rainy season. Researchers and farmers have recorded about 1000–1500 kg/ha yield each in the main and ratoon crop (Saxena *et al.*, 1992c).

Post-rainy Season Crop

In areas with higher total annual rainfall (e.g., > 1000 mm), wider seasonal distribution of rainfall (e.g., bimodal distribution), moderate to warm winters, and deep moisture-retentive soils it is possible to plant pigeonpea well after the longest day and still obtain good crops. Roy Sharma *et al.* (1981) in Bihar, India, obtained 3.4 t./ha of seed yield from a sole crop of pigeonpea planted after the rainy season in comparison with 1.6 t./ha obtained with pigeonpea planted as a traditional intercrop with maize in May or June. Similarly, Sengupta (1982) reported three-year mean yields of October-planted sole crops of pigeonpea in the order of 2.5 t/ha; these crops matured in about 160 days. There is a large potential for growing pigeonpea in rice fallows. However, suitable genotypes need to be identified which would be more productive than other short season grain legumes (lathyrus, blackgram, greengram, lentils, etc.) usually grown in this situation.

With the availability of relatively less photoperiod-thermosensitive short-duration genotypes, the area of adaptation of pigeonpea can be further extended to spring plantings. In February plantings at Bangalore (India), the short-duration genotype ICPL 8332 yielded 2.8 t/ha. This crop can be harvested in early June to permit normal rice planting. During the crop growth, mean minimum and maximum temperatures ranged respectively from 12° to 22°C and 29° to 36°C (Yadahalli and Balakrishna, 1987).

Perennial Crop

Since pigeonpea remains basically perennial in growth habit it can be used essentially as a tree crop for agroforestry systems and alley cropping. It is widely grown as perennial backyard or garden crop in Africa and the Caribbean and also in tribal areas of India for dry grain, green peas, fodder, construction material, and fuel wood.

Genetic Studies

In pigeonpea, relatively few genetic studies of qualitative and quantitative traits have been carried out. A summary of the qualitative traits and their inheritance has recently been compiled by Saxena and Sharma (1990).

For genetic improvement in the yielding ability of the crop basic information on the genetic parameters of yield and related traits such as maturity, pods/plant, and seed size is required to determine the most efficient breeding approach. In pigeonpea such information is rather limited

Table 4
Summary of gene action for various economic traits in pigeonpea as reported in literature

Character	Gene action		
	Additive	Non-additive	Additive + Non-additive
Plant height	*	*	*
Plant width	*		*
Days to mature	*	*	*
Pods/plant		*	*
Seeds/pod	*		*
Days to flower	*		
100-seed weight	*	*	*
Seed yield	*	*	
Protein %		*	*

and research on this aspect has only recently received attention. Various genetic studies on quantitative traits of pigeonpea reported in literature are abstracted in Table 4. Sharma and Green (1977) concluded that the important agronomic characters in general are primarily controlled by genes with additive effects. Dominance and non-additive effects were detected in the cases of yield, plant height, and protein content. Saxena *et al.* (1981b) studied the genetic nature of yield and its components in short-duration pigeonpeas. They opined that in quantitative breeding of any crop in which phenology is sensitive to environmental influences, a basic problem is that the interpretation of the results of mating designs is complicated by physiological changes associated with phenological differences. They further concluded that the inheritance of yield and associated characters is confounded with the pleiotropic effects of genes influencing phenology.

Genetic Improvement

Although efforts to improve the productivity of pigeonpea started quite early in India (Howard *et al.*, 1919) and Hawaii (Krauss, 1932), land races still dominate the acreage in most pigeonpea-growing areas throughout the world. The lack of well-directed efforts for the genetic improvement of pigeonpea is one reason for its poor productivity. Most of the breeding programmes in the past were based on selection in the land races and limited hybridization with narrow genetic base. Hence, there has been little improvement in yielding ability and the national productivity levels are low. Lately, large-scale multidisciplinary efforts have been channelized for pigeonpea improvement and stability in India.

Unlike in other economic crops, little information has been generated on the relative efficiency of classical breeding methods in pigeonpea. The choice of breeding methods depends on the floral biology (mating systems) of the crop, gene action controlling economic traits, and desired and product (cultivar, composite, or hybrid). In pigeonpea, the presence of both additive and non-additive genetic variation has been reported and its floral morphology permits the exploitation of both. The additive genetic variation can be used in pedigree selection, mass selection, and population improvement, while in hybrids non-additive genetic variation can be exploited. Pedigree and backcrosses can be used for specific purposes such as incorporation of disease resistance, increased seed size, and greater number of seeds/pod. There is no consensus regarding the most appropriate breeding approach for pigeonpea. However, at present, most of the programmes are deigned to exploit its tendency towards self-pollination (Byth, 1981).

Germplasm Utilization

Plant breeders throughout the world have now very well realized the potential values of genetic diversity in the economic crops as well as their wild relatives. In many crops of economic importance there is an international effort for collection and preservation of the available genetic diversity for the use of present and future plant breeders. ICRISAT has the responsibility for the collection, maintenance, preservation, and distribution of pigeonpea and *Atylosia* germplasm.

The world collection of pigeonpea germplasm maintained at ICRISAT is over 11,171 and represents 52 countries. Besides this, 270 accessions of 47 wild species belonging to 6 genera are being maintained. In this programme each accession is examined for 53 traits and the data are stored in computer and can be retrieved as and when necessary at a short

Table 5
Range of variability in pigeonpea germplasm for various economic traits

Character	Minimum	Maximum
50% flowering (days)	55.0	210.0
75% maturity (days)	97.0	260.0
Plant height (cm)	39.0	385.0
Primary branches (no.)	2.3	66.0
Secondary branches (no.)	0.3	145.3
Racemes (no.)	6.0	915.0
Seeds/pod	1.6	7.6
100-seed weight (g)	2.8	22.4
Harvest index (%)	0.6	62.7
Shelling ratio (%)	5.8	86.6
Protein %	12.4	29.5

Source: Pigeonpea Germplasm Catalogue, Genetic Resources Unit, ICRISAT

notice. Table 5 shows the range of variation available in the pigeonpea germplasm for important characters at ICRISAT Centre.

Natural outcrossing poses problems in maintaining pigeonpea germplasm. Each year a certain number of accessions are grown in field for seed multiplication. For maintenance of the genotypes within each accession about 30 plants are selfed with muslin cloth bags and the selfed seed is composited for the next generation. The open-pollinated seed is generally used for seed supply to cater for the needs of scientists of different disciplines throughout the world. The germplasm is stored in short- and long-term storages. In the former, it is kept at 4°C at 30–40 per cent RH, while in the latter the temperature is kept around –18°C. In the long-term storage the seeds can be kept in viable state for 30 to 100 years (van der Maesen, 1986).

In recent years, pigeonpea breeders in India have successfully exploited the natural genetic variability available in the land races by selecting high-yielding, widely adapted genotypes in short-, medium-, and long-duration groups (Table 6). On the other hand, pathologists have identified genotypes having resistance to one or more diseases (Table 7) that have been used in breeding programmes to develop disease-resistant high-yielding cultivars.

Table 6
List of cultivars developed by selection from land races in different states of India

Cultivars	Origin	Maturity
LRG 30, LRG 36, C 11, ST-1	Local selections from Andhra Pradesh	Medium to mid-late
Bahar, Basant BR 183, BR60, BR65	Local selections from Bihar	Medium and late
T-15-15	Local selections from Gujarat	Late
GS 1, PT 221, C28	Local selections from Karnataka and Maharashtra	Late
Gwalior 3, Khargone 2, AS 31-37	Local selections from Madhya Pradesh	Medium and late
BDN 1, BDN 2, No. 148	Local selections from Maharashtra	Medium
SAI, Co 1, Co 4	Local selections from Tamil Nadu, Andhra Pradesh	Medium
T7, T17, UPAS 120, Pant A 3	Local selections from Uttar Pradesh	Early and late
B 7, B 517	Local selections from West Bengal	Late

Source: Pigeonpea cultivation in India—A package of practices, Directorate of Pulses Research, Kanpur, 1983.

Table 7
Pigeonpea lines with resistance or tolerance to one or more diseases maintained at ICRISAT Centre

ICP no.	Wilt	Sterility mosaic	Phytophthora	Alternaria blight	Root knot
8861	R	R		R	R
8862	R	R		R	R
5097	R	R	R		
8094	R	R	R		
8863	R			R	
8867	R	R		R	
2376			R	R	
2719			R	R	
2630		R		R	
4725		R		R	
7188		R		R	
8460		R	R		
11300		R	R		
11302		R	R		
8859	R			R	
11286	R			R	
11299	R			R	
8868	R		R		
10958	R		R		
11287	R		R		
11289	R	R			
11297	R	R			
11298	R	R			
6997		R			
7065			R		
8858	R				

R = Resistance reaction.
Source: Pulse Pathology Unit, ICRISAT.

Pure-line Breeding

Choice of parents and mating schemes. The importance of genetic diversity for improving yield potential *per se* has been emphasized by several workers and reviewed by Frey (1971). In pigeonpea, a large variation exists for most of the economic traits (Table 5). The parents for hybridization should be chosen keeping in view the breeding objectives and the special characteristics of the lines such as seed size, pod size, branching, disease and insect resistance, and/or yield *per se*. For determining the value of parents with respect to their combining ability in single-cross combinations, diallel or line × tester mating schemes may be used. Hartwig (1972) suggested that one or more backcrosses to the adapted parent were useful in retaining the productivity and adaptation to specific environment. Thorne and Fehr (1970) and Frey (1972) observed that triple crosses were superior to two-way crosses for developing high-yielding progenies.

Hybridization. In pigeonpea, crosses for developing breeding populations are made by hand emasculation and pollination. Sharma and Green (1980) have given a detailed description of the hybridization technique in pigeonpea. Pathak (1970) reported that emasculation of pigeonpea buds cannot be carried out after 10 a.m. because at that time the anthers begin to dehisce. Since the pigeonpea stigma becomes receptive about 24 hours before the flower opens, the younger bud, where anthers are immature, should be selected for emasculation to avoid the possibility of self-pollination. The pollination can be done immediately after emasculation.

The success in hybridization is greatly influenced by weather conditions, particularly temperature and sunshine. Veeraswamy and Sheriff (1973) made artificial cross-pollinations from 9 a.m. to 5 p.m. at intervals of every 30 minutes. They found that the crosses made between 10 and 10:30 a.m. gave maximum success and a very low pod set was recorded between 1 and 2 p.m. Under Hyderabad conditions successful pollinations are made from 8 a.m. to 5 p.m. in the months of September to February. The pod setting was found to reduce considerably in the months of March and April when the maximum temperature rises to 35–40°C. Dutta and Deb (1970) reported that in pigeonpea fertilization occurs 5 hours after pollination, which results in poor pod set. The final pod set following artificial hybridization can also be influenced by a number of other factors such as parents involved in crosses, efficiency of the pollinator, and insect-pest damage. Saxena *et al.* (1976) reported 19.5 per cent average success with a range of 5–40 per cent from 283, 560 hand pollinations made at ICRISAT during 1974. Singh *et al.* (1980) also recorded 20 per cent success in hybridization programme at Pantnagar. They also reported that pod setting can be increased by the use of plant hormones such as gibberellic acid and indole acetic acid. Saxena *et al.* (1987a) reported cross-incompatibility in some pigeonpea crosses and postulated that the cross-incompatibility was genetically controlled. Singh *et al.* (1980) also suggested genetic control of crossability in pigeonpea.

Selection among crosses. Information on potential of crosses at an early stage of a breeding programme helps in efficient use of resources. Results of a number of experiments conducted in various crops to identify potential value of crosses are not consistent.

In pigeonpea, Saxena and Sharma (1983) studied the value of early generation testing in eight crosses involving diverse parents. They concluded that for practical purposes, low-yielding crosses can be safely rejected on the basis of their F_1 performance. They further suggested that the crosses that record high yield in F_1 generation should be tested again in the F_2 for further selection, since the F_2 performance was consistently related to the cross performance in succeeding generations. An F_2 multilocation test will be more useful in reducing the bias caused by genotype and environment interaction in selecting crosses based on single location performance. According to Allard (1960), in early generation, selection for yield among crosses could be made but selection of lines within a cross was not possible. Lupton (1961) concluded that the value of early generation testing is rather limited.

Pedigree selection. The effectiveness of pedigree selection mainly depends upon closeness of a phenotypic value to its genotype. Therefore, it is a more powerful breeding tool for the traits having high heritability. Besides environmental variation, non-additive gene effects in heterozygotes and high level of genotype-environment interaction contribute to low heritability in early generations for quantitative characters such as yield. Selections based on single plant yield in early segregating generations have been highly ineffective in cereals (McGinnis and Shebeski, 1968) as well as in legumes (Byth *et al.,* 1980). For a successful pedigree breeding, besides high mean performance, the cross in which selection is being exercised should also have high variance. Green *et al.* (1981) observed that in single crosses of pigeonpea the variance of individual plant yield was similar in the parents and their F_2s, indicating high level of environmental influence on the expression of individual plant yield. In spite of these limitations pedigree selection has resulted in the development of some high-yielding varieties such as ICPL 87, ICPL 151, Prabhat, T.21, Pusa Ageti, CO 5, and JA 3 in the breeding programmes at ICRISAT and Indian Research Centres.

Breeding for disease resistance. About 50 pigeonpea diseases caused by a number of agents have been documented in literature. However, only a few of them cause serious economic losses to this crop. Nene *et al.* (1981) reviewed disease situation in different regions of pigeonpea-growing areas and concluded that in the Indian subcontinent and eastern Africa wilt is the most important disease; sterility mosaic and *Phytophthora* blight are two other important diseases of the Indian subcontinent. In Central America, besides rust, witches broom causes serious losses to pigeonpea. Disease-resistant cultivars are an integral component of crop disease management. Success in breeding high-yielding, stable, and disease-resistant genotypes mainly depends on the availability and genetic understanding of resistant gene(s), disease screening facilities, and appropriate mating and selection schemes. Besides the desired level of disease resistance, the varieties should be capable of producing high yield levels in farmers' fields under natural disease pressure.

The first description of *wilt* disease and its causal organism was given by Butler as early as 1906. The causal fungus *Fusarium* sp. *udum* is a soil-borne vascular pathogen. Wilting of the infected plants is caused by blockage of the vascular tissues and before wilting the plants appear chlorotic. The wilting may be complete or partial. The expression of this disease is high during the reproductive stage of the plants. Sheldrake *et al.* (1978) observed that wilt development is influenced by the physiological state of the plant and the depletion of the assimilates in the roots may facilitate the entry or subsequent growth of the fungus. Butler (1906) noted that the intensity of wilt disease in field is affected by the crop rotation being followed in a particular field. Continuous cultivation of pigeonpea in a field can increase wilt incidence. On the contrary, the disease incidence can be minimized by practising long-term crop rotations (Butler, 1906; Natarajan *et al.*, 1985).

The inheritance of wilt resistance is not fully understood. Pal (1934) reported that the resistance to wilt in pigeonpea was controlled by multiple factors, while Shaw (1936) suggested that wilt resistance was conditioned by two complementary genes. Joshi (1957) indicated that resistance to wilt was controlled by a single dominant gene. Pathak (1970) observed the presence of two complementary genes in determining the resistance to fusarium wilt. Inheritance studies of wilt resistance at ICRISAT have shown that in some crosses the resistance reaction was governed by a dominant gene (Laxman Singh, personal communication).

Breeding of wilt-resistant genotypes in pigeonpea was initiated in India in the first decade of the 20th century (Butler, 1908, 1910). Subsequently, selection of resistant plants in wilt-sick plots was carried out at several other research stations. These breeding efforts have resulted in the identification of a number of high-yielding wilt-resistant genotypes (Table 8).

Table 8
Performance of sterility mosaic and wilt resistant lines at ICRISAT Centre, 1982–1985

Line	Sterility mosaic* (%)	Wilt* (%)	Yield (kg/ha)
ICPL 335	8.2 (4)	0.0 (2)	1969 (4)
ICPL 227	7.3 (4)	1.6 (2)	2054 (4)
ICPL 8363	5.8 (4)	0.9 (2)	1881 (4)
ICPL 8362	7.5 (4)	0.0 (2)	1858 (4)
ICPL 343	99.0 (2)	2.4 (4)	2435 (3)
ICPL 8357	6.8 (4)	100.0 (2)	2218 (4)
ICPL 345	67.0 (2)	0.0 (4)	2162 (3)
C 11 (control)	100	60	2127 (4)

Number of years tested is given in parentheses.
*Disease reaction recorded in sick plot.

Sterility mosaic is the second most important disease of pigeonpea and its occurrence was first reported from Pusa, Bihar, in India (Alam, 1931; Mitra, 1931). The infected plants show mosaic symptoms on leaves and excessive proliferation. The affected plants also show either complete or partial sterility. The disease is known to occur in almost all the pigeonpea-growing areas of India (Kannaiyan *et al.*, 1981b), and its incidence in farmers' field varies from 0 to 100 per cent. Capoor (1952) and Narayana-swamy and Jagnathan (1973) considered the disease to be caused by a virus. According to Oldfield *et al.* (1981), the causal agent has not been identified but is believed to be a virus. Seth (1962) and Nene (1972) reported that the causal organism is transmitted by eriophyid mite *Aceria cajani* Channabasavanna (Channa Basavanna, 1963). The perennial nature of pigeonpea plants helps in the survival of viruliferous eriophyid mites (Seth, 1965). The disease was reported to be transmitted through graft (Capoor, 1952). Nene and Reddy (1976) developed a simple 'leaf stapling' technique for transmitting the mites and consequently the disease. The viruliferous mites can be airborne and can spread up to 2000 m in the favourable wind direction (Reddy *et al.*, 1986)

Singh *et al.*, (1983) reported that the resistance to sterility mosaic in pigeonpea was controlled by four independent non-allelic genes. Of these, two were dominant and the remaining two recessive and the presence of at least one dominant and one recessive gene was essential for imparting resistance to sterility mosaic. Sharma *et al.* (1984) reported that sterility mosaic reaction in pigeonpea was controlled by four alleles at two loci. Two alleles control immunity, one of which is dominant and the other recessive to the allele for tolerance. One allele conditions tolerance. The allele responsible for susceptibility is dominant over the other three alleles.

Though some cultural control methods for this disease have been suggested, the use of resistant varieties is the best way to control it. Seth (1965) recommended the removal of perennial and volunteer plants of pigeonpea in the vicinity of pigeonpea field well before sowing. According to Raychaudhary and Nariani (1977) both the mite and the causal organism have narrow host range and, therefore, removal of perennial and self-sown plants can be a very effective control measure. But these measures are not practicable. Bhatnagar *et al.* (1984) reported higher sterility mosaic incidence in sorghum/pigeonpea intercrop in comparison to sole pigeonpea plots. Gurha *et al.* (1983) reported low sterility mosaic incidence at higher plant population. Since the disease is mite-transmitted, the control measures should aim to eradicate or at least reduce the mite population. This could easily be achieved by the use of appropriate pesticides but it may not be economical.

With the availability of easy and effective sterility mosaic screening method, the breeding for resistant genotypes started with the identification of resistant germplasm accessions (Nene and Reddy, 1976). Some of the resistant sources have been used in developing high-yielding disease-resistant cultivars (Table 8).

Phytophthora stem blight, an important disease of pigeonpea, was first reported in India by Williams *et al.* (1968). Since then, it has been reported from most pigeonpea-growing areas in India, particularly during long wet spells (Nene *et al.,* 1981), resulting in severe losses (Williams *et al.,* 1975). This disease is caused by fungus *Phytophthora drachsleri* f. sp. *cajani* (Kannaiyan *et al.,* 1980). The symptoms of the stem blight, confined to the aerial parts of the plants, are characterized by the appearance of dark brown to purple lesions on stem, petioles, and leaves (Nene *et al.,* 1981). These lesions coalesce to girdle the stem and lead to a rapid wilting and drying of the foliage.

The disease is a problem when pigeonpea is grown in fields subjected to waterlogging, though avoiding waterlogged fields is the best way to avoid the disease. Kannaiyan *et al.* (1981a) identified sources of resistance to P2 isolate of the fungus, but most of them are susceptible to P3 isolate. Efforts are being made to identify lines with resistance to P2 and P3 isolates.

Sharma *et al.* (1982) studied the inheritance of resistance to P2 isolate and reported a single dominant gene for resistance. Screening of advanced breeding lines and adapted cultivars indicated that several high-yielding lines had resistance to P2 isolate.

Alternaria blight is of minor importance in the main pigeonpea crop planted with the onset of monsoon. However, in the post-rainy season crop in parts of Bihar and West Bengal the disease causes severe losses. This disease is caused by *Alternaria tenuissima* (Kunze ex. pers. Wittshire; Pavgi and Singh, 1971). The symptoms are generally confined to the older leaves (Kannaiyan and Nene, 1977). Dark lesions with concentric rings, typical of *Alternaria* diseases, are formed on the leaves, branches, and stem resulting in complete defoliation of the plants. Inheritance studies have shown that the resistance to *Alternaria* blight is controlled by a single recessive gene (Sharma *et al.,* 1987). Several high-yielding *Alternaria*-resistant lines have been identified at ICRISAT and at Varanasi (Venkateswarlu *et al.,* 1981). At present, no serious efforts are being made to breed for the resistant cultivars.

Fifty-nine species of *plant-parasitic nematodes* have been reported to be associated with pigeonpeas from different pigeonpea-growing areas in the world. Of these, pigeonpea cyst nematode (*Heterodera cajani*), the root-knot nematodes (*Meloidogyne incognita,* and *M. javanica*), and the reniform nematode (*Rotylenchulus reniformis*) are very important.

The pigeonpea cyst nematode is fairly widespread in the pigeonpea-growing states of India. Nematode attack can easily be detected by careful observations of the young plant roots. Many white to brown pearl-like structures (cysts) can be seen attached to the roots that contain numerous nematode eggs and larvae. The infection retards the root and vegetative growth and the plants become stunted and weak. The root-knot disease of pigeonpea is prevalent in India and Malawi. The nematode attack produces typical galls on the root system and the infected plants appear weak, yellowish, and stunted. The reniform nematode affects the pigeonpeas in many areas. The infected plants have stunted growth.

Currently no economically effective control measures are available against these diseases and keeping in view the high cost of the nematicides, host resistance appears to be the most promising method of controlling these diseases (S.B. Sharma, personal communication).

Insect pest resistance. About 200 insect species have been recorded feeding upon pigeonpea crop during its life cycle (ICRISAT, 1985) but most of them are of minor importance, sporadic, or localized. The insect complex feeding on pods and seeds causes maximum losses. Under favourable growing conditions the insect damage caused to the foliage, floral buds, and flowers can be fully compensated by extended growth of the crop (Reed *et al.,* 1981). The surveillance of insect pest of pigeonpea conducted by ICRISAT in 15 major pigeonpea growing states of India for a period extending from 1975 to 1981 revealed that in the southern region pod borer *Helicoverpa armigera* (Hub.) causes maximum damage, while in northern India pod fly *Melanogromyza obtusa* (Mall.) is the most serious pest (Reed *et al.,* 1981).

Helicoverpa armigera has a wide range of plant hosts including tobacco, chickpea, green gram, soybean, and cotton (Davis and Lateef, 1975). A female moth of *H. armigera* lays numerous tiny white eggs singly, mostly on flower buds or developing pods. On hatching, the larva bores into the pods and feeds on seeds. In general, it has been observed that determinate cultivars are more prone to this damage than indeterminate cultivars mainly because of the clustering of pods in the former.

Pod fly damage is generally underestimated because it is difficult to detect pod fly infestation since the eggs are laid inside the pods and there are no visible symptoms on the pods until the adult emerges from the pod wall. According to Ahmad (1937), a female pod fly lays eggs through the pod wall and the eggs project into the lumen. The wound at the point heals. Initially, the larva searches for a soft seed and bores into the epidermis. In the second or third stage it bores into the cotyledons and in most cases one seed is enough for the larva to complete its development. The final instar leaves the seed and 'windows' the pod prior to pupation.

In search of the sources of resistance to pod borer and pod fly, a large number of pigeonpea accessions have been screened under unsprayed open field trials at ICRISAT Centre. Although no genotype with sufficient level of resistance to withstand heavy infestation of insects is available, there were clear differences in susceptibility. These tolerant lines have been used by pigeonpea breeders to develop high-yielding lines that have moderate levels of resistance or tolerance to pod borer (K.C. Jain, personal communication). The performance of breeding lines that combine high yield and *Helicoverpa* tolerance are given in Table 9.

Table 9
Performance of *Helicoverpa*-tolerant lines on large (81 m^2) insecticide-free plots at ICRISAT Centre, 1989

Line	Pod damage (%)	Yield (kg/ha)
ICP 1903	30.7	1010
ICPL 87088	26.0	910
ICPL 84060	32.3	840
ICP 10531	33.5	880
ICPL 87089	36.2	770
Control		
BDN-1	57.9	500
SE	± 5.2	± 57
Mean	34.7	810

Source: Legume Programme Annual Report, ICRISAT, 1990.

Breeding vegetable pigeonpeas. In India pigeonpea is generally consumed in the form of cooked *dhal* (decorticated dry split mature seed) along with cereals. However, in some parts of India immature pigeonpea seeds are also consumed as green vegetable. Vegetable pigeonpea is also a very important crop in eastern Africa and the Caribbean. For use as vegetable, pigeonpea pods are picked when the seeds have reached physiological maturity, i.e., when they are fully grown but just before they lose their green colour. The best vegetable varieties have long pods and large, sweet seeds. Harvest of pigeonpea for use as vegetable is more common near cities where the green pods can be quickly transported to market. Vegetable pigeonpeas are, in general, late-maturing types and their cultivation has been limited to kitchen backyard, hedge crop, or small plots. In Gujarat, large-podded and large-seeded varieties are cultivated commercially and the immature pods are sold and consumed as vegetable while the remainder of the crop is then harvested in the usual manner to produce a crop of dry seeds. Singh *et al.* (1986) observed that pigeonpea seeds, when used green as a vegetable, are nutritionally better than mature seeds as the former have higher protein, sugar, and fat content (Table 10). In addition, the protein of vegetable pigeonpea is easily digestible. Compared on a fresh weight basis with garden pea, vegetable pigeonpea has a greater edible portion and more protein, carbohydrate, crude fibre, and fat (Faris *et al.,* 1987). Vegetable pigeonpea also contains more minerals and generally much more of many vitamins than garden pea. The pods can be shelled by hand like green garden pea and the seeds can be substituted in recipes directly for green garden pea. Vegetable pigeonpeas are also canned (Mansfield, 1981), frozen (Sanchev Nieva *et al.,* 1961), and dehydrated (Rahman, 1961) for storage and marketing.

Genetic improvement work on vegetable pigeonpea is being carried out

Table 10
Mean values of various nutritional constituents of green and mature seed from nine pigeonpea genotypes

Green/ mature	Protein (%)	Starch (%)	Soluble sugars (%)	Seed coat	Dietary fibre (%)	Cal (mg/100)	Iron (g)
Green	21.0	48.4	5.1	22.3	23.1	94.6	4.6
Mature	18.8	53.0	3.1	13.6	20.1	120.8	3.9
SD	0.50	0.78	0.18	0.54	0.60	4.52	0.16

Source: Singh *et al.,* 1986.

at the University of West Indies and a few lines have been released for cultivation. However, in India no systematic breeding efforts were directed towards developing vegetable types until a special project was initiated at ICRISAT in 1976. At present, the vegetable breeding programme at ICRISAT is aimed towards developing extra short-duration and near photoperiod-insensitive cultivars so that a continuous supply of fresh immature pods for the market and the canneries can be assured over a longer period of time. The breeding lines developed at ICRISAT for this purpose, having high yield potential and large white seeds, are under testing in multilocational tests. One such line, ICPL 87091, is becoming popular among the growers and consumers.

Breeding hybrid pigeonpeas. Being predominantly a self-pollinated crop, breeding methods designed to exploit additive genetic variance have been adopted by pigeonpea breeders in the past for developing high-yielding pure lines. The substantial amount of non-additive genetic variance (Sharma *et al.,* 1973; Reddy *et al.,* 1979; Reddy *et al.,* 1981; Saxena *et al.,* 1981b) and hybrid vigour (Solomon *et al.,* 1957) available in this crop have not been profitably tapped for its genetic improvement of yield and stability. Unlike other legumes, the presence of natural outcrossing coupled with the discovery of stable genetic male-sterility systems has opened up the possibility of commercial exploitation of heterosis in pigeonpeas. At ICRISAT, considerable research efforts have been devoted since 1977 for identifying heterotic cross-combinations and to develop economic hybrid seed production systems.

Saxena *et al.* (1986b) summarized the performance of several medium-duration experimental hybrids made and tested at ICRISAT between 1977 and 1981. The range of standard heterosis (heterosis over the control cultivar) for seed yield in different crosses was from –38 to 39 per cent. Of the 106 hand-made combinations tested, 9 showed heterosis of more than 20 per cent. The best hybrid (ICPH 2) recorded 32 per cent superiority over the control cultivar.

Some of the newly developed short-duration experimental hybrids have shown very high (4–5 t/ha) yield potential (Saxena *et al.*, 1989b). Of these, ICPH 8 (Fig. 4) has been found to be the most promising. In 1991, the world's first pigeonpea hybrid (ICPH 8) was released for general cultivation in India. This is an indeterminate hybrid developed by crossing a male-sterile line MS Prabhat (DT) and an indeterminate breeding line ICPL 161. Since 1981, ICPH 8 has been tested in a range of diverse environments and its performance in 100 such yield tests is summarized in Table 11. In

Fig. 4 World's first pigeonpea hybrid (Photo: ICRISAT)

Table 11
Seed yield of short-duration pigeonpea hybrid ICPH 8 and control cultivars in different zones in India over years

Zone	Trial[1]	Year	No. of trials	Yield (kg/ha) ICPH 8	Yield (kg/ha) UPAS 120	Yield (kg/ha) Manak	% increase over UPAS 120	% increase over Manak
Northwestern plains	EACT	1987	5	1889	1486	—	27.1	—
	EPHYT	1984	4	2626	2239	2416	17.3	8.7
	EPHYT	1985	2	3152	1866	2689	68.9	17.2
	EPHYT	1987	4	2392	1913	1811	25.0	32.1
	Station	1981	1	3900	2225	—	75.3	—
	trials	1983	1	3560	2569	2715	38.6	31.1
		1984	2	3952	2738	2717	44.3	45.4
		1985	2	1742	1463	—	19.1	—
		1986	7	2953	2114	2137	39.7	38.2
		1987	8	3411	2479	2523	37.6	35.2
Zonal weighted mean			36	2851	2100	2342	35.0	31.0
Central	EACT	1986	10	829	640	609	29.5	36.1
	EACT	1987	4	2127	1556	—	36.7	—
	EPHYT	1984	8	2315	1730	1284	33.8	80.3
	EPHYT	1985	4	1193	929	841	28.4	41.9
	EPHYT	1987	3	1500	1147	1047	30.8	43.3
	EPHYT	1986	1	2147	1270	1312	69.1	63.6
Zonal weighted mean			30	1558	1163	930	32.9	52.5
Southern	EACT	1986	4	695	435	347	58.3	100.3
	EACT	1987	2	861	650	—	32.5	—
	EPHYT	1984	3	1421	1035	824	37.3	72.5
	EPHYT	1985	3	1529	1305	1377	13.5	2.1
	EPHYT	1987	3	1311	1182	1272	10.9	3.1
	Station	1985	1	1493	1301	—	14.8	—
	trials	1986	3	1813	1079	1510	68.0	20.1
		1987	11	1676	1666	1613	0.6	3.9
Zonal weighted mean			30	1422	1216	1262	23.6	27.3
Northwestern hill	EACT	1986	2	1564	1499	1194	4.3	31.0
Northeastern hill	EACT	1987	1	1679	1153	—	45.6	—
Western	EACT	1986	1	2057	1413	1588	45.6	29.5
Overall weighted mean			100	1989	1525	1354	30.5	34.2

1. EACT = Extra-early Arhar All India Coordinated Trial; EPHYT = Early Pigeonpea Hybrid Trial.

the northwestern plains of India, ICPH 8 yielded 2851 kg/ha, reflecting 35 per cent superiority over control UPAS 120 (2100 kg/ha). Similarly, this hybrid recorded 33 and 24 per cent superiority in yield over UPAS 120 in central and southern India respectively (Saxena *et al.*, 1989a). Besides high yield, the hybrid ICPH 8 has shown higher stability and drought tolerant (Saxena *et al.*, 1989b).

The degree of heterosis expressed in the hybrids shows that a fairly good level of heterosis is present in this crop that could be exploited through heterosis breeding. Various aspects of hybrid seed production systems have been discussed in detail by Saxena *et al.* (1986b). A few private and public seed companies in India are presently involved in production and research of hybrid pigeonpea. This non-traditional breeding approach in pigeonpea is likely to enhance yielding ability, stability, and general productivity of pigeonpea in the semi-arid tropics.

Mutation Breeding

Mutation breeding has been successfully used in the genetic improvement of a number of economic crop plants. In pigeonpea a number of mutants, both induced as well as spontaneous, for various qualitative characters have been reported from time to time. This information is available in *An Annotated Bibliography of Pigeonpea* (Dahiya, 1980) and for the sake of brevity does not find a place in this chapter.

A few pigeonpea cultivars have been developed through the use of induced quantitative variation. A high-yielding and large-seeded cultivar T6

Table 12
Performance of selected M_4 lines during 1987 rainy season at ICRISAT Centre

Line	Days to flower	Plant height (cm)	100-seed weight (g)	Plant stand	Yield (kg/ha)	% Wilt in sick nursery
25 Kr-B-WB	128	197	9.9	37	2004	0
25 Kr-33-WB	128	197	9.8	35	1993	4
10 Kr-17-WB	112	174	10.2	35	1846	7
20 Kr-20-WB	134	227	8.8	35	1769	0
Controls						
LRG 30 (susceptible)	129	205	8.1	38	1700	66
ICPL 270 (resistant)	111	178	12.1	35	1691	10
SE	± 0.5	± 4.0	± 0.24	± 1.7	± 176.8	
Mean (n = 25)	121	191	9.0	36	1656	
CV (%)	0.8	4	4	8	18	

Source: Dwivedi *et al.*, 1989.

was developed by irradiating a small-seeded culture T 21 at BARC Trombay (Pawar *et al.*,1982). This variety has performed well in the coordinated trials and has about 35 per cent larger seeds. Another high-yielding pigeonpea cultivar Co-2 was developed at Coimbatore through mutation breeding. This cultivar is suitable for both rainfed and irrigated conditions (Veeraswamy *et al.*, 1975). Dwivedi *et al.* (1989) developed high-yielding, wilt-resistant lines from the irradiated populations of a highly susceptible cultivar, LRG 30 (Table 12).

Population Breeding

The level of natural outcrossing in pigeonpea appears to be insufficient to allow effective use of different breeding methods designed for outcrossed crops. Despite this, population improvement methods have been advocated (Khan, 1973; Byth *et al.*, 1981) and used (Onim, 1981) with limited success in pigeonpea. The identification of usable genetic male sterility allowed the identification of 100 per cent outcrossed plants in the field and meant that pigeonpea breeders could take a fresh look at using the population breeding methods suited to cross-pollinated crops (ICRISAT, 1977, 1980).

To increase recombination by intermating of genotypes, a population breeding programme based on the dual population system (Rachie and Gardner, 1975) and use of a male sterility gene was initiated at ICRISAT Centre. In the dual population method a parent with an easily observable recessive marker is used. The F_2 is grown in isolation and only plants with the recessive marker are harvested. In F_3 only plants with the dominant marker gene are harvested and in the F_4 only plants with the recessive marker are taken. This alteration ensures that only cross-pollinated plants are advanced. At ICRISAT this method did not produce good results, since the finished products from these populations did not show any significant gain in yield levels. Onim (1981) tested two population improvement methods—stratified mass selection and mass selection with progeny testing—in marginal rainfall areas. He reported 2 and 4 per cent yield gain per cycle under stratified mass selection and mass selection with progeny testing respectively. At present, pigeonpea population improvement programmes based on genetic male sterility have been initiated at ICRISAT to develop high-yielding, disease-resistant short- and medium-duration populations and to enhance the gene frequency of favourable alleles in the population for insect tolerance.

Intergeneric Hybridization

The genetic potential of wild relatives in crop improvement is now well established. The primitive cultivars and related wild species of a crop are known to harbour diverse genes for various economic traits. Lawrence (1974) studied the introgression effects of wild germplasm into cultivated oats. He concluded that genes for quantitatively inherited traits in cultivated species may be found in strains of their weedy species. He also pointed out that in oats it is necessary to introgress less than 50 per cent of wild germplasm into the breeding population to recover the already present background of desirable genes. The genetic diversity of the wild gene-pool of *Cajanus* has been discussed by Remanandan (1981). Saxena *et al.* (1990b) reported a considerable intraspecies variation for a number of important traits among the accessions of *Atylosia scarabaeoides*.

The intergeneric hybridization has been successful only with some *Atylosia* species. Most of the *Atylosia* spp. are rich in protein and this useful character has been incorporated into pigeonpea. Saxena *et al.* (1987d) have now combined high yield and high protein characters in pigeonpea (Table 13). It has also been suggested that seed size and protein can be improved simultaneously in pigeonpea (Saxena *et al.* 1987b). The high protein lines derived from *Cajanus-Atylosia* crosses have high protein content and give yield similar to those of standard cultivars. Two such lines (HPL 40 and HPL 8) yield about 100 kg/ha additional protein in comparison to the check cultivars (Table 13). These results have important implications particularly in countries of tropical Asia such as Thailand and Indonesia, which import soybean as protein supplement for animal feed. Pigeonpea can produce a good crop in these countries where soybean is unadapted and feeding trials show that pigeonpea can substitute for much of the soybean in hog and sheep rations and can completely replace soybean in poultry feed. High-protein lines also can improve the human diet where

Table 13
Performance of high protein pigeonpea selections at ICRISAT Centre during 1985 and 1986

Line	Days to mature	Plant height (cm)	100-seet wt. (g)	Grain yield (kg/ha)	Protein	
					%	kg/ha
1986						
HPL 8-10	163	102	10.5	1663	26.5	353
HPL 8-16	162	108	10.5	1569	27.4	344
ICPL 211 (check)	162	112	14.3	1457	21.6	251
SE ±	1.1	3.9	0.15	195.7	0.21	38.5
CV (%)	2.2	7.1	2.51	27.0	1.66	25.8
1985						
HPL 40-5	169	170	9.6	2096	26.9	452
HPL 40-17	169	161	8.5	2074	26.5	440
BDN 1 (check)	168	168	9.6	2022	23.2	373
SE ±	0.9	4.5	0.18	181.0	0.46	37.3
CV (%)	0.9	4.7	3.43	17.3	3.03	17.0

pigeonpea is already grown and used, as the protein quality, measured by the proportion of sulphur-bearing amino acids, is similar in *Atylosia* spp., high-protein lines, and normal cultivars (Saxena *et al.,* 1986a).

International Research Network

India has a large, well-organized national programme for pigeonpea improvement under All India Coordinated Pulse Improvement Project of the Indian Council of Agricultural Research (ICAR). ICRISAT, located at Hyderabad (India), has the worldwide responsibility for overall improvement of pigeonpea. ICRISAT cooperates with ICAR and agricultural universities in India and aboard in exchange of germplasm, multilocation varietal testing, and screening for disease and pest resistances through an excellent network of experimental stations located all over the country. Training programmes, workshops, and conferences on pigeonpea research and development

Table 14
Performance of some pigeonpea lines in various countries in ICRISAT's international trials

Region	Country	Promising entries
Caribbean and others	St Kitts	ICPL 151 (1.5)
	Antigua	ICPL 151 (1.5), ICPL 270 (2.5), ICPL 304 (2.5)
	Surinam	ICPL 151, ICPL 147
	Trinidad	ICPL 151
	Argentina	ICPL 87 (4.8), ICPL 8313 (4.8), ICPL 8310 (4.4)
Southeast	Myanmar	ICPL 151 (2.0), ICPL 8308 (3.2), ICPL 8313 (2.4)
	Nepal	ICP 7363 (2.2), ICP 8865 (1.7)
	Thailand	ICPL 151 (2.1), ICPL 148 (2.5), ICP 8863 (5.0)
	Philippines	ICPL 87 (2.3), ICP 8863 (3.5), ICPL 304 (2.5)
	Pakistan	ICPL 82 (1.8), ICPL 142 (1.7), ICPL 1 (2.4)
	Sri Lanka	ICPL 87 (2.0), ICPL 161 (1.8), ICPL 2 (1.7)
Africa	Burkina Faso	ICPL 151 (1.2)
	Ethiopia	ICP 9180 (2.9), NP (WR) 15 (2.8), ICPL 304 (2.0)
	Mali	ICPL 148, ICPL 154
	Sudan	BDN 1 (2.1), ICPL 270 (1.8)
	Tanzania	ICPL 87 (2.2)

Yield in t/ha is given in parentheses.

are organized by national, regional, and international organizations to provide a forum for information and material exchange.

The Asian Grain Legume Network has been operating at ICRISAT to promote research and development of pigeonpea, chickpea, and groundnut in Asia. ICRISAT has started a collaborative network for testing the adaptation of newly developed pigeonpea cultivars in eastern, central, and southern Africa.

The University of Queensland in Australia has successfully collaborated with ICRISAT to develop a mechanized commercial production system of pigeonpea using short-duration genotypes. The cultivars 'Hunt' and 'Quantum' developed in this programme have been successful in Thailand, the Philippines, and Indonesia also. Argentina, Sri Lanka, Surinam, and South Korea, where pigeonpea is not grown presently, have found it feasible to grow this crop successfully using short-duration cultivars. Similar efforts are being made in the United States.

The international linkages for pigeonpea research have led to increased efforts in the development of this crop in recent years. International testing of pigeonpea lines in various countries has resulted in the identification of high-yielding genotypes (Table 14). This can help in widening the adaption of pigeonpea in new areas. Although short-duration pigeonpea as a monocrop opens up vast potential of commercial mechanized production of the crop in new environments, one would not like to neglect intercropping systems with longer-duration types because of their continued role in subsistence farming as well as their potential role in commercial farming. Research on the genetic improvement of pigeonpea would be more meaningful if an attempt is made to keep in mind total cropping systems into which the various types of pigeonpea could fit. This includes enhancement of traditional systems as well as creation of new areas such as short-duration monocrops, long-term sustainable crop, and agroforestry-based production systems.

REFERENCES

Abrams, A. 1967. Studies on natural cross-pollination in pigeonpea (*Cajanus cajan*). *J. Agric. Univ. Puerto Rico* 51: 1–3.

Abrams, R., and Julia, F.J. 1973. Effect of planting time, plant population and row spacing on yield and other characters of pigeonpea, *Cajanus cajan* (L.) Millsp. *J. Agric. Univ. Puerto Rico* 57: 275–285.

Ahmad, T.Y. 1937. The Tur-podfly *Agromyza obtusa* Mall, a pest of *Cajanus cajan*. *Indian J. Agric. Sci.* 8: 63–76.

Akinola, J.O., Pritchard, A.J., and Whiteman, P.C. 1972. Chromosome number in pigeonpea (*Cajanus cajan* (L.) Millsp.) *J. Aust. Inst. Agric. Sci.* 38: 305–306.

Akinola, J.O., and Whiteman, P.C. 1974. Agronomic studies on pigeonpea (*Cajanus cajan* (L.) Millsp.). Responses to sowing density. *Aust. J. Agric. Res.* 26: 57–66.

Alam, M. 1931. Administration report of the botanical section for the year ending 31st March 1931. Dept. Agric. Bihar, Orissa for the period from 1st April 1930 to 31st March 1931, pp. 42–65.

Allard, R.W. 1960. *Principles of Plant Breeding*. John Wiley and Sons, New York.

Ariyanayagam, R.P. 1976. Outcrossing and isolation in pigeonpea. *Trop. Grain Legume Bull.* 5: 14–17.

Ariyanayagam, R.P., Jain, K.C., Gupta, S.C., and Saxena, K.B. 1991. New approaches to pigeonpea breeding. Paper presented at the Golden Jubilee Symposium on Genetic Research and Education. Current Trends and the Next Fifty Years. 12–15 Feb. 1991, New Delhi.

Bajaj, Y.P.S., Singh, H., and Gosal, S.S. 1980. Haploid embryogenesis in anther culture of pigeonpea (*Cajanus cajan*). *Theo. Appl. Genet.* 58: 157–159.

Baker, J.G. 1876. In *Flora of British India*. L. Hooker, ed. London, Royal Botonic Gardens, Kew, U.K., pp. 212–217.

Bhatia, G.K., Gupta, S.C., Green, J.M., and Sharma, D. 1981. Estimates of natural cross pollination in *Cajanus cajan* (L.) Millsp. Several experimental approaches. *Proc. Int. Workshop on Pigeonpea,* 15–19 Dec. 1980. ICRISAT, Patancheru, A.P., India. Vol. 2, pp. 129–136.

Bhatnagar, V.S., Nene, Y.L., and Jadhav, D.R. 1984. Sterility mosaic disease of pigeonpea in a sorghum-based cropping system. *Int. Pigeonpea Newslett.* 3 : 37–38.

Bhattacharjee, S.K. 1956. Study of autotetraploid *Cajanus cajan* (Linn.) Millsp. *Caryologia* 9: 149–159.

Brucher, M. 1977. *Tropische Nutzpflanzen*. Springer-Verlag, Berlin, Heidelberg, New York, pp. 166–167.

Burman, J. 1737. *Thesaurus Zeylanicus*. Amstelae damimi.

Butler, E.J. 1906. The wilt disease of pigeonpea and pepper. *Agric. India* 1: 25–36.

Butler, E.J. 1908. Selection of pigeonpea for wilt disease. *Agric. India* 3: 182–183.

Butler, E.J. 1910. The wilt disease of pigeonpea and the parasitism of *Necosmospora vasinfecta* Smith. *Mem. Dept. Agric. India* (Bot. Ser.) 2: 1–62.

Byth, D.E. 1968. Comparative photoperiod responses for several soybean varieties of tropical and temperate origin. *Aust. J. Agric. Res.* 19: 879–880.

Byth, D.E. 1981. Breeding. *Proc. Int. Workshop on Pigeonpea,* 15–19 Dec. 1980. ICRISAT, Patancheru, A.P., India. Vol. 1, pp. 487–495.

Byth, D.E., Green, J.M., and Hawtin, G.C. 1980. ICRISAT/ICARDA chickpea breeding strategies. *Proc. Int. Workshop on Chickpea Improvement,* 28 Feb. to 2 March 1979. ICRISAT, Hyderabad, India, pp, 11–27.

Byth, D.E., Wallis, E.S., and Saxena, K.B. 1981. Adaptation and breeding strategies for pigeonpea. *Proc. Int. Workshop on Pigeonpea,* 15–19 Dec. 1980. ICRISAT, Patancheru, AP, India. Vol. 1, pp. 450–465.

Byth, D.E., Saxena, K.B., and Wallis, E.S. 1982. A mechanism for inhibiting cross fertilization in pigeonpea (*Cajanus cajan* (L.) Millsp). *Euphytica* 31 : 405–408.

Capoor, S.P. 1952. Observations on the sterility disease of pigeonpea. *Indian J. Agric. Sci.* 22: 271–274.

Channa Basavanna, G.P. 1963. Ph.D. thesis, Division of Entomology, IARI, New Delhi, India.

Chauhan, Y.S., Venkataratnam, N., and Sheldrake, A.R. 1984. The high yield potential of extra-early pigeonpeas in peninsular India. *Int. Pigeonpea Newslett.* 3: 28–31.

Chauhan, Y.S., Rathnam, N.V., and Sheldrake, A.R. 1987. Factors affecting growth and yield of short duration pigeonpea and its potential for multiple harvest. *J. Agric. Sci. Camb.* 109: 519–529.

Dahiya, B.S. 1980. *An Annotated Bibliography of Pigeonpea 1900–1977.* ICRISAT, 183 pp.

Davis, J.C., and Lateef, S.S. 1975. Insect pests of chickpea and pigeonpea in India and prospects for control. *Proc. Int. Workshop on Grain Legumes*, 13–16 Jan. 1975. ICRISAT, Hyderabad, India, pp. 319–331.

De, D.N. 1974. Pigeonpea. In *Evolutionary studies in World Crops*. J. Hutchinson, ed. Cornbridge University Press, London, pp. 79–87.

de Candolle, A.P. 1813. *Catalogus Hortus Monspeliensis*, pp. 85–86.

de Candolle, A.P. 1884. *Origin of Cultivated Plants*. Kegan Paul, Trench and Co., London.

D'Cruz, R., and Jadav, A.S. 1972. Aneuploidy in tur (*Cajanus cajan* (L.) Mills). *Mahatma Phule Agric. Univ. Res. J.* 3: 61–62.

Deodikar, G.B., and Thakar, C.V. 1956. Cyto-taxonomic evidence for the affinity between *Cajanus indicus* Spreng and certain erect species of *Atylosia* W. & A. *Proc. Ind. Acad. Sci.* 43: 37–45.

Deshmukh, N.Y., and Rekhi, S.S. 1963. Study of natural crossing in pigeonpea (*Cajanus cajan* (L.) Millsp.) *Proc. Bihar Acad. of Agric. Sci.* 8–9: 135–139.

Dundas, I.S. 1990. Pigeonpea: Cytology and cytogenetics—perspectives and prospects. In *The Pigeonpea*. Y.L. Nene, S.D. Hall and V.K. Sheila eds. CAB International, Wallingford, UK, pp. 117–136.

Dundas, I.S., Saxena, K.B., and Byth, D.E. 1981. Microsporogenesis and anther wall development in male sterile and fertile lines of pigeonpea (*Cajanus cajan* (L.) Millsp.). *Euphytica* 30: 431–435.

Dundas, I.S., Saxena, K.B., and Byth, D.E. 1982. Pollen mother cell and anther wall development in a photoperiod insensitive male sterile mutant of pigeonpea (*Cajanus cajan* (L.) Millsp). *Euphytica* 31: 371–375.

Dutta, P.C., and Deb, A. 1970. Floral biology of *Cajanus cajan* (Linn.) *Bull. Bot. Soc. Bengal* 24: 135–145.

Dwivedi, S., Faris, D.G., and Jain, K.C. 1989. Gamma ray induced *Fusarium* wilt resistance in pigeonpea (*Cajanus cajan* (L.) Millsp). *Proc. 6th Int. Cong.* SABRAO 781–784.

Faris, D.G., Saxena, K.B., Mazumder, S., and Singh, U. 1987. Vegetable pigeonpeas: a promising crop for India. Information Bull. ICRISAT, Patancheru, AP, India, 13 pp.

Frey, K.J. 1971. Improving crop yields through plant breeding. In *Moving Off the Yield Plateau*. Amer. Soc. of Agron. Special Publication 20, Madison, Wisconsin, pp. 15–58.

Frey, K.J. 1972. Stability indices for isolines of oats (*Avena sativa* (L.)). *Crop Sci.* 12: 809–812.

Green, J.M., Sharma, D., Saxena, K.B., Reddy, L.J., and Gupta, S.C. 1979. Pigeonpea breeding at ICRISAT. Presented at the Regional Workshop on Tropical Grain Legumes, 18–22 June, 1979. Univ. of the West Indies, St. Augustine, Trinidad.

Green, J.M., Sharma, D., Reddy, L.J., Saxena, K.B., Gupta, S.C., Jain, K.C., Reddy, B.V.S., and Rao, M.R. 1981. Methodology and progress in the ICRISAT pigeonpea breeding program. In *Proc. Int. Workshop on Pigeonpea*, 15–19 Dec. 1980. ICRISAT, Patancheru, AP, India, Vol. 1, pp. 437–449.

Gupta, S.C., Reddy L.J., Sharma, D., Green, J.M., Murthi, A.N., and Saxena, K.B. 1981. Maintenance of pigeonpea cultivars. *Proc. Int. Workshop on Pigeonpea*, 15–19 Dec. 1980 ICRISAT, Patancheru, AP, India, Vol. 1, pp. 295–302.

Gurha, S.N., Singh, D.N. and Gangal, L.K. 1983. Effect of population density on the incidence of sterility and yellow mosaic diseases in pigeonpea. *Int. Pigeonpea Newslett.* 3: 47–48.

Hains, H.H. 1920. In *The Botany of Bihar and Orissa* 3. Botanical Survey of India, Calcutta, pp. 287.

Hartwig, E.E. 1972. Utilization of soybean germplasm strains in a soybean improvement program. *Crop Sci.* 12: 856–859.

Howard, A., Howard, G.L.C., and Khan, A.R. 1919. Studies in the population of Indian Crops. I. *Mem. Dept. Agric. India* (Bot. Ser.) 10: 195–200.

Hutchinson, J., and Dalziel, J.M. 1958. *Flora of West Tropical Africa*, 2nd edn., revised by R.W.J. Keay. Crown Agents for Overseas Govt. and Admin., London.

ICRISAT. 1975. ICRISAT Annual Report 1974-1975. ICRISAT, Patancheru, India.

ICRISAT. 1977. ICRISAT Annual Report 1976-77. ICRISAT, Patancheru, India.

ICRISAT. 1980. ICRISAT Annual Report 1979-80. ICRISAT, Patancheru, India.

ICRISAT. 1985. ICRISAT Annual Report 1984-85. ICRISAT, Patancheru, India.

Jacquin, N.J. 1772. In *Hortus botanicus vindobonensis seu plantarum rariorum* 2.

Joshi, A.B. 1957. Genetics of resistance to diseases and pests. *Indian J. Genet.* 17: 305–317.

Kannaiyan, J., and Nene, Y.L. 1977. Alternaria leafspot of pigeonpea. *Trop. Grain Legume Bull.* 9: 34.

Kannaiyan, J., Ribeiro, O.K., Erwin, D.C., and Nene, Y.L. 1980. Phytophthora blight of pigeonpea in India. *Mycologia* 72: 169–181.

Kannaiyan, J., Nene, Y.L., Raju, T.N., and Sheila, V.K. 1981a. Screening for resistance to Phytophthora blight of pigeonpea. *Plant Dis.* 65: 61–62.

Kannaiyan, J., Reddy, M.V., and Nene, Y.L. 1981b. Survey of pigeonpea diseases with special reference to wilt and sterility mosaic in India. *Proc. Int. Workshop on Pigeonpea,* 15–19 Dec. 1980. ICRISAT, Patancheru, AP, India, Vol. 2, pp. 297–304.

Khan, T.N. 1973. A new approach to the breeding of pigeonpea (*Cajanus cajan* (L.) Millsp.). Formation of composites. *Euphytica* 22: 373–377.

Krauss, G.G. 1932. The pigeonpea (*Cajanus indicus*) its improvement, culture and utilization in Hawaii. *Hawaii Agric. Exp. Sta. Bull.* 64.

Krishnaswamy, N., and Ayyangar, G.N.R. 1935. Chromosome number in *Cajanus indicus* Spreng. *Curr. Sci.* 3: 614–615.

Kumar, L.S.S., Abraham, A., and Srinivasa, V.K. 1945. Preliminary note on autotetraploidy in *Cajanus indicus* Spreng. *Proc. Indian Acad. Sci.* (Sec. B) 21: 301–306.

Kumar, L.S.S., Thombre, M.V., and D'Cruz, R. 1958. Cytological studies on an intergeneric hybrid of *Cajanus cajan* (L.) Millsp. and *Atylosia lineata* W and A. *Proc. Indian Acad. Sci.* (Sec. B.) 47: 252–262.

Lawrence, P.K. 1974. Introgression of exotic germplasm into oat breeding populations. Ph.D. thesis, Iowa State University, Ames, Iowa, 120 pp.

Linnaeus, C. 1737. *Hortus Cliffortianus.* Amstelaedami, 354.

Linnaeus, C. 1753. *Species plantarum Holmiae* 2. 739.

Lupton, F.G.H. 1961. Studies in the breeding of self-pollinated cereals 3. Further studies in cross prediction. *Euphytica* 10: 209–224.

Mansfield, G.R. 1981. Processing and marketing of green pigeonpeas, the case of the Dominican Republic. *Proc. Int. Workshop on Pigeonpea,* 15–19 Dec. 1980 ICRISAT, Patancheru, India, Vol. 1, pp. 355–350.

McGinnis, R.C., and Shebeski, L.H. 1968. The reliability of single plant selection for yield in F_2. *Proc. Third Int. Wheat Genetic Symp.* Aust. Acad. Sciences, Canberra, pp. 410–415.

McPherson, H. 1985. Optimum temperature and daylight for rapid generation turnover in pigeonpea. Unpublished report. DSIR, Palmerston North New Zealand.

Mehta, T.R. 1970. General. In *Pulse Crops of India.* P. Kachroo, ed. Indian Council of Agric. Research, New Delhi, pp. 1–13.

Mitra, M. 1931. Report of the Imperial Mycologist. *Sci. Rep. Agric. Res. Inst. Pusa* 1929–30: 58–71.

Naithani, S.P. 1941. Cytological studies on Indian pulses, Part I. The somatic chromosomes and the prochromosomes of *Cajanus. Proc. Nat. Acad. Sci. India* 11: 67–73.

Narayanaswamy, P., and Jagnathan, T. 1973. In *Annual Report 1972-73.* TNAU, Coimbatore, India, 134 pp.

Natrajan, M., Kannaiyan, J., Willey, R.W., and Nene, Y.L. 1985. Studies on the effect of cropping system on Fusarium wilt of pigeonpea. *Field Crop Res.* 10: 333–346.

Nene, Y.L. 1972. Viral disease of pulse crops in Uttar Pradesh. *G.B. Pant Univ. of Agric. and Tech. Res. Bull.* 4: 191.

Nene, Y.L., and Reddy, M.V. 1976. A new technique to screen pigeonpea for resistance to sterility mosaic. *Trop. Grain Legume Bull.* 5: 23.

Nene, Y.L., Kannaiyan, J., and Reddy, M.V. 1981. Resistance to major pigeonpea diseases. *Proc. Int. Workshop on Pigeonpea,* 15–19 Dec. 1980. Patancheru, A.P., India. Vol. 1, pp. 121–128.

Oldfield, G.N., Reddy, M.V., Nene, Y.L., and Reed, W. 1981. Préliminary studies on the eriophyid vector of sterility mosaic. *Int. Pigeonpea Newslett.* 1: 25–27.

Onim, J.F.M. 1981. Pigeonpea improvement in Kenya. *Proc. Int. Workshop on Pigeonpea,* 15–19 Dec. 1980. ICRISAT, Patancheru, AP, India, Vol. 1, pp. 427–436.

Oza, G.M. 1972. What is the native home of the pigeonpea? *Ind. Forester* 98(8): 477–478.

Pal, B.P. 1934. Recent progress in plant breeding at Pusa, Bihar. *Agric. Livestock India* 4(5): 511–512.

Pandey, R.K., and Singh, V.B. 1981. Influence of source and sink size on flower drop and seed yield of pigeonpea. *Indian J. Agric. Sci.* 51: 185–188.

Panwar, K.S., and Yadav, H.L. 1981. Response of short-duration pigeonpea (Cv. T 21) to early planting and phosphorus levels in different cropping systems. *Proc. Int. Workshop on Pigeonpea,* 15–19 Dec. 1980. ICRISAT, Patancheru, A.P., India. Vol. 1, pp. 37–44.

Pathak, G.N. 1970. Red gram. In *Pluse Crops of India.* Indian Council of Agricultural Research, New Delhi, India pp. 14–53.

Pathak, G.S., and Yadav, R.S. 1951. Spontaneously originated hexaploid and tetraploid plant in *Cajanus cajan* Millsp. *Curr. Sci.* 20: 304.

Pavgi, M.S., and Singh, R.A. 1971. Parasitic fungi from North India VIII. *Mycopathd. Mycologia Applicata* 43: 117–125.

Pawar, S.E., Thakare, R.G., Mitra, R., Krishna, T.S., and Bhatia, C.R., 1982. In *Pulse Production Constraints and Opportunities.* H.C. Srivastava, S. Bhaskaran, K.K.G. Menon, S. Ramanujam and M.V. Rao, eds. Oxford and IBH Pub. Co., New Delhi, pp. 361–367.

Plukenet, L. 1692. *Phytographia* 3, London.

Purseglove, J.W. 1968. *Tropical Crops. Dicotyledons,* Vol. 1. Longmans Green, London, pp. 236–241.

Rachie, K.O., and Gardner, C.O. 1975. Increasing efficiency in breeding partially outcrossing grain legumes. *Proc. Int. Workshop on Grain Legumes,* 13–16 Jan. 1975. ICRISAT Hyderabad, pp. 285–297.

Rahman, A.R. 1961. The effect of chemical pretreatment on the quality of dehydrated pigeonpeas. *J. Agric. Univ. Puerto Rico* 45 (3): 172–191.

Rao, M.R., and Willey, R.W. 1980. Preliminary studies of intercropping combinations based on pigeonpea and sorghum, *Exp. Agric.* 16: 29–39.

Rawson, H.M., and Constable, G.A. 1981. Gas exchange of pigeonpea: a comparison with other crops and model of carbon production and its distribution within the plant. *Proc. Int. Workshop on Pigeonpea,* 15–19. Dec. 1980. ICRISAT, Patancheru. Vol. 1, pp. 175–189.

Raychaudhary, S.P., and Naraini, T.K. 1977. *Virus and Mycoplasma Diseases of Plants in India.* Oxford & IBH Pub. Co., New Delhi, pp. 19–20.

Reddy, B.V.S., Reddy, L.J., and Murthi, A.N. 1977. Reproductive variants in *Cajanus cajan* (L.) Millsp. *Trop. Grain Legume Bull.* 7: 11.

Reddy, B.V.S., Green, J.M., and Bisan, S.S. 1978. Genetic male sterility in pigeonpea. *Crop Sci.* 18: 362–364.

Reddy, L.J. 1973. Interrelationship of *Cajanus* and *Atylosia* species as revealed by hybridization and pachytene analysis. Unpublished Ph.D. thesis, Indian Institute of Technology, Kharagpur, India.

Reddy, L.J. 1981a. Pachytene analysis in *Cajanus cajan, Atylosia lineata* and their hybrid. *Cytologia* 46: 397–412.

Reddy, L.J. 1981b. Pachytene analysis in *Atylosia sericea* and *Cajanus cajan* × *A. sericea* hybrid. *Cytologia* 46: 576–577.

Reddy, L.J. 1981c. Pachytene analysis in *Atylosia scarabaeoides* and *Cajanus cajan* × *A. scarabaeoides* hybrid. *Cytologia* 46: 579–589.

Reddy, L.J., and De, D.N. 1983. Cytomorphological studies in *Cajanus cajan* × *Atylosia lineata*. *Ind. J. Genet.* 43: 96–103.

Reddy, L.J., Saxena, K.B., Sharma, D., and Green, J.M. 1981. Some combining ability analyses in pigeonpea (*Cajanus cajan* (L.) Millsp.). In *Proc. Int. Workshop* on pigeonpea 15–19 Dec. 1980. ICRISAT, Patancheru, AP, India, Vol. 2, pp. 93–104.

Reddy, M.V., Beniwal, S.P.S., Sheila, V.K., Sithanantham, S., and Nene, Y.L. 1986. Role of eriophyid mite *Aceria cajani* (Order: Acariformes; Family: Eriophidae) in transmission and spread of sterility mosaic of pigeonpea. Paper presented at Int. Cong. Acarology 1986, Bangalore, India.

Reddy, P.R., Rao, V.K., and Rao, N.G.P. 1979. Heterosis and combining ability in pigeonpea. *Indian J. Genet.* 39: 240–246.

Reddy, S.J., and Virmani, S.M. 1981. Pigeonpea and climatic environment. *Proc. Int. Workshop on Pigeonpea.* 15–19 Dec. 1980. ICRISAT, Patancheru, AP, India. Vol. 1, pp. 259–270.

Reed, W., Lateef, S.S., and Sithanantham, S. 1981. Pest management in low-input pigeonpea. *Proc. Int. Workshop on Pigeonpea,* 15–19 Dec. 1980. ICRISAT, Patancheru, AP, India. Vol. 1, pp. 99–105.

Remanandan, P. 1981. The wild gene pool of *Cajanus* at ICRISAT, Present and Future. *Proc. Int. Workshop on Pigeonpea,* 15–19 Dec. 1980. Patancheru, AP, India. Vol. 2, pp. 29–38.

Roy, A., and De, D.N. 1965. Intergeneric hybridization of *Cajanus* and *Atylosia*. *Sci. Cult.* 31–32: 93–95.

Roy, B. 1933. Studies in the development of female gametophyte in some leguminous crop plants in India. *Indian J. Agric. Sci.* 3: 1098–1107.

Royes, V.W. 1976. Pigeonpea *Cajanus cajan* (Leguminosae; Papilionaceae): In *Evolution of Crop Plants.* N.W. Simmonds, Longman, London, pp. 154–156.

Roy Sharma, R.P., Thakur, H.C., and Sharma, H.M. 1981. Pigeonpea as a *Rabi* crop in India. *Proc. Int. Workshop on Pigeonpea,* 15–19 Dec. 1980. ICRISAT, Patancheru, AP, India. Vol. 1, pp. 26–36.

Rumphius, G.E. 1747. *Herbarium Amboinense,* Paris 5 Amstelaedami.

Sanchez Nieva, M, Gonzalez, A., and Benero, J.R. 1961. The freezing of pigeonpea for market. *J. Agric. Unvi. Puerto Rico* 45: 205–216.

Saraf, C.S., and Amarchand. 1980. Agronomic and economic aspects of intercropping systems with grain legumes in relation to phosphate levels. *Indian J. Genet.* 41A: 31–38.

Saxena, K.B. 1981. Report of research work carried out by K.B. Saxena at the Department of Agriculture, University of Queensland, Brisbane, Australia, 40 pp.

Saxena, K.B., and Sharma, D. 1981. Variation for leaf characteristics and relationship with grain yield and seed size in pigeonpea. *Indian J. Genet.* 41: 340–345.

Saxena, K.B., and Sharma, D. 1983. Early generation testing in pigeonpea *Cajanus cajan* (L.) Millsp. *Trop. Plant Sci. Res.* 1: 309–313.

Saxena, K.B., and Sharma, D. 1990. Pigeonpea genetics. In *The Pigeonpea.* Y.L. Nene, S.D. Hall, and V.K. Sheila, eds. CAB International, Wallingford, UK, pp. 137–158.

Saxena, K.B., Sharma, D., and Green, J.M. 1976. Pigeonpea ratooning—an aid to breeders. *Trop. Grain Legume Bull.* 4: 21.

Saxena, K.B., Byth, D.E., Dundas, I.S., and Wallis, E.S. 1981a. Genetic control of sparse pollen production in pigeonpea. *Int. Pigeonpea Newslett.* 1: 17–18.

Saxena, K.B., Byth, D.E., Wallis, E.S., and Delacy, I.H. 1981b. Genetic analysis of diallel cross of early flowering pigeonpea lines. *Proc. Int. Workshop of Pigeonpea,* ICRISAT, Patancheru, AP, India, Vol. 2, pp. 81–92.

Saxena, K.B., Dundas, I.S., Byth, D.E., and Wallis, E.S. 1982. Effect of spontaneous chromosome doubling on some vegetative and reproductive characteristics of pigeonpea (*Cajanus cajan* (L.) Millsp.). *Legume Res.* 5: 83–86.

Saxena, K.B., Wallis, E.S., and Byth, D.E. 1983. A new gene for male sterility in pigeonpeas. *Heredity* 51: 419–421.

Saxena, K.B., Gupta, M.D., and Faris, D.G. 1984. A non-flowering mutant of pigeonpea (*Cajanus cajan* (L.) Millsp.). *Int. Pigeonpea Newslett.* 3: 11–12.

Saxena, K.B., Gupta, M.D., Chauhan, Y.S., and Faris, D.G. 1985. Low temperature induced floral abscission in pigeonpea. *Int. Pigeonpea Newslett.* 4: 27–28.

Saxena, K.B., Faris, D.G., and Kumar, R.V. 1986a. Breeding for special traits. Progress Report, 7, June 1983–1984, Pigeonpea Breeding. ICRISAT, Patancheru, India, 93 pp.

Saxena, K.B., Faris, D.G,. Reddy, L.J., Sharma, D., Reddy, B.V.S., Gupta, S.C., and Green, J.M. 1986b. Prospects for hybrid pigeonpea. In *New Frontiers in Breeding Researches,* Proc. Fifth Int. Cong. SABRAO, pp. 379–388.

Saxena, K.B., Durga, B.K., and Singh, L. 1987a. A case of cross-incompatibility in pigeonpea. *Int. Pigeonpea Newslett.* 6: 33–34.

Saxena, K.B., Faris, D.G., Singh, U., and Kumar, R.V. 1987b. Relationship between seed size and protein content in high selections from crosses between pigeonpea and three *Atylosia* spp. *Plant Food Hum. Nutr.* 36: 335–340.

Saxena, K.B., Sharma, D., and Faris, D.G. 1987c. Ineffectiveness of wrapped flower in inhibiting cross fertilization in pigeonpea. *Euphytica* 36: 295–297.

Saxena, K.B., Kumar, R.V., Faris, D.G., and Singh, L. 1987d. Breeding for special traits. Pigeonpea Breeding Progress Report 9. ICRISAT, Patancheru, India, 156 pp.

Saxena, K.B., Singh, A.K., and Faris, D.G. 1988. The morphogenesis and inheritance of an open carpel mutant *cd* 1 in pigeonpea. *J. Hered.* 79: 75–76.

Saxena, K.B., Gupta, S.C., Kumar, R.V., Reddy, L.J., Singh, L., and Faris, D.G. 1989a. ICPH 8 — The first promising short-duration pigeonpea hybrid. *Int. Pigeonpea Newslett.* 9: 8–10.

Saxena, K.B., Singh, L., Gupta, S.C., Chauhan, Y.S., and Reddy, M.V. 1989b. Hybrid pigeonpea development and prospects. Paper presented at a seminar on Strategies for Increasing Pulses Production, 3–4 Jan. 1989, Agricultural Research Station, Badnapur (India), 23 pp.

Saxena, K.B., Singh, L., and Gupta, M.D. 1990a. Variation for natural outcrossing in pigeonpea. *Euphytica* 46: 143–148.

Saxena, K.B., Singh, L., Reddy, M.V., Lateef, S.S., Sharma, S.B., and Remanandan, P. 1990b. Intraspecies variation in *Atylosia scarabaeoides* (L.) Benth., a wild relative of pigeonpea (*Cajanus cajan* (L.) Millsp.). *Euphytica* 49: 185–191.

Saxena, K.B., Singh, L., and Reddy, L.J. 1990c. Low natural outcrossing in cleistogamous pigeonpea mutant. *Int. Pigeonpea Newslett.* 11: 9–10.

Saxena, K.B., Chauhan, Y.S., Ariyanayagam, R.P., and Johansen, C. 1992a. Annual pigeonpea—prospects for a new plant type. *Int. Pigeonpea Newslett.* 15: 8–9.

Saxena, K.B., Ariyanayagam, R.P., and Reddy, L.J. 1992b. Genetics of a high selfing trait in pigeonpea. *Euphytics* 59: 125–127.

Saxena, K.B., Jayasekera, S.J.B.A., and Ariyaratne, H.P. 1992c. Pigconpea varietal adaptation and production studies in Sri Lanka. Report of work Phase I. ICRISAT, 45 pp.

Sengupta, J.C. 1955. The physiology of growth and development of some crop plants. *Bull. Bot. Soc. Bengal* 9: 62–80.

Sengupta, K. 1982. Promise of cultivation of Arhar (*Cajanus cajan*) in *rabi* season. In *Pulses and Oilseeds in West Bengal.* D.K. Mukherji, ed. Directorate of Agric., Calcutta, India.

Seth, M.L. 1962. Transmission of pigeonpea sterility by an eriophyid mite. *Ind. Phytopathol.* 15: 225–227.

Seth, M.L. 1965. Further observations and studies on pigeonpea sterility. *Ind. Phytopathol.* 18: 317–319.

Sharma, D., and Green, J.M. 1977. Problems and prospects of upgrading yield levels in pigeonpea. *Proc. Third Int. Cong. SABRAO,* Vol. 2, pp. 6–10.

Sharma, D., and Green, J.M. 1980. Pigeonpea. In *Hybridization of Crop Plants* Amer. Soc. Agron. Crop Sci. Society of America, Wisconsin, pp. 471–481.

Sharma, D., Kannaiyan, J., and Reddy, L.J. 1982. Inheritance of resistance to blight in pigeonpea. *Plant Dis.* 66: 22–25.

Sharma, D., Gupta, S.C., Rai, G.S., and Reddy, M.V. 1984. Inheritance of resistance to sterility mosaic disease in pigeonpea I. *Ind. J. Genet.* 44: 84–90.

Sharma, D., Kannaiyan, J., and Saxena, K.B. 1987. Sources and inheritance to *Alternaria* blight in pigeonpea. *SABRAO J.* 19: 109–114.

Sharma, H.K., Singh, L., and Sharma, D. 1973. Genetic analysis of flower initiation in pigeonpea. *Indian J. Genet.* 33: 393–397.

Shaw, F.J.F. 1936. Studies in Indian pulses: The inheritance of morphological characters and wilt resistance in arhar (*Cajanus indicus* Spreng.). *Indian J. Agric. Sci.* 6: 139–187.

Sheldrake, A.R. 1979. A hydrodynamical model of pod set in pigeonpea. *Ind. J. Plant Phys.* 22: 137–143.

Sheldrake, A.R. 1984. Pigeonpea. In *The Physiology of Tropical Field Crops.* P.R. Goldsworthy and N.M. Fisher, eds. John Wiley and Sons, Ltd., Chichester, U.K., pp. 385–417.

Sheldrake, A.R., and Narayanan, A. 1979a. Growth development, and nutrient uptake in pigeonpea (*Cajanus cajan*). *J. Agric. Sci. Camb.* 92: 513–526.

Sheldrake, A.R., and Narayanan, A. 1979b. Comparison of earlier and later formed pods in pigeonpea (*Cajanus cajan* (L.) Millsp.). *Ann. Bot.* 43: 459–466.

Sheldrake, A.R., Narayanan, A., and Kannaiyan, J. 1978. Some effects of physiological state of pigeonpea on the incidence of wilt disease. *Trop. Grain Legume Bull.* 11/12: 24–25.

Singh, B.V., Pandey, M.P., and Pandya, B.P. 1980. Pod set in pigeonpea. *Indian J. Genet.* 40: 568–572.

Singh, B.V., Pandya, B.P., Gautam, P.L., Beniwal, S.P.S., and Pandey, M.P. 1983. Inheritance of resistance to sterility mosaic virus in pigeonpea. *Indian J. Genet.* 43: 487–493.

Singh, L. 1980. Prevalent practices of mixed cropping and cropping systems of pulse crops in India. *Indian J. Genet.* 40A: 11–15.

Singh, L. and Shrivastava, M.P. 1976. Cultivation systems and varietal adaptations of pigeonpea (*Cajanus cajan* (L.) Millsp.) in Madhya Pradesh. *Indian J. Genet.* 36: 293–300.

Singh, L., Maheshwari, S.K., and Sharma, D. 1971. Effect of date of planting and plant population on growth, yield, and yield components and protein content of pigeonpea (*Cajanus cajan* (L.) Millsp.). *Indian J. Agric. Sci.* 41(6): 535–538.

Singh, U., Jambunathan, R., Saxena, K.B., and Faris, D.G. 1986. Nutritive value of green and mature seed of pigeonpeas. Paper presented at ACIAR Workshop on Grain Legume, Khon Kaen, Thailand, 1986.

Sinha, S.S.N., and Kumar, P. 1979. Mitotic analysis in thirteen varieties of *Cajanus cajan* (L.) Millsp. *Cytologia* 44: 571–580.

Sloane, H. 1725. A voyage to the islands Madera. Barbados, Nieves, S. Christopher and Jamaica. London, Vol. 2, 31: 383.

Solomon, S., Argikar, G.P., Salanki, M.S., and Morbad, I.R. 1957. A study of heterosis in *Cajanus cajan* (L.) Millsp. *Indian J. Genet.* 17: 90–95.

Spence, J.A., and Williams, S.J.A. 1972. Use of photoperiod to change plant design. *Crop Sci.* 12: 121–122.

Thompson, J.A., Kumar Rao, J.D.V.K., and Dart, P.J. 1981. Measurement of inoculation responses in pigeonpea. In *Proc. Int. Workshop on Pigeonpea,* 15–19 Dec. 1980. ICRISAT, Patancheru, AP, India, Vol. 1, pp. 249–253.

Thorne, J.C., and Fehr, W.R. 1970. Exotic germplasm for yield improvement in 2-way and 3-way soybean crosses. *Crop Sci.* 10: 677–678.

Thothathri, K., and Jain, S.K. 1981. Taxonomy of the genus *Cajanus* DC. In *Proc. Int. Workshop on Pigeonpea,* 15–19 Dec. 1980. ICRISAT, Patancheru, AP, India, Vol. 2, 3–7.

Turnbull, L.V., Whiteman, P.C., and Byth, D.E. 1981. The influence of temperature and photoperiod on floral development of early flowering pigeonpea. In *Proc. Int. Workshop on Pigeonpea,* 15–19 Dec. 1980. ICRISAT, AP, India. Vol. 2, pp. 217–222.

van der Maesen, L.J.G. 1980. India is the native home of the pigeonpea. In *Liber Gratulatorius in honerem H.C.D. de Wit.* Agric. Univ. miscellaneous paper, 19. Wageningen, the Netherlands, pp. 257–262.

van der Maesen, L.J.G. 1981. Taxonomy of *Cajanus. Proc. Int. Workshop on Pigeonpea,* 15–19 Dec. 1980. ICRISAT, Patancheru, AP, India. Vol. 2, pp. 9–13.

van der Maesen, L.J.G. 1986. *Cajanus* DC and *Atylosia* W. & A. (Leguminosae). *Agric. Univ.*, Wageningen Papers, 85–89 (1985), the Netherlands.

Vavilov, N.I. 1939. The new systematics of cultivated plants. In *The New Systematics.* J. Huxley, ed. Oxford University Press, London, pp. 549–566.

van Rheede, D.H. 1686. *Hortus Indicus Malabaricus,* 6, 13: 23–24.

Veeraswamy, R., and Sheriff, N.M. 1973. Artificial crossing in *Cajanus cajan* (L.) Millsp. *Madras Agric. J.* 60(9–12): 1826–1827.

Veeraswamy, R., Rangasamy, P. and Sherieff, N.M. 1975. CO-2 red gram—a new strain with early maturity and improved plant type. *Madras Agric. J.* 62: 541–543.

Venkateswarlu, S., Reddy, A.R., Singh, O.N., and Chauhan, V.B. 1981. *Alternaria blight* : The most serious disease of post-rainy season (Rabi) pigeonpea. *Int. Pigeonpea Newslett.* 1: 28–29.

Wallis, E.S., Byth, D.E., and Saxena, K.B. 1981a. Flowering responses of thirty-seven early maturing lines of pigeonpea. *Proc. Int. Workshop on Pigeonpea,* 15–19 Dec. 1980. ICRISAT, Patancheru, AP, India. Vol. 2, pp. 143–150.

Wallis, E.S., Saxena, K.B., and Byth, D.E. 1981b. A new source of genetic male sterility in pigeonpea. *Proc. Int. Workshop on Pigeonpea*, 15–19 Dec. 1980. ICRISAT, Patancheru, AP, India, Vol. 2, pp. 105–108.

Whiteman, P.C., Byth, D.E., and Wallis, E.S. 1985. Pigeonpea (*Cajanus cajan* (L.) Millsp.) In *Grain Legume Crops.* R.J. Sumerfield and E.H. Roberts, eds. Collins Professional and Technical Books, London, 658–698.

Williams, F.J., Grewal, J.S., and Amin, K.S. 1968. Serious and new disease of pulse crops in India in 1966. *Plant Dis. Rep.* 52: 300–304.

Williams, F.J., Amin, K.S., and Baldev, B. 1975. Phytophthora stem blight of *Cajanus cajan. Phytopathol.* 65: 1029–1030.

Williams, I.H. 1977. Behaviour of insects foraging on pigeonpea (*Cajanus cajan* (L.) Millsp.) in India. *Trop. Agric.* 54: 353–363.

Willey, R.W. 1985. Evaluation and presentation of intercropping advantages. *Exp. Agric.* 21: 119–133.

Wilsie, C.P., and Takahashi, M. 1934. Natural crossing in pigeonpea. *J. Agric. Res.* 49(10): 923–925.

Yadahalli, Y.H., and Balakrishna, P.C. 1987. Performance of short-duration pigeonpea genotypes during summer. *Int. Pigeonpea Newslett.* 6: 39–41.

Zeven, A.C., and Zhukosky, P.M. 1975. Dictionary of cultivated plants. In *The New Systematics.* J. Huxley, ed. Oxford Univ., London, pp. 549–566.

3. LENTIL

F.J. Muehlbauer
Reserch Geneticist, U.S. Department of Agriculture, Agricultural Research Service, Grain Legume Genetics and Physiology Research, Washington State University, Pullman, Washington 99164-6421, U.S.A.

N.I. Haddad
Agronomist and Regional Coordinator, West Asia Program, ICARDA, P.O. Box 950764, Amman, Jordan

A.E. Slinkard
Agronomist and food legume breeder, Crop Development Centre, University of Saskatchewan, Saskatoon, Saskatchewan, Canada S7N 0W0

B. Sakr
Graduate research assistant, Department of Crop and Soil Sciences, Washington State University, Pullman, Washington 99163-6420, U.S.A.

Introduction

Lentil (*Lens culinaris* Medic.) is one of the principal food legume crops cultivated in the Indian subcontinent and throughout the developing countries of the Middle East and northern Africa, where they are an important source of protein and calories for human consumption. In dry areas, lentils have a reputation as a crop that will produce at least some yield where other crops are likely to fail completely under the prevailing conditions of poor fertility and limited moisture. Crop residues are also valuable as livestock feed in many regions where grazing is limited. The crop is important in dryland farming systems; it offers farmers an alternative to cereal grains and, by fixing dinitrogen in symbiosis with *Rhizobium*, can contribute to the nitrogen budget of infertile soil.

World production of lentil in 1990 was estimated at nearly 2.7 million t, which is an increase of 140 per cent over production recorded in 1974 (FAO, 1991). About 88 per cent of the total production was from developing countries; and, as in many grain legumes, production increases reflected greater areas harvested far more than gains in productivity (FAO, 1974, 1984, 1991). Significant shifts in lentil production have taken place throughout the last 25 years. For example, the area sown declined sharply in Syria, Jordan, Iraq, Ethiopia, and countries of the former USSR. In most of those countries, lentil production is costly because the crop is harvested entirely by hand (Khayrallah, 1981; Haddad and Arabiat, 1985). Lentil production has remained more or less constant in India, Pakistan, and Bangladesh, countries where the crop is a dietary mainstay and where labour is more readily available.

A dramatic increase in lentil production (to more than 900,000 t annually) has taken place in Turkey, where output has nearly doubled between 1984 and 1990 (FAO, 1991). Turkey is now the world's largest lentil exporter. The increased production was brought about mainly by extending the crop to replace summer fallowing in cereal production systems in those areas of the country that receive sufficient rainfall to permit annual cropping (Acikgoz *et al.,* 1992).

In North and South America, where mechanized agricultural systems are used, the area sown to lentil has remained more or less constant. Yearly fluctuations in production in the United States are the result of the wheat-legume-fallow cropping system followed and responses to export market demands (Summerfield *et al.,* 1982). An average of 85 per cent of the US production is exported; Chile and Argentina are also large exporters. Canada has recently become a significant producer of lentil, and of the nearly 230,000 t produced in that country annually (FAO, 1991), nearly all are exported. Canada is the world's second largest exporter after Turkey.

Yields of lentils vary widely between years and countries. Unfortunately, no significant yield increases have been reported in any of the major lentil-producing countries during the last decade. Furthermore, the smallest yields are recorded in those countries where the crop is most important and where nearly all the production is consumed domestically—as in the case of India, Syria, Ethiopia, and Morocco. The factors responsible for these poor yields include primitive production systems, non-optimal planting dates and seeding rates, lack of water and fertilizers, and inadequate pest and weed control. Moreover, with the exception of India, improved cultivars are seldom available, and farmers rely on traditional and indigenous land races. In India, an increasing number of improved cultivars of lentil have been developed and released for cultivation (Malik *et al.,* 1988). On the other hand, yields are still relatively small in those countries where the crop is grown with modern, mechanized agricultural methods and inputs, but where the crop still depends on natural rainfall, as in the United States, Canada, and Argentina. The yields typical of those countries (990 kg/ha) are below what would be expected from improved cultivars grown in what is considered to be a better production system (1800–3000 kg/ha). The world record farm yield is 4200 kg/ha from New Zealand.

Many of the problems related to poor yield potential can be addressed in suitably designed crop improvement programmes. And so, the better understanding of genetics, cytogenetics, and breeding of the crop and issues related to crop improvement in lentil are important objectives. They are the principal subjects of this chapter.

Origin of Lentil

The Near East arc and Asia Minor are believed to be the centres of origin of cultivated lentils (Zohary, 1972; Williams *et al.,* 1974). Most points of view favour *L. orientalis* as the progenitor of *L. culinaris* but some still favour *L. nigricans.* The various conclusions are based on discoveries of carbonized remains of cultivated lentils in the same region over which *L. orientalis* is distributed, together with the appearance of lentils in early Neolithic settlements that date back to 7000–6000 BC (Helbaek, 1959). The conclusion is that the Near East is the most probable area where lentils were domesticated. (A review of the evidence and points of view for the place of domestication of lentils is available in Ladizinsky, 1979a.)

Botany

Lentil plants are herbaceous annuals with slender stems and branches. Plant height usually ranges from 25 to 30 cm for the majority of genotypes, but may vary from extremes of 15 to 75 cm depending on genotype and environmental conditions (Saxena and Hawtin, 1981).

Plants have a slender taproot with fibrous lateral roots. Several rooting patterns, which range from a much-branched shallow root system to intermediate types that are less branched and more deeply rooted, have been observed on similar soil types (Nezamuddin, 1970). The taproot and lateral roots in surface layers of the soil have numerous, indeterminate nodules that vary in shape from round to elongate (Saxena and Hawtin, 1981).

The herbaceous stems of lentil plants are square and ribbed and usually thin and weak. Primary branches arise directly from the main stem and can emerge from the cotyledonary node below ground. Many secondary branches usually arise from the primary branches but the habit of plants is plastic and so the number of primary and secondary branches depend on the genotype and on stand density and environmental conditions (Wilson and Teare, 1972; Malhotra *et al.,* 1974).

The leaves are relatively small compared to those of other large-seeded food legumes. They are described as pinnate or imparipinnate and comprise as many as 14 sessile, ovate or elliptic, or abovate or lanceolate leaflets. They vary in length from 1 to 3 cm. Each leaf is subtended by two small stipules that may or may not terminate in a tendril. The entire stipules are oblong-lanceolate and unappendaged (Davis and Plitmann, 1970; Summerfield *et al.,* 1982).

The flowers are borne singly or in multiples on peduncles that originate from the upper nodes of the plant. Each peduncle normally bears from one to three, but rarely four flowers, although seven flowers per peduncle have been reported for plants grown in a glasshouse (Hawtin, 1977).

The individual flower is complete and has a typical papilionaceous ('butterflylike') structure (Fig. 1A). Flowers are small (4 to 8 mm long) and white, pale purple, or purple-blue. The flower has a calyx comprising five, equally elongated sepals that equal or exceed the length of the corolla of the unopened flower. The corolla has a standard, two wings, and two lower petals that lie internal to the wings and are united at their lower margin to form the keel (Muehlbauer *et al.*, 1980; Summerfield *et al.*, 1982). The stamens are diadelphous (9 + 1) with the upper vexillary stamen free (Fig. 1B). The ovary is flat and glabrous; it normally contains one or two ovules that alternate along the margin and terminate in a short curved style. The style is pilose on the inner side; it usually develops at a right angle to the ovary, and is flattened on the outer side (Muehlbauer *et al.*, 1980; Saxena and Hawtin, 1981; Summerfield *et al.*, 1985).

The fruits (referred to as pods) are oblong, laterally compressed, 6–20 mm long and 3.5–11 mm wide, and usually contain one or two, but rarely three seeds (Saxena and Hawtin, 1981).

Seeds are lens-shaped and weigh between 20 and 80 mg. Their diameter ranges from 2 to 9 mm and the testa may be light red, green or greenish-red, grey, brown, or black. Purple and black mottling and speckling of seeds is also common in some cultivars and accessions (Duke, 1981; Saxena and Hawtin, 1981). Seed size differs according to genotype, and researchers frequently follow the classification of Barulina (1930), who grouped lentils as '*macrosperma*', with large seeds that range from 6 to 9 mm in diameter, and '*microsperma*', with smaller seeds that range from 2 to 6 mm in diameter. The *macrosperma* types are common to the mediterranean basin and in the western hemisphere, while the *microsperma* predominate throughout the Indian subcontinent and in parts of the Near East. Other groupings of cultivated lentils have been described and include *europeae, asiaticae, intermediate, subspontaneae, aethiopicae,* and *pilosae.* Detailed descriptions of these groups are available in Muehlbauer *et al.* (1985).

Taxonomy

Cultivated lentils belong to the genus *Lens*, which is associated with other genera of the tribe Vicieae (Kupicha, 1981). The Vicieae tribe has been reported to contain, in addition to *Lens, Vicia* L., *Pisum* L., *Lathyrus* L., and *Vavilovia* A. Fed. *Cicer* L. has in the past been considered part of the Vicieae but anatomical and morphological evidence indicates that it is quite

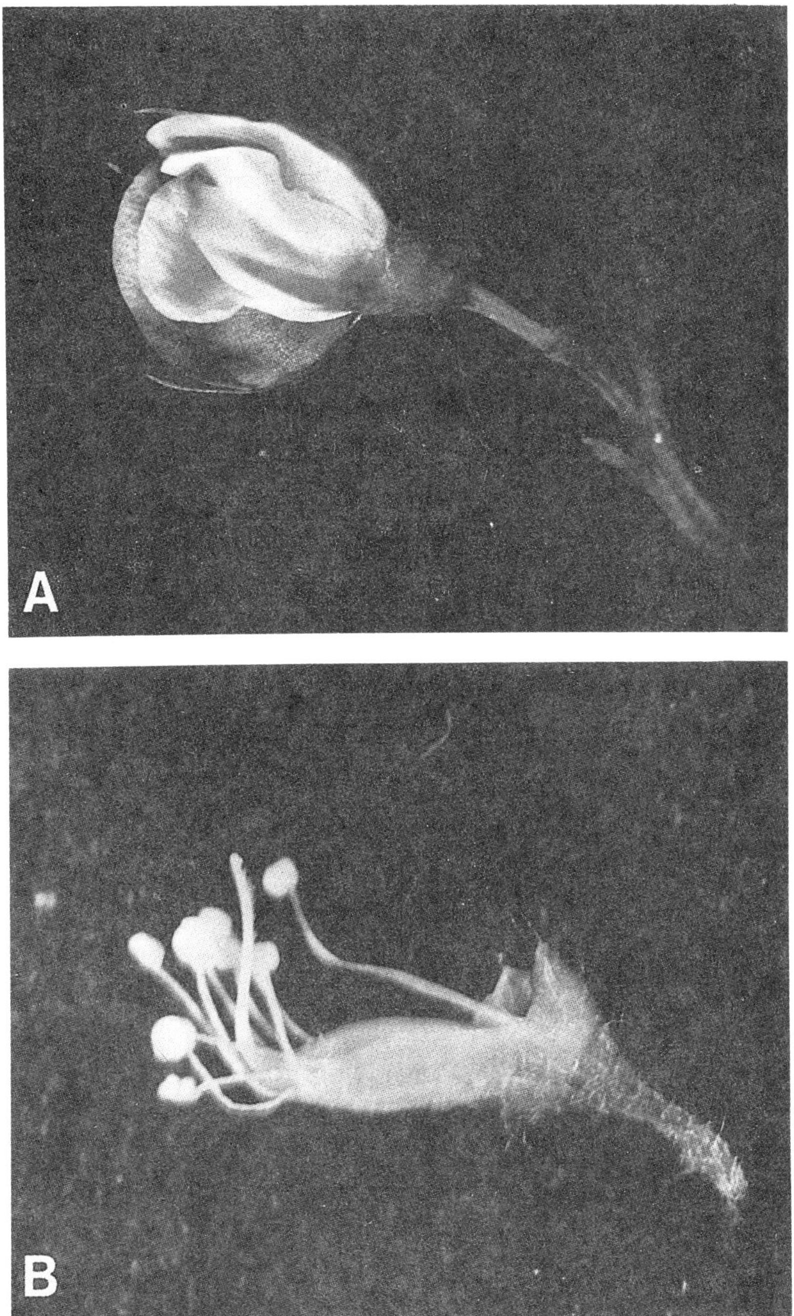

Fig. 1. Typical lentil flower (A) and staminal column and stigma (B)

different from other members of the tribe and it is now more appropriately classified in the monogeneric tribe Cicereae Alef. (Kupicha, 1975; Clarke and Kupicha, 1976; Kupicha, 1977).

In the Vicieae, *Lens* is described by Davis and Plitmann (1970) as holding a position that is intermediate between *Vicia* and *Lathyrus*, but closer to *Vicia* sec. Ervum. *Lens* is distinguished from *Vicia* by clayx morphology, stylar characters, and pod and seed shape. The calyx tube in *Lens* is subequal, while in *Vicia* the calyx tube is oblique. Other calyx traits that distinguish the two genera include the lengths of the calyx teeth relative to the corolla tube.

Medikus is considered the authority for cultivated lentils (*L. culinaris*) because the publication of *L. culinaris* Medikus predates that of *L. esculenta* Moench. (Slinkard, 1974). *Lens montbretii* was recently shown by Ladizinsky and Sakar (1982) not to be a *Lens* species. They used morphological and karyological information to show considerable differences between the two species. Their foremost observation was that *L. montbretii* has 2n = 12 chromosomes whereas the other *Lens* species have 2n = 14. Based on that information, it is clear that *L. montbretii* is more appropriately reclassified as *V. montbretii*. Characteristics of the *Lens* species are shown in Fig. 2.

Until the work of Ladizinsky *et al.* (1984) on the biological species of *Lens*, the genus was considered to be composed of the cultigen, *L. culinaris*, and four indigenous species: *L. orientalis* (Boiss) Hand. Mazz., *L. nigricans* (M. Bieb.) Godron, *L. ervoides* (Brign.) Grande, and *L. montbretii* (Fischer and C. Meyer) Davis and Plitm. Synonyms were used by Barulina (1930) for these species and included: *L. esculenta* Moench for *L. culinaris*; *L. lenticula* (Schreb.) Alefed. for *L. ervoides*; and *L. kotschyana* (Boiss.) Alefed. for *L. montbretii*. A more complete list of synonyms can be found in Barulina (1930) and in Cubero (1981).

Ladizinsky *et al.* (1984), using cytological evidence and morphological distinctions, have suggested the following classification within *Lens*:

1)	Stipules lanceolate, entire if semi-hastate, situated at right angles to the stem, and calyx teeth ± as long as corolla	*L. culinaris*
	stipules semi-hastate, mature fruits dehisce when dry	ssp. *odemensis*
	stipules lanceolate, mature fruits dehisce when dry	ssp. *orientalis*
	mature fruits do not dehisce, or only slightly so, cultivated	ssp. *culinaris*
2)	Stipules semi-hastate	*L. nigricans*
	stipules entire or slightly dentate, calyx teeth much shorter than corolla	ssp. *ervoides*
	stipules pointing upward and dentate, calyx teeth as long as or longer than corolla	ssp. *nigricans*

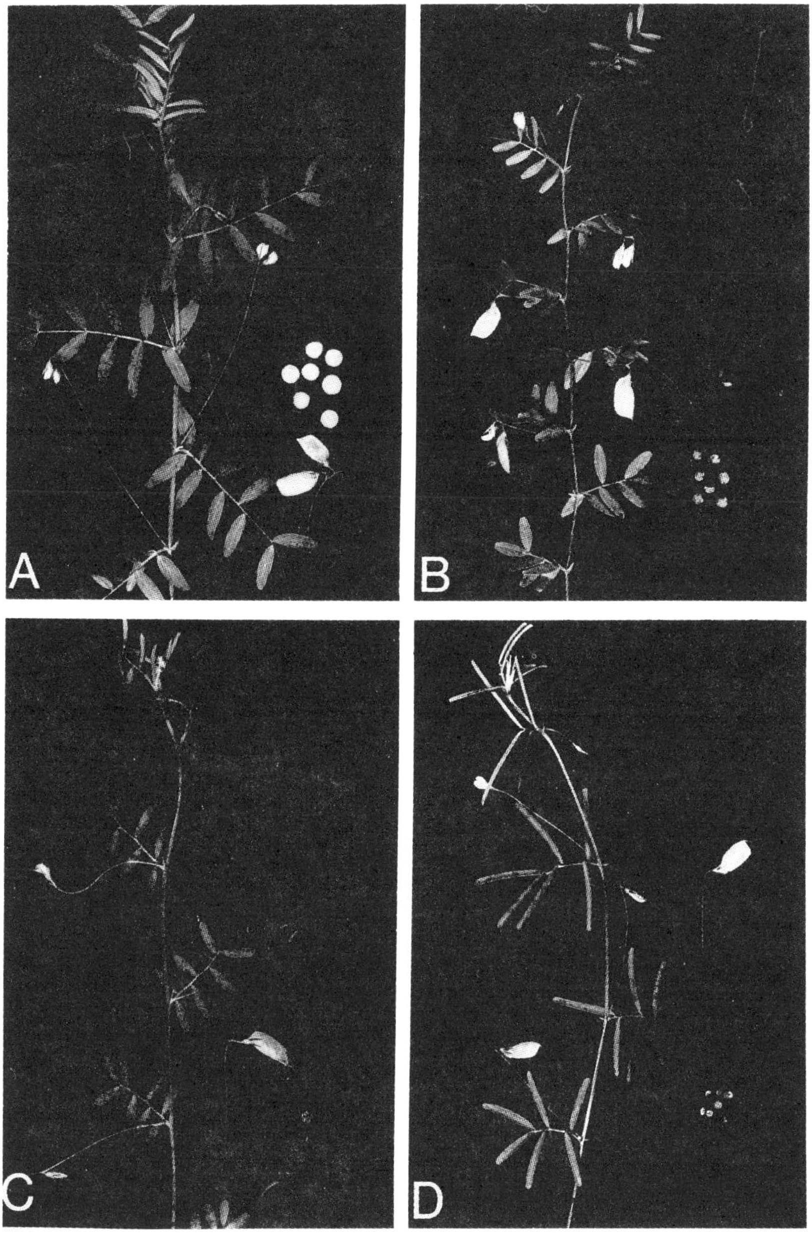

Fig. 2. Typical plants, pods, seeds, and stipules of *Lens culinaris* ssp. *culinaris* (A); *L. culinaris* ssp. *orientalis* (B); *L. nigricans* ssp. *nigricans* (C); and *L. nigricans* ssp. *ervoides* (D). Lanceolate stipules of ssp. *orientalis* and semi-hastate stipules of ssp. *nigricans* are shown.

The cytogenetic basis and crossability evidence for this classification is discussed in the section on cytogenetics.

Genetics

Qualitatively Inherited Traits

Knowledge of lentil genetics has progressed significantly in recent years with the identification of numerous genes that determine the inheritance of morphological traits and the identification of various isozyme markers. It is expected that the number of genes identified in lentils will continue to increase and that a detailed gene map will be constructed that will serve to facilitate the orderly accumulation of genetic information in the species.

The following discussion covers the genetics of the cotyledon, flower, seed coat (testa) and epicotyl colour, the number of flowers per inflorescence, growth habit, pod dehiscence, and virus resistance. A list of described genes is presented in Table 1. The use of isozyme markers is

Table 1. List of gene symbols in lentils

Symbol—allele	Character	Reference
Fn	Number of flowers per inflorescence	Gill and Malhotra (1980)
Gh	Plant growth habit	Ladizinsky (1979b)
Gs	epicotyl colour	Ladizinsky (1979b)
I	cotyledon colour inhibitor	Slinkard (1978b)
O	cotyledon colour (synonymous with *Yc)*	Singh (1978)
Ggc	grey seed coat ground colour	Vandenberg and Slinkard (1990)
Tgc	tan seed coat ground colour	Vandenberg and Slinkard (1990)
P	flower colour	Lal and Srivastava (1975)
Pi	pod indehiscence	Ladizinsky (1979b)
Sbv	resistance to pea seed-borne mosaic virus	Haddad *et al.* (1978)
Scp	seed coat spotting	Ladizinsky (1979b)
V	flower colour	Lal and Srivastava (1975) Wilson and Hudson (1978)
W	flower colour	Wilson and Hudson (1978)
Yc	cotyledon colour	Slinkard (1978b)
Glp	glabrous pod	Vandenberg and Slinkard (1989)
Grp	green pod colour	Vandeberg and Slinkard (1989)
Tnl	tendrilless leaf	Vandenberg and Slinkard (1989)
Chl	chlorina chlorophyll mutant	Vandenberg and Slinkard (1989)

discussed and a standardized system of genetic nomenclature for lentil is suggested.

Cotyledon colour. The first research reports on genetics of lentil were the studies of cotyledon colour by Tschermak (1928) and by Wilson *et al.* (1970). Observations of F_1s from reciprocal crosses indicated that red cotyledon colour was completely dominant to yellow. The F_2 seeds segregated 3 red: 1 yellow, indicating that the trait is controlled by a single dominant gene. Wilson (1972) noted that red cotyledon F_1 seeds resulting from crosses between yellow cotyledon maternal parents and red cotyledon paternal parents were readily distinguishable from selfed maternal seeds with yellow cotyledons. This observation suggested the use of cotyledon colour (yellow cotyledon maternal parent) as a genetic marker in hybridization programmes to ensure that resulting seeds were indeed hybrids.

Both Singh (1978) and Slinkard (1978b) confirmed the mode of inheritance of cotyledon colour. Singh referred to the red cotyledons as orange and proposed the gene symbols *O* for orange and *o* for yellow cotyledons. He also noted that the intensity of orange cotyledon colour did not vary in reciprocal crosses and so concluded that there was no cytoplasmic effect on gene expression. In addition to red and yellow cotyledons, Slinkard (1978b) reported the inheritance of green cotyledons. The F_2 populations of crosses between red cotyledon parents and green cotyledon parents revealed that red was completely dominant to both yellow and green, and that yellow was completely dominant to green. He obtained an F_2 ratio of 9 red : 3 yellow : 4 green that indicated a dihybrid ratio with a recessive colour inhibitor gene. Based on those data, Slinkard (1978b) proposed the following gene symbols: *yc* = yellow cotyledons; *Yc* = red cotyledons; and *i* = the inhibitor of cotyledon colour alleles such that dominant *I* does not inhibit either *Yc* or *yc,* but homozygous recessive *i* in the presence of either *Yc* or *yc* gives green cotyledons. These symbols describe the inheritance of cotyledon colour in lentils and it is suggested that they be used in preference to the *O* symbol proposed by Singh (1978).

Flower colour. Two reports of the inheritance of flower colour in lentil have postulated that two genes are involved. However, these reports differ in the F_2 ratios obtained, which suggests the possibility that three or more genes control the trait. Two genes, each with complete dominance, were suggested by Lal and Srivastava (1975), who obtained an F_2 ratio of 3 violet : 1 white when they crossed parents with violet and pink flowers. In addition, when pink- and white-flowered parents were crossed, an F_2 ratio of 9 violet : 3 white : 3 pink : 1 rose was obtained that further indicated the presence of two independent genes with complete dominance and no epistasis. Several gene symbols were proposed for the genes that control flower colour: *V* for violet flower with colour expressed only in the presence of *P;*

p produces pink flowers in the presence of *v;* and the double recessive *vv pp* produces rose-coloured flowers. Thus, the homozygous genotypes and phenotypes are *VV PP* (violet), *VV pp* (white), *vv PP* (pink), and *vv pp* (rose).

Wilson and Hudson (1978) studied the progenies of crosses between white- and violet-flowered parents and obtained an F_2 ratio of 9 violet : 6 intermediate : 1 white; they suggested that the effects of the genes were additive. They noted that the intermediate class consisted of two types that were difficult to distinguish from each other; one was darker and had pale-violet standards and white wings and keels, while the other type was lighter with pale-violet standard and white wings and keel. The white flowers often had pale-pink or pale-violet veins deep within the throat of the standard. They proposed *V* as the symbol for violet and *ww* for white. The homozygous genotypes and phenotypes were suggested as follows: *VV WW* (violet), *VV ww* and *vv WW* (intermediate), and *vv ww* (white).

Gill and Malhotra (1980) also obtained a 3 violet : 1 white flower colour ratio in F_2 progenies. In a concurrent study of linkage between genes for flower colour, presumably *V*, and the gene for number of flowers per inflorescence (*Fn*), they found no linkage between the two genes.

The differing results between the studies of Lal and Srivastava (1975) and those of Wilson and Hudson (1978) suggest that additional work is needed on the genetics of flower colour in lentils. The complexity of flower colour inheritance in other grain legumes suggests that a similar complex system may exist in lentils. Some of the recent work of Ladizinsky (1979b), using interspecific crosses of *L. culinaris* ssp. *culinaris* with the closely related *L. culinaris* ssp. *orientalis*, indicated that genes in addition to the *V* gene may be involved in the inheritance of flower colour. In those interspecific crosses, he found what appeared to be an additional gene, a conclusion he based on segregation for bluish flower colour, which he presumed was different from the two genes previously reported (Lal and Srivastava, 1975).

Seed coat colour. Seed coat colour is controlled by several genes and pleiotropic action is likely for epicotyl and flower colour because of the similar variability for the traits. Pleiotropic action of genes for colouration is expected and has numerous parallels in other genera including *Cicer* and *Pisum.* A single gene for seed coat spotting was proposed by Ladizinsky (1979b), who found a 3 : 1 ratio of F_2 plants with spotted and non-spotted seed coats and proposed the symbol *Scp* for the gene controlling this trait. Non-spotted individuals would have the *scp scp* genotype. Five alleles of the *Scp* gene have been reported (Vandenberg and Slinkard, 1990).

Seed coat colour was also studied by Wilson and Hudson (1978) but their results are difficult to interpret. In F_3 families used to identify individual

F_2 plant genotypes, they found a 1 black : 14 mottled : 1 beige segregation ratio. These results were interpreted as indicating two genes without dominance. Within the mottled class there were differing degrees of mottling.

Background colour of lentil seed coats is reportedly (Vandenberg and Slinkard, 1990) controlled by two genes. Dominant *Ggc* determines grey ground colour while the dominant *Tgc* gene produces tan ground colour. When both dominant genes are present *(Ggc Tgc)*, brown seed coat colour is produced. The double recessive *(ggc tgc)* has green seed coat colour.

Epicotyl colour. Epicotyl colour was shown by Ladizinsky (1979b) to be simply inherited with purple epicotyl dominant to green epicotyl. The trait has proved to be very useful as a genetic marker for the identification of F_1 hybrids in a breeding programme. The gene symbol *Gs* has been proposed with purple epicotyl having the *Gs GS* genotype and green epicotyl having *gs gs*. The intensity of the pigmentation is suspected to be influenced by environmental conditions (e.g., temperature, irradiance, and soil fertility). These environmental effects can make classification of plants difficult.

Growth habit. Growth habit has received a great deal of attention from breeders attempting to develop cultivars with more upright stature that are also lodging resistant and adaptable to mechanized harvest. In F_2s from crosses of *L. culinaris* ssp. *culinaris* with *L. culinaris* ssp. *orientalis*, Ladizinsky (1979b) found erect, intermediate, and prostrate types in a ratio that indicated a single gene with incomplete dominance. Based on that study, the prostrate type was designated as *gh gh* and the erect type as *Gh Gh.*

Leaves without tendrils are reportedly controlled by a single recessive gene designated as *tnl* by Vandenberg and Slinkard (1990).

Flower number. In crosses of parents with two-flowered racemes with parents with three-flowered racemes, Gill and Malhotra (1980) obtained an F_2 ratio of 3 two-flowered : 1 three-flowered plants. They proposed that the gene symbol *Fn* be used for the trait with *Fn Fn* being two-flowered and *fn fn* being the three-flowered genotype.

Pod indehiscence. Genes that are expressed in the pod include *Glp* and *Grp* (Vandenberg and Slinkard, 1990). The dominant allele of *Glp* produces pod pubescence, while the homozygous recessive allele *(glp)* produces glabrous pods. The dominant *Grp* gene produces red pods, while the homozygous recessive *grp* allele produces green pods.

Pod indehiscence (tendency to shatter) is the major distinction between wild and cultivated species of *Lens* and is considered to be the gene most critical to domestication of the species. In crosses of the putative progenitor, *L. culinaris* ssp. *orientalis,* with *L. culinaris* ssp. *culinaris,* Ladizinsky (1979b) found that pod indehiscence was controlled by a single recessive gene and assigned the symbol *pi.* Therefore, the designation of indehiscent cultivated types was *pi pi* and that of the wild progenitor, *L. culinaris* ssp. *orientalis,* was *Pi Pi.* Differences in shattering susceptibility within cultivated germplasm, even in the presence of the recessive *pi* gene for indehiscence, indicate the presence of modifying genes for the indehiscence trait. It is those modifiers that breeders of the crop now need to work with in order to improve shatter resistance, which can still prove troublesome in some circumstances.

Virus resistance. Resistance to pea seed-borne mosaic virus (PSbMV) and resistance to pea enation mosaic virus (PEMV) has been found in lentil germplasm (Haddad *et al.,* 1978; Aydin *et al.,* 1987), but the genetics of only the former has been studied. Haddad *et al.* (1978) found that the resistance was controlled by a single recessive gene, which they designated as *Sbv* for susceptibility and *sbv* for the recessive allele for immunity. The resistance was found in PI 368648 and 212610 obtained from the USDA world lentil collection. In crosses of the resistant accessions with known susceptible cultivars, they obtained an F_2 ratio of 3 susceptible : 1 immune. Several other PI lines have subsequently been shown to carry the gene for resistance. The resistance to PEMV does not exhibit an immune response to the virus; however, symptoms on infected plants are extremely mild, indicating a very useful degree of tolerance. Studies of the inheritance of the resistance to PEMV have not yet been completed.

Other traits. Vaillancourt *et al.* (1986) reported a recessive gene *tan* for zero tannin in the seed coat. Vandenberg and Slinkard (1987) reported a recessive gene *xan* for a xantha mutant. The *xan* gene is epistatic to and independent of the two loci conditioning cotyledon colour. Thus, cotyledons of seeds homozygous for xantha are yellow regardless of the genotype for cotyledon colour. A chlorina chlorophyll mutant was reported to be controlled by a single recessive gene designated as *chl* (Vandendberg and Slinkard, 1990) and produces pale yellow foliage.

Isozymes. Enzymic gene variation is becoming as increasingly important tool that enables geneticists to study gene organization from a molecular and population perspective. Besides being important in genetical analysis, isozymes can be an important plant breeding tool by providing a means of 'tagging' economically important traits. Also, isozyme variation in conjunction with morphological genes can be used to establish linkage groups in the genus that can then be used by breeders. The more intense gene maps that result may also be exploited in the future to study the genetics of quantitative traits. Useful descriptions of this technique are found in Tanksley and Orton (1983).

Linkage of enzymic and morphological genes in lentil were first reported by Zamir and Ladizinsky (1984). Eight isozyme loci and two linkage groups involving five loci, one of which included epicotyl colour (*Gs*) were observed. Linkage was shown between *Gs* and an isozyme of glutamate-oxaloacetate-transminase (*Got-2*) (14 map units apart) and between *Gs* and an isozyme of malic enzyme (*Me-1*) (25 map units apart). Linkage was also found between *Got-3* and an isozyme of alcohol dehydrogenase (*Adh-1*) (21 map units apart). Segregation for epicotyl colour in their hybrids confirmed that epicotyl colour was controlled by a single gene, previously designated *Gs* with purple dominant to green. Codominance was evident for the alleles at the loci involved in the control of the isozymes.

Additional linkage groups and preliminary gene maps of lentil were subsequently reported (Muehlbauer *et al.*, 1989; Havey and Muehlbauer, 1989). Allozymic polymorphisms for 18 loci and the genes for 4 morphological traits were analysed by Muehlbauer *et al.* (1989) to produce a linkage map for lentil consisting of six linkage groups. Havey and Muehlbauer (1989) constructed a more detailed map for lentil by including restriction fragment length polymorphisms (RFLP). This latter map contains 34 loci arranged in 9 linkage groups. More recently, Simon and Muehlbauer (1992) reported a lentil linkage map containing over 100 loci that include morphological and isozymic markers, RFLPs and Random amplified polymorphic DNAs (RAPDs).

Genetic nomenclature. A standardized system of genetic nomenclature is essential for orderly development in understanding the genetics of lentil. Muehlbauer and Slinkard (1981) proposed that the system employed in *Pisum* (Blixt *et al.*, 1977) could serve as the basis for naming genes in lentil. Clearance of new gene symbols is also critically important to prevent the proliferation of different symbols describing the same or even different characters. The task of clearing new symbols is now the responsibility of the Lentil Research Organization, which is currently headed by A.E. Slinkard, Crop Development Centre, University of Saskatchewan, Saskatoon, Saskatchewan, Canada. According to the policy for newly discovered genes, the discoverer who identifies a new symbols submits the evidence and proposed symbol to that authority, who then scrutinizes the evidence and registers the new gene if the symbol is not already in use. If, for some reason, the symbol cannot be validated, an alternative symbol is chosen as mutually agreed, or the evidence is rejected. Approved symbols are then announced in the newsletter *LENS*, published by the International Centre for Agricultural Research in the Dry Areas (ICARDA) at

Aleppo in Syria and the University of Saskatchewan.

Quantitative Genetics

Estimates of genetic parameters for quantitative traits are very useful, since they provide information on the inheritance of such traits and help to identify appropriate breeding methods. Such parameters include the subdivision of genetic variance into proportions resulting from additive, dominance, and epistatic effects of genes. Moreover, parameters such as heritability, expected genetic advance in response to selection, and the degree of association between traits are also important for the design of more effective breeding and selection programmes.

Genetic variance. In lentil, a self-pollinated species, genetic variance is expected to be primarily additive; however, the non-additive proportion of genetic variance could also be important. Singh and Jain (1971) observed heterosis in F_1 hybrids for the respective numbers of branches per plant, pods per plant, pod clusters per plant, and also seed yield, indicating non-additive genetic variance for these traits. Similarly, Sagar and Chandra (1980) found non-additive genetic effects that influenced seed yields in the F_1s and F_2s of nine lentil crosses. On the other hand, Goyal *et al.* (1976, 1977) reported greater variances due to general combining ability effects (GCA) than to specific combining ability effects (SCA) for seed yield, number of seeds per pod, seed size, and number of primary branches, indicating the importance of additive genetic effects for those traits. However, their SCA variances were also significant. GCA and SCA were equally important for the respective number of pods per plant and of secondary branches. Surprisingly, there was no heterosis for plant height.

Haddad *et al.* (1982) estimated the genotypic variances for several traits in each of three lentil crosses and partitioned them into components due to additive, dominance, and additive × additive variances. They advanced the hybrids from F_2 to F_5 using a hierarchic design. Additive genetic variance was a major component of total variance in one of the crosses for all traits examined except plant height and mean seed dry weight. Dominance variance estimates were unexpectedly large in two of the crosses. Where parental means were similar, additive genetic variance estimates were small and, in many cases, negative. Where the parents differed appreciably, additive genetic variance was the major component of total variance for days to flowering and days to maturity (from sowing, in each case), height to the lowest pod, and growth habit. Significant dominance variance estimates were obtained for plant height in all three crosses and for mean seed dry weight in two of them.

Sakar (1983) estimated the genetic effects for several lentil traits by using two lentil cultivars that differed for plant height, maturity, and seed size. By using the F_1, F_2, B_1, and B_2 populations, he found additive genetic effects to be significant for days to first flower and mean seed dry weight. Additive and dominance genetic effects were large and equally important for both plant height and basal pod height with small additive × additive epistasis components. Non-additive genetic effects were important for flowering duration (the number of days from first flower to the last flower), days from first flower to maturity, days from sowing to maturity, and seed yield per plant.

Selection in the F_2 was effective for characters with large additive genetic effects. Traits for which progress was made from early selection included days from sowing to first flower, plant height and basal pod height, and mean seed dry weight. But, in contrast, selection for flowering duration and seed yield per plant in the F_1 was ineffective.

In his study, Sakar (1983) estimated that three genes were responsible for variations in plant height in the cross of two lentil cultivars 'Laird' (Slinkard and Bhatty, 1979) and 'Precoz' (Riva, 1975). Four genes were estimated to be responsible for variations in basal pod height, and at least two genes were responsible for variations in mean seed dry weight. As expected, the same genes were considered to influence plant height and basal pod height (i.e., they influenced node number per stem or internode length, or both).

When the additive genetic variance is the major genotypic component, all systems of selection are useful for developing desirable cultivars (Brim and Cockerham, 1961). Where dominance is important in the control of particular traits, the improvement of those traits will need appropriate combinations of mass, family, and progeny selection (Cockerham, 1961, 1963). Where linkage is suspected to have a major influence, a generation of intercrossing to increase recombination may be necessary.

Current information on the components of genetic variance in lentil suggests that additive genetic effects are the major component of the genetic variance for most traits, and that a considerable non-additive component can often be anticipated. The influence of the non-additive components should be considered in the design of breeding and selection programmes.

Heritability estimates. Heritability estimates can be used to increase the efficiency of breeding systems and to indicate the degree to which selection of genotypes can be based on phenotypic performance (Johnson *et al.*, 1955). However, heritability estimates may vary depending on the method used to calculate them, the breeding material chosen, the nature of the selection units (individual plants, replicated plot, and means of several plots are three common examples), and the relative magnitude of sampling errors. Therefore, heritability estimates should be treated with caution. Broad-

Table 2. Heritability estimates for important traits in lentil

Trait	Range of heritability estimates	
	Broad-sense[1]	Narrow-sense[2]
Yield and yield components		
Seed yield per plant	0.42–0.80	0.07[3]
Weight of 100 seeds	0.59–0.98	0.52–0.68
No. of seeds per pod	0.74[3]	—
No. of pods per plant	0.45–0.80	—
Plant traits		
Plant height	0.16–0.83	0–0.38
Height of lowest pod	0.17–0.67	0.36[3]
Plant type (growth habit)	0.05–0.44	—
No. of secondary branches per plant	0.55[3]	—
Reproductive traits		
Days from sowing to first flowering	0.47–0.78	0.53[3]
Days from sowing to maturity	0.54–0.96	0.42[3]
Quality factors		
Cooking time	0.82[3]	—
Protein concentration (%)	0.71[3]	—

[1]Sources for these broad-sense heritability estimates are Erskine *et al.* (1985), Haddad (1979), Muehlbauer (1974), Sakar (1983), and Sindhu and Misra (1982).

[2]Sources for narrow-sense heritability estimates are Haddad (1979) and Sakar (1983).

[3]Only one estimate considered.

and narrow-sense heritability estimates have been reported for several important traits in lentil and some of these are shown in Table 2.

Most of the heritability estimates reported so far are based on a single experiment that was conducted at only one location and in only one year. Therefore, the estimates of genetic variance may be biased by the inclusion of an environmental variance component. Nevertheless, the estimates may provide an indication of the relative ease or otherwise of making rapid progress through breeding. The estimates for seed yield and its components are moderately large and indicate that good progress can be expected from effective selection. The small narrow-sense estimate of 0.07 of Sakar (1983) indicates some difficulty in selection for yield in the hybrids he used. The heritability of several plant traits was quite variable and sometimes small, indicating that selection progress for traits such as plant height and height of the lowest pod will depend heavily on the variation available in the source material and on the effects of climate and weather. Reproductive traits, such as days from sowing to flowering and to maturity, had moderately large heritabilities indicating that these traits would respond to selection. Quality factors such as seed size and seed thickness seemed to be strongly heritable.

Character association and path analysis. Larger and more stable yields are primary goals in most lentil breeding programmes. However, selection for productive genotypes is not always easy or possible because of the large number of genes involved and the relatively small heritability of yield *per se.* Indirect selection for easily measured traits strongly correlated with yield has been suggested (Gallias, 1984). This approach may allow for the screening of larger populations in early generations but without intense selection pressures.

Correlation coefficients between lentil traits need to be given very careful consideration and interpretation by plant breeders because the degree of association between traits might well differ according to type of correlation measured (phenotypic, genotypic, or environmental), the magnitude of the correlation coefficient, the breeding materials used, and also the environmental conditions where the study was undertaken. It is possible that a large negative correlation between seed yield and maturity could be obtained under conditions of progressively drying soil and with genotypes with short crop durations. However, with adequate moisture and with well-adapted genotypes, such a correlation might not hold. Correlations between seed yield and the numbers of pods per plant, seeds per pod, and secondary branches per plant; plant height and straw yield have been positive and significant (Muehlbauer, 1974; Haddad, 1979; Chauhan and Sinha, 1982; Sarwar *et al.,* 1982; Erskine, 1983; Kumar *et al.,* 1983; Tyagi and Sharma, 1985). Seed yield was negatively correlated with mean seed dry weight, and with times to flowering and maturity in a range of locations (Muehlbauer, 1974; Haddad, 1979; Sakar, 1983; Tyagi and Sharma, 1985).

Path coefficient analysis applied by Chauhan and Sinha (1982) indicated that the number of pods per plant and the number of secondary branches per plant were the most important characters contributing to large seed yield. However, Singh (1977) and Kumar *et al.* (1983) found that pod number per plant and plant height had the greatest direct effects on seed yield. On the other hand, Tikka *et al.* (1977) found that pod length had the largest direct effect on seed yield, followed by plant height, days to flower, and number of secondary branches per plant.

Cytogenetics

All species of *Lens* are diploid annuals with 2n = 14 chromosomes. Also, they appear to have similar karyotypes consisting of three pairs of metacentric or submetacentric chromosomes, three pairs of acrocentric chromosomes, and one satellited pair of chromosomes (Sharma and Mukhopaday, 1963; Ladizinsky, 1979a). Plant morphology of the cultigen and representatives of the wild forms are shown in Fig. 2. A review of the cytology and cytogenetics of lentil is presented by Slinkard (1985).

Crossability among the *Lens* species was used by Ladizinsky *et al.* (1984) to classify the genus *Lens* into two biological species. Several criteria were used to divide the former, *L. nigricans*, into two groups based on stipule morphology and crossability. The two groups are nearly entirely cross-incompatible and, where hybrids have been obtained, they were completely sterile owing to irregular chromosome associations at meiosis. The first biological species includes the cultivated lentil, *L. orientalis,* and those accessions of the former *L. nigricans* with stipules that are oriented horizontally to the stem. This subgroup of the former *L. nigricans* is readily cross-compatible with *L. culinaris* and with *L. orientalis.* Those accessions of the former *L. nigricans* with the horizontal stipule type were then transferred to the biological species *L. culinaris* as ssp. *odemensis* Ladizinsky (Ladizinsky *et al.*, 1984). Two other subspecies proposed by Williams *et al.* (1974) within *L. culinaris* were ssp. *culinaris* and ssp. *orientalis.* With the addition of ssp. *odemensis, L. culinaris* now contains three subspecies.

The second biological species is *L. nigricans* and includes ssp. *nigricans* (those accessions of the former *L. nigricans* with stipules oriented parallel to the stem) and ssp. *ervoides* (Ladizinsky *et al.,* 1984). The subspecies of *L. nigricans* produce partly fertile hybrids when intercrossed.

Based on the crossability studies of Ladizinsky *et al.* (1984), ssp. *orientalis* and ssp. *odemensis* share a common gene-pool with the cultivated lentil. Thus, interspecific hybrids between the cultivated lentil and the two subspecies, *orientalis* and *odemensis,* are easily obtained. Slight chromosomal rearrangements may be the cause of partial sterility but there are still ample opportunities for gene flow and for the use of these wild forms for breeding purposes. The exploitation of wild forms in crop breeding is a common practice in many species because wild germplasm can be a source of much needed variation that may not exist in the cultivated forms. The wild forms of *Lens* may represent important sources of pest resistance or tolerance to environmental stresses that are no longer available in cultivated germplasm.

Lens nigricans should not be overlooked as a source of needed germplasm for breeding cultivated lentils. The infrequency of success in hybridizations with the cultigen is a significant barrier but these difficulties in hybridization can be overcome by embryo rescue (Cohen *et al.,* 1984) as was possible, for example, in the cross of the cultigen with ssp. *ervoides* that yielded partly fertile F_1 hybrids. Thus, all subspecies of *Lens* can be considered as belonging to a common gene-pool with the cultigen.

Genetic Resources and Methods of Lentil Breeding

Lentils are of ancient origin and were probably one of the first plant species to be domesticated (Zohary, 1972; Ladizinsky, 1979c; Muehlbauer *et al.*, 1985). In spite of this antiquity, lentil breeding is a recent endeavour when compared with breeding efforts devoted to cereals, cash or industrial crops, or to some of the other major food legumes. Progress in lentil breeding has been accelerated by increased international awareness of the importance of the crop, and by the strong national and international lentil improvement programmes established during the last decade.

Land races still occupy most of the cultivated land sown to lentil in the major producting countries (Erskine, 1985a). Moreover, most of the cultivars released to date have been derived from selection within heterogenous land races and are not the result of hybridization (Muehlbauer and Slinkard, 1981). Nevertheless, it is expected that new and improved cultivars will become available in many lentil-producing areas owing to the increased efforts in lentil breeding in both national and international programmes.

Genetic Resources

The basic resources for lentil breeding are the germplasm collections maintained by ICARDA and by several national programmes (e.g., those in India and the United States). These genetic resources comprise the land races of the cultivated species and the related wild species (Solh and Erskine, 1981).

The ICARDA collection is by far the largest. It comprises about 5424 accessions from 53 countries (Erskine, 1985b). The US collection numbers over 3500 accessions and includes more than 150 accessions of the related wild species. The collection maintained at the Indian Agricultural Research Institute (IARI) in New Delhi exceeds 3000 accessions. Samples for breeding and other research are available on request to these respective centres.[1]

An evaluation of 4500 accessions from ICARDA's collection suggested the presence of considerable variation for important agronomical traits such as grain yield, straw yield, 100-seed weight, seed number per pod, time to 50 per cent flowering, time to maturity, plant height, height of lowest pod, and pod number per peduncle (Erskine, 1985b). Thus, significant variation for important traits is available for breeding purposes.

Collection and Utilization of Wild Species

A collection of wild lentil species has been assembled and includes ssp.

[1]Addresses for lentil germplasm repositories are as follows:
ACARDA, PO Box 5466, Aleppo, Syria
IARI, New Delhi 110012, India
USDA Regional Plant Introduction Station, Pullman, Washington 99164, USA

orientalis, ssp. *ervoides,* ssp. *nigircans,* and ssp. *odemensis,* all from different areas of origin (Ladizinsky, 1979c; Muehlbauer, 1981). The accessions have been increased and a series of genetic and cytogenetic studies have been made (Ladizinsky, 1979a, b). Further evaluation of the wild species is needed to identify those traits to be incorporated into the cultivated types. It is believed that the wild species will contribute to the improvement of pest resistance, seed quality, and tolerance of environmental stress in cultivars, and to the understanding of the genetics and cytogenetics of lentils.

The variability within lentil wild species has not been fully explored; however, recent efforts have concentrated on collection and preservation to ensure that the wild relatives are available for future breeding. Even though collections have been made in southern Europe, the Middle East, and Asia Minor, wild species are still under-represented in germplasm collections (Muehlbauer, 1981; Erskine, 1985b). The presence of useful genes in wild lentil species has not been thoroughly explored. Nevertheless, the wild species represent a source of diversity for future breeding.

Methods for Lentil Breeding

The methods of breeding lentil are similar to those used in the breeding of other self-pollinated crops and include: selection (including mass and pure-line selection), hybridization (followed by the use of either the bulk method, pedigree method, or single-seed descent), mutation breeding, and polyploidy breeding.

Selection. Mass selection and pure-line selection within introduced germplasm accessions or local land races can lead to improved uniformity and consistent performance. The ability of land races to survive stress environments (albeit giving poor yields), disease and insect depredations, and often primitive production practices is a valuable consideration in the use of these materials. Nevertheless, valuable genes for seed yield, seed quality, and other traits may have been lost from those populations and can only be reincorporated into them by hybridization and selection. Pure-line selection has provided improved cultivars with a life of less than 10 years and genetically uniform material for use in hybridization programmes. The efficacy of selection is dependent upon the presence of genetic variability within the source material. But, as discussed earlier, genetic variability in the germplasm collections worldwide has not been fully exploited to improve important agronomic traits for local circumstances and could be very useful in providing improved cultivars prior to development by hybridization.

Mass selection. Mass selection consists of either the removal of undesirable types from a mixed or variable population followed by harvesting the remaining plants *en masse,* or by the identification of superior types (by tagging or other means) in the population and harvesting those *en masse.* The end result in both cases in an increased frequency of desirable genotypes in the population. Mass selection may involve any source material (e.g., land races or hybrid populations) that has sufficient genetic variability upon which successful selection can be expected to be worthwhile.

Pure-line selection. Pure-line selection within land races or other heterogeneous lentil populations can lead to significant progress because the plants are homozygous and therefore will not segregate upon selfing. This presumption is essentially correct for lentils since Wilson and Law (1972) found little (less than 1 per cent) or no outcrossing under field conditions.

Several steps are involved in pure-line selection. Initially, a large number of selections are made from a genetically diverse population. The number of selections made depends on constraints of time, space, and available monetary and other resources. The selections are then observed in single plant rows, and those with obvious defects are eliminated. Such defects might include undesirable plant characteristics (e.g., too short, too late in flowering, or inappropriate seed size), susceptibility to a particular pest or disease, or even an undesirable quality trait. The greatest reduction in the number of selected lines is made in these preliminary observation nurseries. Promising lines are then chosen on the basis of their relative performance in yield trials at several locations. Based on their range of adaptation, lines are then considered for release to growers.

Mass and pure-line selection for lentil breeding have been valuable in providing improved cultivars. Examples include 'Tekoa', a pure-line selection from the US Department of Agriculture accession, PI 251784 (Wilson *et al.,* 1971), and 'Chilean 78', a cultivar developed by mass selection for large seed size and desirable colour (Muehlbauer *et al.,* 1981). 'Laird', a pure-line selection derived from the US Department of Agriculture accession PI 343028, was developed and licensed in Canada (Slinkard, 1978a). Laird was selected for its good yield and large seed size. In India, the cultivar Pant-L-406, evolved through selection from an Indian genetic stock, P-495, has improved yield potential and is resistant to rust and wilt (Pandya *et al.,* 1980). 'Araucana-INIA' is a new cultivar released in Chile and was developed by mass selection from N-1284, an accession from the Chilean Lentil Germplasm Collection. The cultivar is tall, has a good tolerance to rust, and yields well (Tay *et al.,* 1981). Many cultivars developed by mass and pure-line selection are probably grown today in many lentil-producing countries.

Hybridization. As mentioned earlier, there is a limit to the improvement possible in lentil by use of pure-line selection because of the almost obligatory self-fertilization that takes place, which prevents natural outcrossing and consequently limits gene recombination. Hybridization is now being more widely used in order to generate additional genetic variability.

Hybridization is an essential step in crop improvement. However, hybridization of lentil flowers with hand emasculation and pollination is difficult because they are so small and delicate. But, if special care is given to the selection of flowers at the correct stage of development, and to careful emasculation and pollination procedures, and if environmental conditions are appropriate, then cross-pollination can be very successful. It is essential, therefore, in obtaining hybrid seeds, to understand the reproductive system and to know the steps and conditions that are conducive to success.

Crossing technique. Lentil flowers are cleistogamous and naturally self-pollinated, with extremely infrequent natural cross-pollination (estimated to be less than 0.8 per cent) that is presumed to be caused by small insects such as thrips (Wilson and Law, 1972).

Flowers usually open before 10 a.m. on cloudless days, but when the sky is overcast, they may not open until 5 p.m. The corolla petals fade within three days, and pods are visible three or four days later (Meimandi-Nejad, 1977; Muehlbauer *et al.*, 1985; Summerfield *et al.*, 1985).

The small, delicate flowers prevent rapid and readily successful hybridization in lentil. In addition, environmental factors play a major role in the degree of success. Flowering and seed set, for example, are improved by a photoperiod of 16 hours or longer with good irradiance. Artificial light can be used to extend naturally short photoperiods. Lentils are considered to be either long-day or day-neutral plants (Shukla, 1955; Moursi and Ab El-Gawad, 1963; Saint-Clair, 1972; Summerfield *et al.*, 1985). Under glasshouse and growth chamber conditions, day temperatures of about 25°C and night temperatures of about 18°C are reasonable combinations for good plant growth and seed set. Seed set can also be improved by maintaining relative humidity at about 50 per cent (Wilson, 1972).

Equipment needed. The equipment needed for lentil hybridization is modest and consists of a dissecting microscope or magnifying glasses, a pair of sharp-pointed forceps, 95 per cent alcohol, and a supply of small tags. Viewing of flower parts can be improved with visors giving 5 × to 10 × magnification. Some workers with keen eyesight are able to hybridize lentils without the aid of magnification. The forceps to be used for emasculation and pollination are immersed in the alcohol between crosses to prevent contamination by foreign pollen. The tags are used to record the parents, the date the cross was made, and the identity of the person who made it. They are fixed to the internode directly below the cross-pollinated flower (Muehlbauer and Slinkard, 1981; Muehlbauer *et al.*, 1985).

Emasculation of the female flower. Emasculation is necessary to prevent self-pollination (Wilson, 1972). Flowers to be emasculated should be selected when the corollas have reached a length equivalent to about 75 per cent of that of the sepals (Fig. 3A). At this stage of development only a few flowers will have been pollinated and the stigma will be receptive to the mature pollen grains. Later, when corolla and sepal are of equal length, the flowers will have pollinated.

The selected flower bud is held between the thumb and forefinger with the suture of the keel facing the operator. The flower peduncle is weak and tender and so must be handled carefully. Sharp-pointed forceps are used to remove the sepals closest to the keel. The wings are then removed and

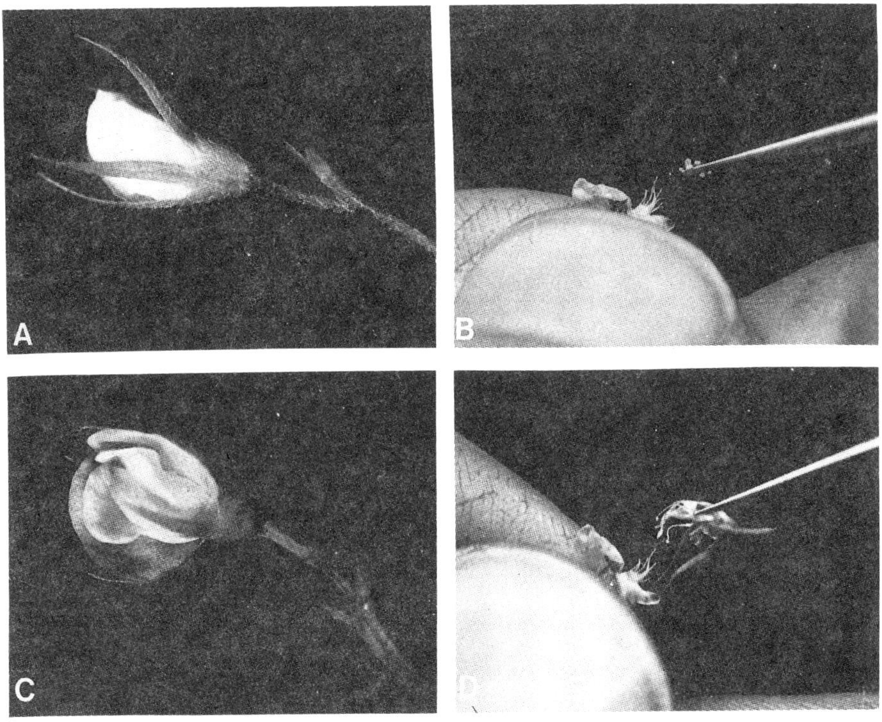

Fig. 3. Emasculation and pollination in lentil crossing: appropriate stage for emasculation (A); technique of emasculation (B); appropriate flower stage for use as a pollen source (C); pollination of an emasculated flower (D)

the standard is folded back. The keel is either folded back or removed to expose the staminal column and stigma. Finally, the stamens are removed by grasping the filaments with the forceps and breaking them free from the staminal column (Fig. 3B). The female flower is now emasculated and ready for pollination.

A different emasculation technique was used by Solh *et al.* (1980), who reported fairly good success (30–50%) under field conditions. They emasculated flowers without removing the petals. Instead, the petals were slit open by inserting sharp-pointed forceps into the base of the flower on the ventral side. With slight pressure on the calyx exerted by the thumb, the anthers became visible and were then removed with the forceps.

Pollination. Partly opened flowers, in which the corollas have elongated to about the same length as the sepals, are selected as sources of pollen (Fig. 3C). The flowers at this stage of development have anthers that have recently dehisced. Viable pollen suitable for transfer is identified by its bright orange-yellow colour and is contained in flowers with obviously turgid keel and wing petals. Pollen retains its viability for several days if flowers are collected immediately after anthesis, placed in petri dishes, and stored in the dark at 4°–10°C.

Pollination directly after emasculation gave the highest percentage of crossed seed (Wilson, 1972; Malhotra *et al.,* 1978; Solh *et al.,* 1980). Under high relative humidity (greater than 50 per cent) and low temperature (18°–20°C), the stigma can remain receptive for at least 24 hours. However, we recommend that pollination proceed immediately after the completion of emasculation, while the emasculated female flower is in hand.

Pollination is accomplished by first removing or folding back the keel and standard of the pollen source flower to expose the 10 dehisced anthers and the pollen-laden stigma. The anthers and stigma are then held with the forceps and brushed lightly against the stigma of the emasculated flower until pollen is visible on that stigma (Fig. 3D). It is important that sufficient pollen is transferred to the stigma to be visible, which may require use of several male flowers. The petals of the pollinated female flower are carefully returned to their original position to protect the stigma. Pollinated flowers seldom need protection from foreign pollen.

After pollination has been completed, a tag is fixed to the internode directly below the pollinated flower to identify the cross. An alternative method of identifying crosses is to use coloured thread looped around the peduncle and then different colours for various parent combinations.

The proportion (%) of successful pollinations and seed setting will vary according to individual skill and prevailing evironmental conditions. Under a glasshouse environment, Wilson (1972) reported up to 82 per cent successful manual pollinations by experienced operators. Success in that study was achieved with adequate irradiance, about 50 per cent relative humidity and temperatures between 15° and 25°C. Pollination success greater than 80 per cent was obtained by Slinkard in controlled environment chambers with day/night temperatures of 21°/15°C (Muehlbauer and Slinkard, 1981).

As expected, the success of pollination decreases under field conditions, especially due to dry air and hot weather at the time of lentil flowering. Malhotra *et al.* (1978), for example, reported 27–43 per cent success under field conditions in India. Also, Solh *et al.* (1980) recorded 47–72 per cent successful pollinations, depending on the method of emasculation and pollination used. Probability of success was greatest when the plants were pollinated in the morning and then covered for three days with a polyethylene bag immediately after crossing. The bag was perforated for aeration, which decreased canopy temperature by 6°–7°C. Similarly, Mera and Erskine (1982) found that enclosing plants in perforated polyethylene bags improved success compared to the unbagged controls. They also discovered that pollinations of early flowers gave 44.8 per cent success compared to only 17.3 per cent when flowers formed later in the reproductive period were used.

Other considerations for crossing. The most reliable genetic marker in use to identify F_1 hybrid seed is red cotyledon colour, a trait controlled by a single dominant gene. Therefore, F_1 hybrid seeds from crosses between yellow cotyledon female parents and red cotyledon male parents can be identified by their red cotyledon colour. Seeds with yellow cotyledons are discarded as selfs. The *gs* gene for stem colouration is also a useful marker to identify F_1 hybrids where the male parent has *Gs* for purple stems and the female parent has *gs* for green stems.

Due to the differences in seed size between lentil accessions, it is advisable to use large-seeded parental lines as the female parents. This will reduce losses from shattering since the pod of the large-seeded parent can accommodate the hybrid seed while that of the small-seeded parent often cannot. Erskine (personal communication) uses the reverse because the smaller-seeded parent will often yield two hybrid seeds when used as the female. However, timely harvest is necessary to prevent losses due to shattering. Pod and seed development can be observed within three days of pollination. When cross-pollination is successful, ovary development is rapid and the swelling seeds in the pod are obvious. When cross-pollination is unsuccessful, the ovary often enlarges but aborts later even when the developing pod may have reached as much as 50 per cent of its potential size.

Seeds are always harvested and separated from the pods by hand and as soon as the pods become yellow-brown in colour, so as to avoid seed losses from pod dehiscence. Harvested pods should be allowed to dry completely in envelopes or bags before the seeds are removed from them.

Hybrid seeds can usually be kept at room temperature provided they are to be sown in a reasonable period of time. For long-term storage, lentil breeding material may be stored at about 10°C with 30 per cent relative humidity or in a freezer at –20°C.

Handling the crosses and progenies. After successful crosses have been made, there are several options available for handling the resultant populations and these are similar to those used in other self-pollinated crops: bulk population, pedigree selection and single-seed descent, and backcrossing.

Bulk population. Bulk population breeding has been the preferred method for lentil because of the ease of application and because of the problems involved with alternative methods. The method is simple, requires minimal record keeping, and is not labour intensive, Simplicity makes it attractive for programmes designed to develop cultivars adapted over a wide geographical area because subsamples of populations can be widely tested. The advantage of wide testing includes the possibility of natural selection favouring genotypes adapted to particular local environments. Caution is needed when using the system in lentil since good seed set on individual plants within a bulk population may not necessarily ensure that the better genotypes will survive successive generations of bulking.

It is to be expected that seed size differences between parents and the larger numbers of seeds produced on small-seeded genotypes compared with those on large-seeded genotypes may cause rapid shifts in a bulk population toward small-seeded genotypes. The populations can be subjected to mass selection either on the basis of seed size or colour, or plant traits such as, for example, flowering time, plant height, branching characters, or disease resistance. Selection in early generations, such as in the F_2 to F_4, might be effective in eliminating many undesirable genotypes. Selected plants can then be grown in bulk for several generations, followed by reselection after the populations have reached homozygosity.

Because of the problems of genetic shifts during generation advance, modifications of the bulk population method have been devised. These include mass-pedigree, modified bulk, single-seed descent, and other schemes designed to control genetic shifts, or at least to channel them in the desired direction.

ICARDA uses a bulk-pedigree method in which crosses are advanced in bulk to the F_4, after which the pedigree method is used. The generations of advance by bulking allow an early evaluation and selection of bulks on which to concentrate efforts. Selection of plants in the F_4 is based only on highly heritable plant characters, and thereafter the progenies are managed by pedigree procedures. It is assumed that visual selection of F_4 plants in lentil with the selection criteria currently available will lead to improved types.

Pedigree selection. The pedigree method of breeding is not generally applied to early generations of lentil hybrid populations. However, if the method is used, between 5 and 15 F_1 plants would be sufficient to provide from 200 to 2000 F_2 seeds. Pedigree selection is imposed as early as the F_2 where selection may be practised for strongly heritable traits. To allow for successful selection, lentil plants need to be widely spaced so that individuals can be observed for possible selection. The plastic branching habit of lentil plants is a disadvantage of this method because their performance when widely spaced for observation may be entirely different from that in more densely sown stands as used in commercial production. Also, for pedigree selection to succeed, readily identifiable traits need to be available, which is not the case in lentil. However, selection of the F_2 plants for traits with large heritability estimates (such as flowering date, relative maturity, and seed size) is likely to be successful.

In the F_3, sufficient plants are grown in progeny (family) rows to give the breeder some indication of the variability and degree of heterogeneity within and between families; between 20 and 50 plants for each F_3 family will usually suffice. Since the breeding material at this stage will begin to take on family characteristics, selection is usually made on the outstanding plants within the better-looking families. As the emphasis is still on individual plant selection, it is necessary that plants be widely spaced so that plant characters are evident. But, as mentioned earlier, this may be a disadvantage for lentil because of its plastic growth habit and ability to branch profusely if space is available.

In the F_4, there is a shift in emphasis from selection within families to selection between them. Characteristics of F_4 families should be readily apparent, and successful selection among families should be possible. Characteristics that are considered desirable in lentil, such as upright growth habit, greater branch number, earlier flowering, and suitable maturity dates, should be readily distinguishable among F_4 families. Although the emphasis is on family selection, individual plant selection in heterogeneous families is not only still possible but also extremely desirable, because recombination is probably still taking place in such families.

The F_5 gives the first real opportunity to observe selections in comparison to nominated standard cultivars. Selected F_5 lines are usually sown in multi-row plots that have within- and between-row spacings similar to those used in farmers' fields. Preliminary yield trials may also be conducted.

With this approach, line characteristics can be observed in solid sowings and yield potential gauged. Lines that meet the selection criteria of plant type, relative earliness, degree of branching, seed size and colour, and yield are then retained and entered into advanced yield trials in succeeding generations.

Slinkard, in Canada, used a modification of the mass-pedigree method in which individual F_2 plant selections are maintained and evaluated as individual bulk populations, i.e., F_2-derived bulks or families (Muehlbauer and Slinkard, 1985).

Single-seed descent. Single-seed descent, in contrast to bulk population breeding, is not affected by method of plant culture since it does not depend on the numbers of seeds produced by the genotypes involved. Therefore, the method is suitable for rapid generation advance in glasshouses and growth chambers and, since only a small population is needed, less time and space for advancing generations are required. This, in turn, reduces costs.

Haddad and Muehlbauer (1981) found that more genetic variability was maintained by the single-seed descent method than by the bulk method, and that natural selection was operating in the bulk method against less competitive, short-statured lentil genotypes. The populations derived by single-seed descent had 10, 9, and 13 per cent more erect lines in three hybrids, when compared with the same hybrids advanced by the bulk methods. They also found that the time for generation advance was similar, but that single-seed descent used only 15 to 20 per cent of the space used by the bulk method and so required much less labour and money.

The genetic makeup of populations derived by single-seed descent can be affected in each generation by failure of that single seed to germinate, by seedling mortality, and by the inability of each plant to produce at least one seed. In the study by Haddad and Muehlbauer (1981), the maximum loss in three different single-seed descent populations was 29 per cent after four generations of advance, which did not decrease the efficiency of the method, as determined by population mean performance and genetic variances. They found the method convenient for regional programmes where selection criteria are precisely defined.

The backcross method. The backcross method gives predictable and repeatable results in a relatively short time and should be considered for short-term goals. It is especially useful for transferring simply inherited traits with strong heritability, such as disease resistance and mean seed dry weight, into well-adapted genotypic backgrounds. The main consideration in using the backcross method includes the presence of a recurrent parent that is acceptable except for a single undesirable trait; the trait to be transferred must have high heritability and be easily identified within segregating populations. Resistance to pea seed-borne mosaic virus in lentil is well suited to the backcross method because the resistance is controlled by a single recessive gene, and resistance plants are easily identified.

Other traits of strong heritability might also be transferred by backcrossing. Seed weight is strongly heritable in lentil, indicating that selection within segregating populations would be very effective in identifying large-seeded genotypes. The other yield components have considerably smaller heritabilities, indicating some difficulty in selection. The difficulty with poor heritability in this case can be circumvented by using large-yielding small-seeded lines as the recurrent parents in a backcross programme for improvement of large-seeded lentil.

Cyclical recurrent selection (CRS) and diallel selective mating (DSM). Breeding methods such as CRS and DSM can be used where the objective is long-term and the character is controlled by many genes with cumulative effects (Jensen, 1978). The usual approach in using such methods involves choosing a series of parents, intercrossing those parents to create a genetically variable population, screening that population, and intercrossing selected plants to form the base population for the next cycle of selection.

The method has merit for those quantitative traits that have relatively poor heritability, where the objective is to recombine into one population minor genes with cumulative effects on a particular character. The main considerations in using these two methods are to make gradual improvements in mean performance while maintaining genetic variability within the breeding population. Mean performance for each cycle must be monitored but, more importantly, a measure of within-population variation should also be made. Within-population variation is important since there must be variation on which to practise selection. When variation is exhausted, progress form additional cycles of intercrossing and selection is impossible.

Currently conceived objectives of lentil breeding programmes hardly seem to warrant the use of CRS and DSM. In any event, the relatively large number of crosses required for such methods to succeed would prohibit their use because of the difficulties generally encountered in lentil crossing. Then again, the necessary prerequisite for the application of these methods in lentil breeding is the availability of a reliable source of male sterility in order to facilitate the large number of crosses required. Even though male sterility is known in lentil, the small percentage of outcrossing on the male-sterile plants has severely restricted its usefulness (Muehlbauer and Ladizinsky, 1986).

Mutation breeding. The use of mutations resulting from irradiation or chemical mutagens has not received much attention as a breeding method in lentil. Uhlik (1972a, b, 1973) used fast neutrons and chemicals (N-ethyl-N-nitrosourea, N-methyl-N-nitro-N-nitrosoguanidine, and ethyl-methane sulphonate) to produce progenies that segregated for chlorophyll mutants and chimeric plants. N-ethyl-N-nitrosourea at a concentration of 0.005 per cent, and ethyl-methane sulphonate at a concentration of 0.125 per cent each applied for 20 hours at 19°C, were effective in producing mutant plants. Sharma and Kant (1975) used gamma rays and N-nitrosoethyl urea to induce mutations and were more successful with the chemical.

Mutation breeding has merit if the character being sought is simply inherited and suitable screening methods are also available to identify that character within large populations of mutagenized material. The technique is more adaptable for inducing recessive genes than dominant ones. After treatment, the material should be handled in similar way as in other breeding methods.

Polyploid breeding. The use of polyploids has received little attention in lentil breeding. However, the method offers the possibility of developing radically changed, vigorous, large-seeded, and well-adapted types. Polyploidy has been brought in other crop species through the use of cholchicine, temperature shocks, wounds irradiation, and twin seedlings, and has also arisen spontaneously. Gupta and Singh (1982) produced autotetraploid lentil plants by placing cotton plugs saturated with colchicine solution (0.5 per cent) on the apical bud for 10 hours. The method most often used is to soak seeds for several days in a 0.001–1.6 per cent solution of colchicine before germinating them and observing the plants for abnormal characters (Tawar and Gour, 1984). Abnormal growth would include improved vigour, large leaves, and partial sterility. Such plants are likely to be autopolyploids. Reduced fertility is likely to be troublesome, but selection for increased seeds set will improve fertility.

The original induced polyploids should be considered to be starting material that needs to be intercrossed with other polyploids of diverse origin. The progenies must then be selected for fertility and handled in the breeding programme similarly to other progenies.

Breeding Objectives

The major objectives of lentil breeding programmes are similar, with larger and more stable seed yield being the most important. Breeding for adaptation to stress environments, especially to drought, and resistance to diseases and insects are also major objectives. Priorities and breeding goals usually differ between regions depending on the critical location, specific problem, and special considerations related to farmers' needs and consumer demands. In the developing countries, for example, one of the major breeding goals is the development of genotypes suitable for mechanical harvesting. Moreover, improving straw yield is also important because of the value placed on lentil straw as animal feed. On the other hand, increased seed yield, improved disease resistance, and improved seed quality are principal breeding goals in the major exporting countries.

A description follows of the current objectives of lentil breeding in the major producing areas.

Seed and Straw Yields

Increased seed and straw yields with acceptable respective quality are principal objectives in lentil breeding, but strategies for improvement differ. In North America, Muehlbauer (1981) has emphasized the development of cultivars appropriately adapted to the Palouse region of the Pacific Northwest, whereas Slinkard (1985) has also emphasized the importance of environmental adaptation and disease resistance. In the Middle East, where erratic and limited rainfall prevails in the lentil producing areas, genotypes better adapted to drying soil and hot weather are desired. Erskine (1985a) suggested that improved yield could be achieved through the exploitation of genotype × environment interactions to identify genotypes for specific local environments rather than relying on fewer genotypes that are more widely adapted.

Selection for improved yield and wider adaptation can be practised within land races, seeking suitable parental materials for hybridization; however, little progress in yield can be anticipated when compared to locally adopted land races. The introgression of *microsperma* with *macrosperma* types is viewed as a potentially productive strategy for hybridization because the two types evolved from and became important in different ecological regions and, therefore, are likely to possess different genes and adaptive complexes. Summerfield (1981) pointed out that no single factor has been or is likely to be identified to explain relative adaptation to environments and that well-adapted genotypes would probably be endowed with several, individually unspectacular, traits the best combinations of which are difficult to predict.

Larger straw and seed yields are often emphasized in those developing countries where the straw is important for feed. The correlation between seed and straw yields is strong and positive and so simultaneous selection for both traits should be possible.

Diseases

The diseases of lentil are, in general, relatively less damaging than those of most other food legume crops. However, there are some important and potentially devastating diseases that include the wilt/root rot complex in the

Indian subcontinent, rust in India and South America, and Ascochyta (*Ascochyta lentis*) blight and viruses in North America. Lentil genotypes resistant to various races of *Fusarium oxysporum* f. sp. *lentis* have been identified and can be used in breeding programmes (Kannaiyan *et al.*,1978; Khare, 1980). Screening for resistance under field conditions can best be accomplished in wilt-sick plots.

Lentil rust (*Uromyces fabae*) is an important disease in India, Morocco, Pakistan, Ethiopia, Argentina, and Chile. Infection of susceptible cultivars has caused up to 70 per cent yield losses in Chile (Sepulveda, 1985) and total field losses were observed in Morocco (Sakr, 1989b). Sources of resistance have been identified in cultivars such as 'Tekoa', 'Laird', and 'Arancana-INIA', which are now in use in South America, as is 'Pant- L-406' in India and 'Precoz' in Morocco (Pandya *et al.*, 1980; Tay *et al.*, 1981; Muehlbauer and Slinkard, 1985; Sakr, 1989a). Several other lines found to be resistant to rust in India and Morocco have been used in breeding programme (Agrawal *et al.*, 1976; Khare *et al.*, 1979). 'Hot spots' for rust, such as Debre zeit in Ethiopia and Pantnagar in India, were suggested by Erskine (1985a) to be useful locations for establishing screening nurseries for rust resistance. Chemical control of rust was very effective using Dithane-M45 as a foliar spray (Singh *et al.*, 1985; Sakr, 1989b).

Lentil Ascochyta blight attacks the leaves, stems, and pods and is an important disease in parts of western Canada where frequent rains occur between flowering and harvest. According to Gossen (1985), the fungus can be found in Argentina, Brazil, Syria, Greece, Chile, and Pakistan. When seeds from 30 countries were screened for Ascochyta infection, the fungus was isolated from seeds of 16 countries including Australia, Canada, Ethiopia, Hungary, India, Italy, Morocco, Russia, Spain, Turkey, and Yugoslavia (Kaiser and Hannan, 1986). In some cases, infection of seeds is so severe that lentils are unmarketable (Kaiser, 1981; Gossen and Morral, 1983). Resistance sources have been identified in North America; 'Laird', ILL 5588, and ILL 5684 have good resistance to prevailing race(s) and are used as resistant parents in crosses. Resistance in Laird was found to be controlled by a single recessive gene, ral_1 while that of ILL 5588 and 5684 was due to two dominant genes, Ral_2 and Ral_3. ILL 5588 also carries the ral_1 gene (Tay, 1989). Natural infection that can be obtained in cooler, moister parts of Saskatchewan is used for making selections. The mode of inheritance of resistance is not yet known.

Several viruses are reported to infect lentils (Kaiser and Eskandari, 1970; Kaiser, 1972; Haddad *et al.*, 1978) and pea seed-borne mosaic virus (PSbMV) is potentially serious because it can be seed-borne and is transmissible by aphids. Screening of the USDA collection indicated that PI lines 212610, 151786, 297745 and 368648 were immune to the virus. Jermyn (1980), in New Zealand, confirmed PI 212610 as resistant to aphids,

PSbMV, and other viruses. Even though PSbMV has not been detected in farmers' lentil fields in the United States, Muehlbauer has begun to develop breeding material immune to the virus. Incorporation of multiple disease resistance into breeding material and acceptable cultivars is possible with the use of resistant germplasm already identified. Pea enation mosaic virus (PEMV), a natural pathogen of lentil (Volvas and Rana, 1972), became a serious problem in lentil production in the United States during the late 1980s (Aydin *et al.*, 1987). PI 472547 and 472609 were identified as tolerant (Aydin *et al.*, 1987).

Root Rot/Wilt Complex

Root rot/wilt caused by *Fusarium oxysporum* f. sp. *lentis, Rhizoctonia solani* and *Sclerotium rolfsii* is an important disease complex of lentil in India where several resistant lines have been identified (Pandey *et al.*, 1988). Resistance to *F oxysporum* in India was found to be controlled by two dominant genes with duplicate interaction in one line (L234) and complementary effects in two other lines (ILL446 and LP286). The genes in L234 were not allelic to those found in either ILL446 or LP286 (Kamboj *et al.*, 1990).

Root rot caused by *Thielaviopsis basicola* was first noticed in 1984 under field conditions in eastern Washington and northern Idaho by Bowden *et al.* (1985). Sources of partial resistance were identified in lentil breeding lines.

Orobanche

Lentil is susceptible to several species of *Orobanche* including *O. crenata, and O. ramosa* (Basler, 1981). *Orobanche crenata* is the most important species, especially in Mediterranean countries (Erskine,1985a). Control of *Orobanche* is difficult because of the large number of wind-blown seeds that can be produced each year and that may remain dormant in soil for several years thereafter.

Research at ICARDA has shown that lentil accessions can have different susceptibilities to the various *Orobanche* species. This finding suggests that selection for resistance or tolerance should be possible (Basler and Haddad, 1979). Sauerborn *et al.* (1987) have developed a rapid test to screen lentil for resistance to *Orobanche* under laboratory conditions. After 35 days of incubation of lentil seeds at 20°–25°C in attapulgus clay-filled petri dishes, *Orobanche* attachments to lentil roots can be counted directly. Several lentil genotypes from India have shown satisfactory tolerance to *O. crenata* at ICARDA but were poorly adapted to the cooler weather of the Mediterranean region (Erskine, 1985a). Therefore, this resistance source and others that might come to be identified in the future should be recombined with locally adapted material.

Insects

Many insects are attracted to lentil and include aphids, moths, weevils, seed beetles, and agromyzed flies (Hariri, 1981; Tahhan and Hariri, 1982). The most important of these are weevils of the genera *Sitona* and *Bruchus* (Hariri, 1979, 1981; Tahhan and Hariri, 1982). The *Sitona* adults feed on lentil leaves and cause considerable losses in area that have detrimental effects on productivity. Larval feeding on the nodules and the consequent loss of nitrogen-fixing capacity is also a serious problem.

Differences in susceptibility found between lentil genotypes to *Sitona* have been studied but not found repeatable. Progress in this area may well require the development of reliable screening techniques to assess the degree of damage to nodules.

Breeding for insect host-plant resistance in lentil depends on the economic importance of the pest in the target area. It is expected, however, that more insect problems will occur with the spread of newly developed cultivars and, therefore, the need for resistant sources might soon be urgent.

Quality

A major concern of many breeding programmes is to monitor their breeding material so that cultivars for release are acceptable to farmers and consumers. Lentil quality is either related to obvious seed characters such as seed size and testa and cotyledon colour, or to the nutritional quality of seeds, such as their protein and methionine concentrations. Breeders in the Americas are concerned about the development of large-seeded (macrosperma) lentils with yellow cotyledons and light-coloured seed-coats (Muehlbauer and Slinkard, 1985) because of the export market demand. However, small-seeded red cotyledon lentils are often desired elsewhere.

Evaluation of germplasm for nutritional quality and seed decortication has begun at ICARDA. Variability for each of these characters is available in the germplasm collection (Solh and Erskine, 1981).

Erskine *et al.* (1985) studied 24 small-seeded lentils and found a small range in protein concentration of between 25.5 and 28.9 per cent and a negative correlation of protein concentration with seed yield ($r = -0.94$). However, increased seed yield could be found without a significant decrease in protein concentration to values smaller than those in land races. In the same study, they found that cooking time is more related to seed size and less to environment, with a positive genetic correlation ($r = 0.92$) between the two traits.

Suitability for Mechanical Harvesting

The development of genotypes suitable for mechanical harvesting is a prime goal for many breeders in national programmes within the Middle East and southwest Asia, and at ICARDA. Several traits are considered to be important for the success of mechanical harvesting and include increased plant height, pods borne well above the soil surface, erect growth habit, improved standing ability, reduced pod dehiscence, and reduced pod drop.

A clearance of about 15 cm between the soil surface and the lowest pod is required for successful mechanical cutting or pulling of lentil plants (Khayrallah, 1981). This leads to the view that mechanical harvesting of lentil would be facilitated by the introduction of tall cultivars with the lowest pods borne well above the soil surface (Solh and Erskine, 1984). Considerable genetic variability for plant height and lowest pod height was found in the ICARDA collection with ranges from 10 to 45 cm and 6 to 30 cm, respectively (Solh and Erskine, 1981). It was also found that the two traits are positively correlated (Haddad, 1979, Sakar, 1983), which indicates that selection for both traits is possible. However, tall plants had a tendency to lodge (Haddad, 1979), and both traits are highly affected by the environment (Saxena and Hawtin, 1981).

Relative pod indehiscence has been identified in lentil, and selection was feasible for this trait simply by the delaying of harvest. However, significant variability for pod retention, which accounts for as much as twice the loss caused by pod dehiscence, does not seem to be available (Erskine, 1985c).

Non-lodging lentil cultivars could be a very important development toward the success of mechanical harvesting in stony areas and also to reduce losses in those areas where lentil is mechanically harvested. Erskine (1985a) suggested that stem thickness, stem lignification, and greater tendril production may be important contributions to lodging resistance in lentil.

Tall, erect lentil types considered important for successful mechanized harvest of lentil may have reduced yield potential. In the experience of the authors, erect genotypes with acute branch angles tend to yield relatively poorly and do not compete well with weeds. Their poor competitive ability is the result of a reduced ability of strongly erect genotypes to fill available space with a spreading branch habit. By not covering the soil surface as rapidly as more spreading types, they may permit losses of limited soil moisture. Also, the slower rates of canopy closure in upright types tend to provide an advantage to weeds, which then deplete water even more. Genotypes that rapidly cover the soil surface and develop a full canopy should allow for successful mechanical harvest of acceptable seed and straw yields.

It seems that variability for several traits contributing to successful mechanical harvesting is available in lentil germplasm. However, for traditional farmers in the developing countries, mechanization of lentil harvest is a multidimensional problem that can only be solved by careful attention to the cultivars in use and the local management practices employed. Proper equipment to provide a smooth seed bed, planting with seed drills to ensure good crop establishment, and harvesting equipment that is designed to collect the maximum amount of biomass are all necessary for successful mechanical harvesting of lentil.

Enviornmental Stress

Drought is considered to be the major environmental stress that limits lentil yields. Lentil crops are often grown in marginal areas where limited rainfall and deficient soil moisture are encountered. The relegation of lentils to marginal lands is a consequence of the increased area sown to more remunerative crops such as wheat. It is not surprising then that lentil yields have declined in recent years. In general, the most commonly encountered drought and heat stress on lentil in the Middle East region is the one that is experienced during the reproductive period (Erskine, 1985a). Drought-tolerant cultivars are required to stabilize lentil production and also to extend lentil cultivation into those areas that receive less rain.

As with other crops, success in screening for drought tolerance is not yet possible because of the lack of efficient screening techniques and knowledge of what to screen for. Late planting has been used to simulate drought and heat stress; however, meaningful comparisons between late and normally planted lentil might be difficult because of the difference in plant growth duration. A line source irrigation system was used to create a moisture gradient in an area with otherwise insufficient rainfall (Karaki, 1986). Under this line source system, there was considerable variation among 10 lentil genotypes in response to moisture stress, which gives encouragement for further work.

Another approach to drought resistance screening is to select for drought avoidance through early maturity. This could be a suitable approach when the major moisture stress occurs towards the end of the growing season. However, in seasons that receive adequate rainfall, early genotypes may not have the ability to respond to above-average moisture if and when available.

Agents that simulate moisture stress have recently been tested for use in drought resistance screening. Haddad *et al.* (1987) screened 40 lentil genotypes by growing one-week-old seedlings in different concentrations of polyethylene glycol (PEG) solutions that created different degrees of moisture stress. They were able to group the genotypes into strongly resistant, moderately resistant, and susceptible types. Field tests are now needed to confirm these results.

The land races are potential sources of drought tolerance because they have demonstrated their ability to survive under extremely stressful environments. The wild species such as *L. culinaris* ssp. *orientalis* might also have considerable drought tolerance. These potential sources of drought tolerance need to be considered by breeders.

Hot weather is thought to cause empty pods and so to reduce lentil yields in Brazil (Manara and Manara, 1983). Elsewhere, cold tolerance and salt tolerance, if found, could be used by breeders to extend the range of adaptation of the crop. Jana and Slinkard (1979) found differences in salt tolerance in the cultivated lentil, but the level of tolerance was too little to be of any breeding value. Higher levels of salt tolerance are required.

Other Breeding Objectives

There are several other objectives that are either important for certain areas or have been recently identified by breeders and that might be given more attention in the future. These include photothermal insensitivity, resistance to MCPB herbicide, reduced tannin concentration in testa, cytoplasmic male sterility, development of lentil as an annual green manure crop, improving the seasonal fixation of nitrogen, and understanding the empty pod syndrome (Slinkard, 1980; Vaillancourt and Slinkard, 1983; Erskine, 1985a; Muehlbauer and Slinkard, 1985).

Prospects

The accelerated progress made in recent years towards a better understanding of the genetics of lentil and the relation among wild forms should be the basis for substantial future gains by breeding. Several topics that are in obvious need of attention by breeders include: larger and more stable seed and biological yields; resistance to diseases and insects; and better tolerance to heat and drought. The germplasm pool for lentils has been expanded by the availability of the wild subspecies and by the research that has shown that all the related forms of *Lens* share a common gene pool. This common gene pool has not yet been exploited to any significant extent for lentil crop improvement, but several programmes are actively using the wild species for genetical studies and progenies have been evaluated for important traits.

The limitations on lentil yields brought about in some regions by the crop being grown on progressively poorer land because of the competition imposed by more remunerative crops is a barrier that may be impossible to overcome. However, the yields obtained, albeit small, are still particularly important to local economies and consumers.

Breeding programmes have not yet focussed on improving nitrogen fixation by lentil crops. Estimates of fixation are small and indicate only nominal contributions to the nitrogen status of the soil. However, the ability of lentils to fix some nitrogen in marginal areas, albeit small amounts, may still represent important contributions to the farming system.

Cooperation among lentil researchers should be encouraged in order to take full advantage of often limited resources. Efforts in lentil research and breeding are minimal compared to those devoted to other crops. For this reason, it is essential that lentil researchers share their experiences and exchange data and, especially, germplasm. The Lentil Experimental News Service (LENS) is published by ICARDA to foster the exchange of ideas and information. This service provides an example of the collaboration and cooperation that will become more important in the years to come.

REFERENCES

Acikgoz, N., Er, C., and Meyveci, K. 1992. Chickpea and lentil production in Turkey. *The Second Int. Food Legume Res. Conf.* Abstr., 12–16 April, Cairo.

Agrawal, S.C., Khare, M.N., and Agrawal. P.S. 1976. Field screening of lentil lines for resistance to rust. *Indian Phytopathol.* 29: 208.

Aydin, H., Muehlbauer, F.J., and Kaiser, W.J. 1987. Pea enation mosaic virus resistance in lentil. *Plant Disease* 71: 635–638.

Barulina, H. 1930. Lentils of the U.S.S.R. and other countries. *Bull. Appl. Genet. Plant Breeding* (Leningrad), suppl. 40: 1–319 (in Russian).

Basler, F. 1981. Weeds and their control. In *Lentils.* C. Webb and G. Hawtin, eds. CAB International, Farnham Royal, pp. 143–154.

Basler, R., and Haddad, A. 1979. Selection of *Orobanche* resistant cultivars of broadbean and lentil. *Proc. Second Int. Symp. Parasitic Weeds.* L.J. Musselman, A.D. Worsham and R.E. Eplee, eds. North Carolina State University, Raleigh, pp. 254–259.

Blixt, S., Marx, G.A., and Murfet, I.C. 1977. Rules for genetic symbols. *Pisum Newslett.* 9: 67–70.

Bowden, R.L., Wiese, M.V., Crock, J.E., and Auld, D.L. 1985. Root rot of chickpeas and lentils caused by *Thielaviopsis basicola. Plant Disease* 69: 1089–1092.

Brim, C.A., and Cockerham, C. Clark 1961. Inheritance of quantitative traits in soybeans. *Crop Sci.* 1: 187–190.

Chauhan, V.S., and Sinha, P.K. 1982. Correlation and path analysis in lentils. *LENS* 9: 19–22.

Clarke, G.C.S., and Kupicha, F.K. 1976. The relationships of the genus *Cicer* L. (Leguminosae): The evidence from pollen morphology. *Bot. J. Linn. Soc.* 72: 35–44.

Cockerham, C. Clark. 1961. Implications of genetic variances in a hybrid breeding programme. *Crop. Sci.* 1: 47–52.

Cockerham, C. Clark. 1963. Estimation of genetic variances. In *Statistical Genetics and Plant Breeding.* W.D. Hanson and H.R. Robinson, eds. Publ. 1982. NAS-NRC Publ., Washington, DC, pp. 53–93.

Cubero, J.I. 1981. Origin, taxonomy and domestication. In *Lentils.* C. Webb and G.C. Hawtin, eds. CAB International, Farnham Royal, UK, pp. 15–21.

Cohen, D., Ladizinsky, G., Ziv, M., and Muehlbauer, F.J. 1984. Rescue of interspecific *Lens* hybrids by means of embryo culture. *Plant Cell Tissue and Organ Culture* 3: 343–347.
Davis, P.H., and Plitmann, U. 1970. *Flora of Turkey*. Edinburgh University Press, UK, pp. 325–328.
Duke, J.A. 1981. *Handbook of Legumes of World Economic Importance*. Plenum Press, New York, pp. 110–113.
Erskine, W. 1983. Relationship between the yield of seed and straw in lentils. *Field Crops Res.* 7: 115–121.
Erskine, W. 1985a. Perspectives in lentil breeding. In *Fababeans, Kabuli Chickpeas and Lentils in the 1980s.* M.C. Saxena and S. Varma, eds. ICARDA, Aleppo, Syria, pp. 91–100.
Erskine, W. 1985b. Lentil genetic resources. In *Fababeans, Kabuli Chickpeas and Lentils in the 1980s.* M.C. Saxena, and S. Verma, eds. ICARDA, Aleppo, Syria, pp. 29–33.
Erskine, W. 1985c. Selection for pod retention and pod indehiscene in lentils. *Euphytica* 34: 105–112.
Erskine, W., Williams, P.C., and Nakhoul, H. 1985. Genetic and environmental variation in the seed size, protein, yield and cooking quality of lentils. *Field Crops Res.* 12: 153–161.
FAO. 1974. *FAO Production Yearbook No. 28.* FAO, Rome.
FAO. 1984. *FAO Production Yearbook No. 38.* FAO, Rome.
FAO. 1991. *FAO Production Yearbook No. 45.* FAO, Rome.
Gallias, A. 1984. Use of indirect selection in plant breeding. In *Efficiency in Plant Breeding* W. Lange, A.C. Zeven and W.G. Hogenboom, eds. Pudoc, Wageningen, pp. 45–59.
Gossen, B.D., and Morrall, R.A.A. 1983. Effect of Ascochyta blight on seed yield and quality of lentils. *Can. J. Plant Pathol.* 5: 168–173.
Gill, A.S., and Malhotra, R.S. 1980. Inheritance of flower color and flower number per inflorescence in lentils. *LENS* 7: 15–19.
Gossen, B.D. 1985. Ascochyta blight of lentil in Saskatchewan. Ph.D. thesis, Univ. of Saskatchewan, Saskatoon, Canada, 164 pp.
Goyal, S.N., Jaimini, S.N., and Tikka, S.B. 1976. Heterosis in lentil. *LENS* 3: 1–5.
Goyal, S.N., Jaimini, S.N., and Tikka, S.B. 1977. Combining ability analysis for seed yield and some other biometric characters in lentils. *LENS* 4: 10–13.
Gupta, P.K., and Singh, J. 1982. Induced autotetraploids in lentil. *LENS* 9: 15–16.
Haddad, N. 1979. Breeding methods, breeding behavior, and estimation of genetic parameters in three lentil (*Lens culinaris* Medic.) hybrids. Ph.D. thesis, Washington State Univ.
Haddad, N., and Arabiat, S. 1985. Lentil production in Jordan (A case study). *Dirasat* 12: 51–78.
Haddad, N.I., and Muehlbauer, F.J. 1981. Comparison of random bulk population and single-seed-descent methods for lentil breeding. *Euphytica* 30: 643–651.
Haddad, N., Bogyo, T.P., and Muehlbauer, F.J. 1982. Genetic variance of six agronomic characters in three lentil (*Lens culinaris* Medic.) crosses. *Euphytica* 31: 113–120.
Haddad, N., Campbell, G., and Muehlbauer, F.J. 1987. Screening lentil genotypes under simulated environmental stress using PE solutions. *Agron. Abstr.*, p. 64.
Haddad, N.I., Muehlbauer, F.J., and Hampton, R.O. 1978. Inheritance of resistance to pea seed-borne mosaic virus in lentils. *Crop Sci.* 18: 613–615.
Hariri, G. 1979. Insect pests of chickpea and lentil in countries of the Eastern Mediterranean: A review. In *Food Legume Improvement and Development.* G. Hawtin and G. Chancellor, eds. IDRC, ICARDA, pp. 120–123.
Hariri, G. 1981. Insects and other pests. In *Lentils.* C. Webb and G.C. Hawtin, eds., Farnham Royal, UK, pp. 173–189.
Havey, M.J., and Muehlbauer, F.J. 1989. Linkages between restriction fragment lengths, isozyme and morphological markers in lentils. *Theor. Appl. Genet.* 77: 395–401.
Hawtin, G.C. 1977. The ALAD/ICARDA lentil programme. *LENS* 4: 1–3.

Helbaek, H. 1959. Domestication of food plants in the Old World. *Science* 130: 365–372.

Jana, M.K., and Slinkard, A.E. 1979. Screening for salt tolerance in lentils. *LENS* 6: 25–27.

Jensen, N.F. 1978. Composite breeding methods and the DSM system in cereals. *Crop Sci.* 18: 622–626.

Jermyn, W.A. 1980. P.I. 212610 resistant to aphids and viruses. *LENS* 7: 65.

Johnson, H.W., Robinson, H.F., and Comstock, R.E. 1955. Estimates of genetic and environmental variability in soybeans. *Agron. J.* 47: 314–318.

Kaiser, W.J. 1972. Disease of food legumes caused by pea leaf roll virus in Iran. *FAO Plant Prot. Bull.* 20: 127–132.

Kaiser, W.J. 1973. Etiology and biology of viruses affecting lentils (*Lens esculenta*) in Iran. *Phytopathol. Mediterr.* 12: 7–12.

Kaiser, W.J. 1981. Diseases of chickpea, lentil, pigeonpea, and tepary bean in continental United States and Puerto Rico. *Econ. Bot.* 35: 300–320.

Kaiser, W.J., and Eskandari, F. 1970. Studies with yellow mosaic virus in Iran. *Iran J. Plant Pathol.* 6: 26–27.

Kaiser, W.J., and Hannan, R.M. 1986. Incidence of seed-borne *Ascochyta lentis* in lentil germplasm. *Phytopathology.* 76: 355–360.

Kamboj, R.K., Pandey, M.P., and Chaube, H.S. 1990. Inheritance of resistance to Fusarium wilt in India lentil germplasm (*Lens culinaris* Medik.). *Euphytica* 50: 113–117.

Kannaiyan, J., Nene, Y.L., and Pant, G.B. 1978. Strains of *Fusarium oxysporum* f. sp. *lentis* and their pathogenicity on some lentil lines. *LENS* 5: 8–10.

Karaki, G. 1986. Response of ten lentil genotypes to moisture stress. M.Sc. thesis, Univ. of Jordan, Amman, Jordan.

Khare, M.N. 1980. *Wilt of Lentil.* JNKVV, Jabalpur, Madhya Pradesh, India, 155 pp.

Khare, M.N., Agrawal, S.C., and Jain, A.C. 1979. Diseases of lentil and their control. *Tech. Bull.*, JNKVV, Jabalpur, 46 pp.

Khayrallah, W.A. 1981. The mechanization of lentil harvesting. In *Lentils.* Farnham Royal, C. Webb and G.C. Hawtin, eds. CAB International, UK, pp. 131–141.

Kumar, B., Mehra, K.L., and Sapra, R.L. 1983. An investigation on correlation pattern among yield components in lentil. *LENS* 10: 10–12.

Kupicha, F.K. 1975. Observations on the vascular anatomy of the Vicieae. *J. Linn. Soc. Bot.* 70: 231–242.

Kupicha, F.K. 1977. The delimitation of the tribe Vicieae and the relationships of *Cicer* L. *J. Linn. Soc. Bot.* 74: 131–162.

Kupicha, F.K. 1981. *In Advances in Legume Systematics* Part 1. R.M. Polhill and P.H. Raven, eds. Royal Botanic Gardens, Kew, UK, pp. 377–381.

Ladizinsky, G. 1979a. The origin of lentil and its wild genepool. *Euphytica* 28: 179–187.

Ladizinsky, G. 1979b. The genetics of several characteristics in lentil as indicated by crosses between *Lens culinaris* and *L. orientalis. J. Hered.* 70: 135–137.

Ladizinsky, G. 1979c. The wild genepool of lentil. *LENS* 6: 24.

Ladizinsky, G., and Sakar, D. 1982. Morphological and cytogenetical characterization of *Vicia montbretii* Fisch. and Mey. (synonym: *Lens montbretii* [Fisch. and Mey.] Davis and Plitmann). *J. Linn. Soc. Bot.* 85: 209–212.

Ladizinsky, G., Braun, D., and Muehlbauer, F.J. 1984. The biological species of the genus *Lens. Bot. Gaz.* 145: 253–261.

Lal, S., and Srivastava, R.S. 1975. Inheritance of flower colour in lentils. *Indian J. Genet. et.* 35: 29–30.

Malhotra, R.S., Balyan, H.S., and Gupta, P.K. 1978. Crossing techniques in lentils. *LENS* 5: 7–8.

Malhotra, R.S., Singh, K.B., and Singh, J.K. 1974. Genetic variability and genotype-environmental interaction studies in lentil. *J. Res.* PAU 10: 17–21.

Malik, B.A., Verma, M.M., Rahman, M.M., and Bhattarai, A.N. 1988. Production of chickpea, lentil, pea and faba bean in South-East Asia. *World Crops: Cool Season Food Legumes.* R.J. Summerfield, ed. Kluwer Academic Publishers, Dordrecht, the Netherlands, pp. 1095–1111.

Manara, N.T.F., and Manara, W. 1983. Some climatic elements affecting lentil yields sown at several dates in Santa Maria, Central Depression of the State of Rio Grande do Sol. *LENS* 10: 22–24.

Meimandi-Nejad, M.J. 1977. Crop botany. In *Food Legume Crops: Improvements and Production.* FAO Plant Production Paper no. 9. FAO, Rome, pp. 27–62.

Mera, K.M., and Erskine, W. 1982. Crossing techniques for lentil under field conditions. *LENS* 9: 11–15.

Moursi, M.A., and Abd El-Gawad, A.A. 1963. Lentil photoperiodism. 2. Formative and photoperiodic reaction to light duration. *Ann. Agric. Soc. Univ. Ain. Chams.* 8: 297–304.

Muehlbauer, F.J. 1974. Seed yield components in lentils. *Crop Sci.* 14: 403–406.

Muehlbauer, F.J. 1981. Collection of wild *Lens* species in Southern Europe. *LENS* 8: 5.

Muehlbauer, F.J., and Ladizinsky, G. 1986. A cytoplasmic male sterility restorer system in *Lens.* Abstract. In *Int. Food Legume Res. Conf.* Spokane, Washington.

Muehlbauer, F.J., and Slinkard, A.E. 1981. Genetics and breeding methodology. In *Lentils.* C. Webb and G.C. Hawtin, eds. Farnham Royal, CAB International, UK, pp. 69–90.

Muehlbauer, F.J., and Slinkard, A.E. 1985. Lentil improvement in the Americas. In *Faba Beans, Kabuli Chickpeas and Lentils in the 1980s.* M.C. Saxena and S. Varma, eds. ICARDA, Aleppo, Syria, pp. 351–366.

Muehlbauer, F.J., Slinkard, A.E., and Wilson, V.E. 1980. Lentil. In *Hybridization of Crop Plant.* W.R. Fehr and H.H. Hadley, eds. *Am. Soc. Agron.* Madison, Wisconsin, pp. 417–426.

Muehlbauer, F.J., Short, R.W., Summerfield, R.J., Morrison, K. J., and Swan, D.G. 1981. Description and culture of lentils. Washington State Univ. EB 0957, 8p.

Muehlbauer, F.J. Cubero, J.I., and Summerfield, R.J. 1985. Lentil (*Lens culinaris* Medic.). In *Grain Legume Crops.* R.J. Summerfield, and E.H. Roberts, eds. Collins, 8 Grafton Street, London pp. 262–311.

Muehlbauer, F.J., Weeden, N.F., and Hoffman. D.L. 1989. Inheritance and linkage relationships of morphological and isozyme loci in lentil (*Lens* Miller). *J. Hered.* 80: 298–303.

Nezamuddin, S. 1970. Masur. In *Pulse Crops of India.* ICAR, New Delhi,

Pandey, B.K., Shukla, A., and Pandya, B.P. 1988. Variability of resistance to root rot/wilt complex in lentil germplasm, *Indian Phytopathol.* 41: 359–362.

Pandya, B.P., Pandey, M.P., and Singh, J.P. 1980. Development of Pant L 406 lentil, resistant to rust and wilt. *LENS* 7: 34–37.

Riva, E.A., 1975, 'Precoz' a new lentil cultivar for Argentina. *LENS* 2: 9–10.

Russo, M., Kishtah, A.A., and Tolba, M.A. 1981. A disease caused by bean yellow mosaic virus in Egypt. *Plant Diseases* (USA) 35(1): 132–133.

Sagar, P., and Chandra, S. 1980. Breeding behaviour and genetic variation for yield in crosses of lentil (*L. esculenta* Medik). *Indian J. Agric. Res.* 14: 159–163.

Saint-Clair, P.M. 1972. Responses of *Lens esculenta* Moench to controlled environmental factors. Committee on Agriculture, Univ. of Wageningen, the Netherlands, no. 72–12.

Sakar, D. 1983. Breeding behavior of characters related to yield and adaptation to mechanical harvesting in lentils (*Lens culinaris* Medik.). Ph.D. dissertation, Dept. of Agronomy and Soils, Washington State Univ.

Sakr, B. 1989a. *Lentil Breeding Program in Morocco.* INRA, Morocco.

Sakr, B. 1989b. Assessment of yield loss due to rust in lentil. Annual Report of INRA, Morocco.

Sarwar, D.M., Kaul, A.K., and Quader, M. 1982. Correlation studies in lentils. *LENS* 9: 22–23.

Sauerborn, J., Masri, H., Saxena, M.C., and Erskine, W. 1987. A rapid test to screen lentil under laboratory conditions for susceptibility to *Orobanche. LENS* 14: 15–16.

Saxena, M., and Hawtin, G.C. 1981. Morphology and growth patterns. In *Lentils*. C. Webb and G.C. Hawtin, eds. CAG International, Farnham Royal, UK, pp. 39–52.

Sepulveda, R. Paulina. 1985. Effect of *Uromyces fabae* Pers de Bary on lentil yield. *Agric. Tecnica* (Chile). 45(4): 335–339.

Sharma, B., and Kant, K. 1975. Mutation studies in lentils (*Lens culinaris*). *LENS* 2: 17–20.

Sharma, S.K., and Mukhopaday, S. 1963. Karyotype consistency in different strains of *Lens esculenta* Moench. as worked out through recent techniques. *Indian Agric.* 7: 103–111.

Shukla, T.C. 1955. Photoperiodic induction in *Lens esculenta* Moench. *Sci. Cult.* 20: 389–390.

Simon, C.J., and Muehlbauer, F.J. 1992. Status of the integrated linkage maps of chickpea and lentil. *Agron. Abstr.*, p. 114.

Sindhu, J.S., and Misra, H.O. 1982. Genetic variability in Indian microsperma types of lentil. *LENS* 9: 10–11.

Singh, K.G., and Jain, R.P. 1971. Heterosis in lentil. *Indian J. Agric. Sci.* 41: 54–57.

Singh, S.N., Srivastava, S.K., Agrawal, S.C. 1985. Relative efficacy of fungicides in the control of seedling blight and rust of lentil. *Indian J. Agric. Sci.* 55: 284–286.

Singh, T.P. 1977. Harvest index in lentil (*Lens culinaris* Medik.). *Euphytica* 26: 833–839.

Singh, T.P. 1978. Inheritance of cotyledon color in lentil. *Indian J. Agric. Sci.* 41: 54–57.

Slinkard, A.E. (ed). 1974. The Lentil Letter *LENS*, 1: 4.

Slinkard, A.E. 1978a. Laird lentil licensed in Canada. *LENS* 5: 24.

Slinkard, A.E. 1978b. Inheritance of cotyledon color in lentils. *J. Hered.* 69: 129–130.

Slinkard, A.E. 1980. Resistance to MCPB in P.I. 179310 TR. *LENS* 7: 65.

Slinkard, A.E. 1985. Cytology and cytogenetics of lentils. *LENS* 12: 1–10.

Slinkard, A.E. and Bhatty, R.S. 1979. 'Laird' lentil. *Can. J. Plant Sci.* 59: 503–504.

Solh, M., and Erskine, W. 1981. Genetic resources. In *Lentils*. C. Webb and G.C. Hawtin, eds. CAB International, Farnham Royal, UK, pp. 53–67.

Solh, M., and Erskine, W. 1984. Genetic resources of lentils. In *Genetic Resources and Their Exploitation. Chickpeas, Faba Beans, and Lentils*. J.R. Witcombe and W. Erskine, eds. Martinus Nijhoffs, the Netherlands, pp. 205–221.

Solh, M.G., Paredes, O.M. and Tiwari, A.S. 1980. Crossing techniques in lentils under field conditions. *LENS* 8: 9–14.

Summerfield. R.J. 1981. Adaptation to environments. In *Lentils*. C. Webb and G.C. Hawtin, eds. CAB International, Farnham Royal, UK, pp. 91–110.

Summerfield, R.J., Muehlbauer, F.J., and Roberts, E.H. 1985. *Lens culinaris*. In *Handbook of Flowering*. A.H. Halevy, ed. Vol. 1, pp. 118–124.

Summerfield, R.J., Muehlbauer, F.J., and Short, R.W. 1982. Description and culture of lentils. Production Research Report No. 181. US Department of Agriculture, Washington, DC.

Tahhan, O., and Hariri, G. 1982. Survey of lentil insects in North and Northeastern Syria. *LENS* 9: 34–37.

Tanksley, S.D., and Orton, T.J. (eds). 1983 *Isozymes in Plant Genetics and Breeding*. Elsevier, Amsterdam.

Tawar, M.L., and Gour, V.K. 1984. Effect of colchicine on lentils. *LENS* 11: 15–17.

Tay, J. 1989. Inheritance of resistance to Ascochyta blight in lentil. MS Thesis, Univ. of Saskatchewan, Saskatoon, Canada, 70 pp.

Tay, J., Parades, M., and Kramm, V. 1981. 'Araucana-INIA'. A new large-seeded lentil cultivar. *LENS* 8: 30.

Tikka, S.B.S., Asawa, B.M., Jaimini, S.N., and Goyal, S.N. 1977. Component analysis of seed yield and seed size in lentil. *LENS* 4: 17–20.

Tschermak-seysenegg, E. 1928. Lentil and field bean crosses. *Sityringsber, Akad. Wiss. Wein Math. Nat. K1. I. Abt.* 137: 171–181.

Tyagi, M.C., and Sharma, B. 1985. Association among economic traits in lentil. *LENS* 12: 10–12.

Uhlik, J. 1972a. Mutational efficiency of fast neutrons in *Lens esculenta* (Moench.). *Genetika Slechteni* (Praha) 8: 21–28.

Uhlik, J. 1972b. Mutational efficiency of EMS as compared with gamma irradiation in *Lens esculenta* (Moench.). *Genetika a Slechteni* (Praha) 8: 251–260.

Uhlik, J. 1973. Comparison of the mutagenic activity of N-ethyl- N-nitrosourea and N-methyl-n-nitro-N-nitrosoguanidine in *Lens esculenta. Biol. Plant.* (Praha) 15: 274–279.

Vaillancourt, R.E., and Slinkard, A.E. 1983. Variation in tannin content of lentils. *LENS* 10: 36.

Vaillancourt, R.E., Slinkard, A.E., and Reichert, R.D. 1986. The inheritance of condensed tannin concentration in lentil. *Can. J. Plant Sci.* 66: 241–246.

Vandenberg, A., and Slinkard, A.E. 1987. Intheritance of a Xantha chlorophyll deficiency in lentil. *J. Hered.* 78: 130.

Vandenberg, A., and Slinkard, A.E. 1989. Inheritance of four new qualitative genes in lentil. *J. Hered.* 80: 320–322.

Vandenberg, A., and Slinkard, A.E. 1990. Genetics of seed coat color and pattern in lentil. *J. Herd.* 81: 484–488.

Volvas, C., and Rana, G.L. 1972. Market virus diseases of market garden plants in Apulia VII. *Lens culinaris* M. A natural host of pea enation mosaic virus. *Phytopathol. Mediterr.* 11: 97–102.

Williams, J.T., Sanchez, A.M.C., and Jackson, M.T. 1974. Studies on lentils and their variation. I. The taxonomy of the species. *SABRAO J.* 6: 133–145.

Wilson, V.E. 1972. Morphology and techniques for crossing *Lens esculenta* Moench. *Crop Sci.* 12: 231–232.

Wilson, V.E., and Hudson, L.W. 1978. Inheritance of lentil flower color. *J. Hered.* 69: 139–140.

Wilson, V.E., and Law, A.G. 1972. Natural crossing in *Lens esculenta* Moench. *J. Amer. Soc. Hort. Sci.* 97: 142–143.

Wilson, V.E., and Teare, I.D. 1972. Effect of between and within row spacing on components of lentil yield. *Crop Sci.* 12: 507–510.

Wilson, V.E., Law, A.G., and Warner, R.E. 1970. Inheritance of cotyledon color in *Lens culinaris* (Medik.). *Crop Sci.* 10: 205–207.

Wilson, V.E., Morrison, K.J., and Muehlbauer, F.J. 1971. Tekoa lentil and its culture. Washington State Univ. Circ. 375.

Zamir, D., and Ladizinsky, G. 1984. Genetics of allozyme variants and linkage groups in lentil. *Euphytica* 33: 329–336.

Zohary, D. 1972. The wild progenitor and place of origin of the cultivated lentil, *Lens culinaris*. *Econ. Bot.* 26: 236–332.

4. MUNGBEAN

S. Shanmugasundaram

Plant Breeder, and Director, International Cooperation Program, The Asian Vegetable Research and Development Center, Shanhua, Taiwan, 741, Taiwan, Republic of China

D.H. Kim

Mungbean Breeder, Asian Regional Center-AVRDC, Kasetsart University, Kamphaeng Saen Campus, Nakorn Pathom 73140. Thailand

Introduction

Mungbean, *Vigna radiata* (L.) Wilczek, is a short-duration legume crop cultivated primarily for its dry seeds. While the areas for cereals and other pulses have decreased, that for mungbean has doubled in the last two decades with an annual rate of 2.5 per cent. This growth may very likely continue, since mungbean's short growth duration (60 days) makes it suitable for various cropping systems. Annual mungbean production worldwide is around 2.5 to 3.0 million t harvested from about 5.0 million ha (Poehlman, 1991). Around 45 per cent of the total world mungbean production is in India. In China, mungbean is grown in about 0.5 million ha per year with an average of 2 t/ha yield. In Thailand and the Philippines, mungbean is the most important grain legume; in Sri Lanka it ranks second, while in India, Burma, Bangladesh, and Indonesia it is the third-most important grain legume.

In India, mungbean is used as a pulse in the preparation of *dahl*, a soup porridge eaten with a cereal or other traditional cuisines, and is a main protein source for the vegetarian diet. In other Asian countries, mungbean is used for bean sprouts, starch noodles, mungbean soup, and deep-fried patties of different kinds. Mungbean is known for its easy digestibility and low flatus production compared to other pulses. The green pods and seeds may be used as vegetables.

Since mungbean matures in about 60–70 days after sowing, it is an excellent crop for rotation in different cropping systems. Mungbean may also be sown as an intercrop or as a green manure or cover crop.

Botany

Mungbean belongs to the order Leguminosae and family Papilionoideae and is botanically recognized as *Vigna radiata* (L.) Wilczek syn. *Phaseolus radiatus* L., *P. aureus* Roxb (Wilczek, 1954; Verdcourt, 1970). The taxonomic status of the species is given in Table 1. The genus *Vigna* has been broadened to include about 150 species; 22 species are native to India and 16 to Southeast Asia, but the largest number of species are found in Africa (Polhill and van der Maesen, 1985). A few closely related Asian species of *Vigna* and their common names are listed in Table 2.

Table 1
Taxonomic status of mungbean

Kingdom	Plant kingdom
Division	Spermatophyta
Subdivision	Angiospermae
Class	Dicotyledoneae
Order	Leguminosae
Family	Papilionoideae
Tribe	Phaseoleae
Genus	*Vigna*
Subgenus	Ceratotropis
Species	*radiata*
Variety	*radiata*

Source: Rachie and Roberts (1974), Jain and Mehra (1980).

Table 2
Some wild and cultivated species of *Vigna* in Aisa

Species	Common names
Vigna aconitifolia (Jacq.) Maréchal	Moth bean, mat bean
Vigna angularis (Willd.) Ohwi and Ohashi	Adzuki bean, red bean
Vigna mungo (L.) Hepper	Urd, mash, blackgram, mungo bean
Vigna radiata (L.) Wilczek	Mungbean, moong, greengram, golden gram
Vigna trilobata (L.) Verdc.	Phillipesare bean, jungli bean
Vigna umbellata (Thunb.) Ohwi and Ohashi	Rice bean, red bean
Vigna radiata var. *sublobata* (Roxb). Verdc.	Wild progenitor of mungbean

A key to identify four Asiatic *Vigna* species is given below:

1. Germination epigeal and petiole of primary leaves short
   a. Hilum not concave — *V. radiata* (mungbean)
   b. Hilum concave — *V. mungo* (blackgram)
2. Germination hypogeal and petiole of primary leaves long
   a. Hilum not concave — *V. angularis* (adzuki bean)
   b. Hilum concave — *V. umbellata* (rice bean)

Mungbean is an annual, 0.3–1.5 m tall, erect or suberect plant, sometimes slightly twining at the tips. It is deep-rooted and much branched, with long petioles. The leaves are alternate, trifolilate, and dark or light green, the leaflets are ovate and vary from 5 to 12 cm wide and 2 to 10 cm long.

The inflorescence is an axillary raceme with a peduncle 2 to 13 cm long. The flower is yellow and the keel petal is spirally coiled with a horn-like appendage (Rachie and Roberts, 1974). Pods are 6 to 10 cm long, slender, short, and hairy. Seeds are globose and weigh 15 to 85 mg and are mostly green but sometimes yellow, tawny brown, black, or mottled. The white, flat hilum is not concave; germination is epigeal (Bailey, 1970).

Origin

Mungbean probably originated in India (De Candole, 1886; Zhukovsky, 1950; Bailey, 1970) or the Indo-Burmese region (Vavilov, 1951; Singh *et al.*, 1970; Jain and Mehra, 1980). The primary gene centre of diversity for mungbean was suggested to be the central Asian region (Vavilov, 1951) with India as the gene centre and probable centre of domestication (Smartt, 1985). Lukoki *et al.* (1980) proposed that *V. radiata* var. *sublobata* (Roxb.) Verdc., which occurs wild in India and crosses easily with *V. radiata* var. *radiata,* is the wild ancestor of mungbean.

Genetics, Qualitative Characters

Literature on genetic studies of mungbean largely comes from India, where a vast array of researchers work on the crop. With mungbean, genetic studies are sparse compared to crops like soybean, due to lesser economic importance of the crop and less funding for research.

Early genetic studies of mungbean were conducted in India by Bose (1939). Before 1980, 45 different qualitative characters were identified. Recognizing that there were no rules for gene nomenclature for *Vigna,* Fery (1980) suggested that a set of rules proposed for Cucurbitaceae be adopted for *Vigna.* Thus far, no one has responsibility for maintenance of a genetic type collection of mungbean, although there is a need for this activity. Fery (1980) developed a detailed review of the genetics of *Vigna* including *V. radiata* var. *radiata.* Poehlman (1991) extended the review to include genes identified after 1980.

Disease and Insect Reaction

Mungbean is attacked by many pathogens. These include fungi, bacteria, viruses, and nematodes. For reviews of the genetics of disease resistance, see Rachie and Roberts (1974), Fery (1980), Singh (1981, 1982, 1986), and Poehlman (1991). Many studies report on screening of accessions for resistance but few studies report on inheritance of resistance.

Leaf spots. Resistance to bacterial leaf spot (BLS), a disease incited by *Xanthomonas phaseoli* (Smith) Dowson, and resistance to Cercospora leaf spot (CLS), caused by *Cercospora canescens* Ellis and Martin, are each conditioned by single dominant gene (Singh and Patel, 1977; Thakur *et al.* 1977a, b, c; AVRDC, 1981a, b). However, Yadav *et al.* (1981) reported. resistance to BLS to be governed by a single recessive gene. It was shown that peroxidase polymorphism may be used as a selection index for CLS resistance (Menancio and Ramirez, 1977).

Powdery mildew. Resistance to powdery mildew, caused by *Erysiphe polygoni* DC. ext. St. Amans, is controlled by a single dominant gene (AVRDC, 1979), although it has also been reported to be governed by genes with additive action (Yohe and Poehlman, 1975), and recessive polygenes (AVRDC, 1981a, b). RFLP study also showed that the quantitative genes are controlling powdery mildew resistance (Young *et al.,* 1994).

Viruses. Tolerance or resistance to the mungbean yellow mosaic virus (MYMV) is conferred by a single recessive gene (Singh and Patel, 1977; Thakur *et al.,* 1977b; Malik *et al.,* 1988); complementary recessive genes (Shukla and Pandya, 1985); and a dominant gene and complementary recessive genes (Sandhu *et al.,* 1985). RFLP study also indicated a single recessive gene (Poehlman, 1991). Resistance to MYMV was identified in *V. radiata* var. *sublobata* Roxb. Verdc., progenitor of mungbean, and genes for resistance have been transferred to commercial mungbean types (Singh and Ahuja, 1977). Sittiyos *et al.* (1979) identified a dominant gene conferring resistance to a mungbean strain of cucumber mosaic virus.

Thakur *et al.* (1977a) studied the relationship of genes governing resistance to bacterial leaf spot and Cercospora leaf spot and tolerance to MYMV. They concluded that the genes were independently inherited and that traditional breeding methods could be used to develop multiple disease-resistant mungbean.

Bruchid. Resistance to the storage insect pest bruchid is controlled by a single dominant gene in *V. radiata* var. *sublobata* (Kitamura *et al.,* 1988). RFLP study has also confirmed this (Young *et al.,* 1992).

Morphological Characters

Growth habit. Twining growth was reported to be dominant to erect habit (Sen and Ghosh, 1959). Pathak and Singh (1963) determined that the twining growth habit is due to a single recessive gene and that semispreading growth habit is dominant to erect habit.

Leaf types. Lobed leaflet is governed by a single dominant gene (Singh and Mehta, 1953; Singh, 1980). Sen and Ghosh (1959) reported that the heterozygous genotypes have intermediately lobed leaflets. A dominant gene governs the incised leaflet while its allele produces ovate leaflet (Pokle, 1972). A mutant with four or five leaflets is probably dominant over the normal three leaflets (Veeraswamy and Kunjamma, 1958). Mutation-induced unifoliate and multifoliate leaflets were governed by independent recessive genes over the normal trifoliate leaflets (Santos, 1969). A crinkled lanceolate mutant governed by a single semidominant gene, lethal in homozygous condition, was described by Singh and Saxena (1959). Two trilobate mutants were each found to have a dominant gene over normal monolobate trait. When crossed, an F_2 ratio of 15 : 1 was obtained (Sareen, 1985).

Pubescence. Dense plant pubescence is due to a single dominant gene, while the recessive allele has medium-dense pubescence (Murty and Patel, 1973).

Pigmentation

Genes governing the pigmentation of various plant parts were reviewed by Fery (1980) and gene designations revised to conform to the new nomenclature. (Also, see Poehlman, 1991.)

Seed coat. Inheritance of seed coat colour was studied initially by Bose (1939), who suggested two genes for green seed coat colour. Variations in seed coat colours were found to be due to sap-soluble pigments and the colour and number of chloroplasts by Sen and Ghosh (1959). They reported that blue sap colour was due to a single dominant gene while buff sap colour was due to a recessive gene. Van Rheenen (1965) identified dominant genes conditioning green and spotted seed coat colours. Black spot on the seed coat was due to a single dominant gene (Singh and Singh, 1970). Murty and Patel (1972) described a four-gene model to explain seven different seed coat colours. A three-gene model was proposed by Singh (1973) to describe bottle-green and yellowish-green seed coat colours. Green and spotted seed colour traits are each governed by a single dominant gene (Singh and Patel, 1977).

Flower. Light-yellowish olive flower colour is partly dominant to olive yellow flower colour. The genes governing hypocotyl-epicotyl colour also govern the oilve yellow and plam-leaf yellow flower colours. An allelic series of three genes was reported to govern pale-green yellow, bryta yellow, and naphthalene yellow flower colours, respectively, in that order of dominance

(Murty and Patel, 1973).

Pod. The genes for flower colour are known to affect the unripe pod colour (Bose, 1939). Purple pigmentation in the vertical suture of unripe pods is due to a single dominant gene. Two independent recessive genes govern light popcorn and almond biscuit colour; presence of the two dominant alleles produces black pod colour. Black pods are probably due to a single dominant gene over light brown pods (Pathak and Singh, 1963).

Pubescence. Brown pubescence is recessive to colourless, with two dominant alleles required to produce colourless pubescence (Sen and Ghosh, 1959).

Cotyledon, hypocotyl, epicotyl, stem and leaf rachis. Purple hypocotyl is dominant to green hypocotyl and is governed by a single gene. A multiple allelic series with three alleles was suggested to govern the pigmentation of hypocotyl, epicotyl, stem, and leaf rachis. A number of papers reported that the pigmentation in hypocotyl, epicotyl, stem, petiole, and peduncle is due to single dominant genes (see Fery, 1980, for a list of references).

Red colour of cotyledon, hypocotyl, and top of the leaf stalk is due to a dominant gene that is suspected to be the same as the multiple allelic series identified by Sen and Ghosh (1959) and van Rheenen (1965). A dominant gene was shown to govern purple hypocotyl by Swindell and Poehlman (1978). A mutant with anthocyanin pigmentation in the hypocotyl, stem, petiole, and peduncle was suspected to be due to a recessive gene (Appa Rao and Jana, 1973).

Chlorophyll. Albino seedling is due to a recessive gene with a recessive inhibitor gene (Sen and Ghosh, 1960c). Induced xantha and variegata characters are inherited as monogenic recessives, while a greenish-yellow chlorina behaved as a monogenic recessive in some lines but appeared to be due to two recessive genes in others (Santos, 1969).

Inflorescence, Photoperiod Response and Sterility

Inflorescence. Two dominant genes produce a simple inflorescence; in the absence of one or both of these genes a compound inflorescence arises (Sen and Ghosh, 1959). Single cluster of flowers is dominant to three or more clusters (Singh and Singh, 1970).

Photoperiod response. Photoperiod insensitivity is dominant over photoperiod sensitivity (Verma, 1971; Tiwari and Ramanujam, 1976b). A dominant or partly dominant gene governs the sensitivity to photoperiod in

PI 180311. The gene is expressed in 14 to 16 hour photoperiods but not in 12 hour photoperiods (Swindell and Poehlman, 1978). Early flowering and early maturity is either dominant or partly dominant and probably governed by two genes (Tiwari and Ramanujam, 1976b).

Sterility. Successful induction of male- and female-sterile mutants through various methods have been reported. In all instances sterility appears to be monogenic recessive (Ganguli, 1972, 1973; Singh and Chaturvedi, 1981).

Pod and Seed

Pod tip. Sen and Ghosh (1959) reported that a dominant gene controlled the swolled pod tip over tapering pod tip.

Seed shattering. Seed shattering at maturity is due to a single dominant gene (Verma and Krishi, 1969).

Seed size. Small seed size is dominant to large seed size (Sen and Murty, 1960).

Seed coat surface. Bose (1959), Sen and Ghose (1959), and van Rheenen (1965) reported dull, rough seed surface to be monogenic dominant over glossy, smooth seed surface. Sen and Ghosh (1959) identified a second dominant gene that governs dull, rough seed coat. Dull seeds are covered by an inner pod membrane that renders the seed dull; when this membrane is removed, the seed coat underneath is shiny (Watt *et al.*, 1977). The pod membrane may contain brown or black pigment through which the seed coat colour may not be apparent. Based on this information, inheritance of pigmentation in the seed coat that may have been reported earlier for seeds covered with the membrane needs to be reevaluated. With a translucid membrane the seed coat colour is visible through the membrane layer (see Poehlman, 1991, for discussion).

Genetics, Quantitative Characters

Many traits of economic importance are inherited in a quantitative fashion and their expression may be affected by both genetic and environmental influences. The extent of the genetic variability of a quantitatively inherited trait and the manner of its inheritance determines its usefulness in plant improvement.

Genetic Variability

In mungbean the range of genetic variability for characters of economic importance has been studied for large collections of accessions by Banks (1958), Bhargava *et al.* (1966), Singh and Malhotra (1970a), Yohe and Poehlman (1972), Sandhu *et al.* (1980), Virmani *et al.* (1983), Paroda and Thomas (1988) and He *et al.* (1988). In general, the variability is reported to be sufficiently extensive for progress in mungbean breeding programmes.

Relationship of Plant Characters and Yield

A knowledge of the relationship among plant characters is useful when selecting traits to combine for yield improvement. Correlation coefficients of yield vs. some major yield-associated characters are summarized in Table 3. The yield component with the largest and most consistent correlation coefficient in a wide array of experiments is pods per plant, a close positive relationship to yield being reported in all experiments. Although a larger number of seeds per pod and higher 100-seed weight can compensate for low pod number, the association of seeds per pod or 100-seed weight with yield is inconsistent in different environments and has low reliability as a selection criterion for yield.

Heritability, Combining Ability and Gene Action of Yield and Associated Characters

Grain yield in mungbean, as in other crops, is a complex character and is

Table 3
Estimates of phenotypic and genotypic (in parenthesis) correlation coefficients for yield vs. specific traits in mungbean

Pods per plant	Seeds per pod	Seed weight	Days to flower	Pod length	Reference
0.711 (0.689)	0.073 (0.196)	0.509 (0.720)	−0.131 (−0.158)	0.363 (0.537)	Gupta and Singh, 1969
0.90 (0.950)	0.35 (0.980)	0.0004 (−0.2100)	0.70 (0.80)		Joshi and Kabaria, 1973
0.8512 (0.7095)	0.4438 (0.4516)	0.1600 (0.3300)	0.0323 (0.0967)	0.4026 (0.2676)	Singh and Malhotra, 1907b
0.9512 (0.9772)	0.6903 (0.9077)	−0.4861 (−0.6506)	0.6140 (0.8220)	0.1631 (0.1421)	Malhotra *et al.*, 1974a
0.503 (0.565)	0.282 (0.319)	0.159 (0.468)			Malik *et al.*, 1983
0.847 (0.902)	0.433 (0.455)	−0.100 (−0.008)		0.198 (0.204)	Gupta *et al.*, 1982

influenced by both genetic and environmental factors. Thus, the phenotypic variance of yield or a yield-associated character is composed of genetic and environmental components and genotype × environment interactions. The proportion of phenotypic variance of a character due to genetic component determines the heritability for that character. If calculated from the total genetic component, it is referred to as broad-sense heritability, whereas, if only the additive proportion of the genetic component is used in the calculation, the heritability is known as narrow sense. Genetic improvement by selection largely depends on additive gene action component. Most heritability estimates reported for mungbean characters are broad-sense heritabilities and provide a less accurate prediction of response to selection than would be provided by narrow-sense heritability estimates. A particular heritability estimate is specific only for that population and the environment in which the population was grown. Estimates of broad-sense heritability for yield and some yield parameters as reported by different workers are summarized in Table 4. Heritability for 100-seed weight was highest and most consistent among the yield components, followed by days to flowering and plant height. Heritability of seed yield varied in different experiments.

Yield. Heritability estimates for seed yield varied from 8.6 to 84.2 per cent. Similarly, for pod number, the yield component with the highest correlation to yield (Table 4), heritability ranged from 21.7 to 90.4 per cent. The range in heritability estimates may be attributed to variations in environments in which the studies were conducted, variations in gene action in the different cultivars studied, variations in specific gene combinations present in the varieties used, and/or the genotye × environment interactions.

The inheritance of yield and its components in mungbean has been studied for combining ability and type of gene action through the use of diallel crosses (Singh and Jain, 1971a, b; Singh and Singh, 1971, 1972, 1974a; Yohe and Poehlman, 1975; Ramanujam, 1978; Reddy and Sreeramulu, 1982; Ahuja and Chowdhury, 1984). General combining ability (GCA) values for yield and yield components were generally greater than specific combining ability (SCA) values; therefore, it may be inferred that yield and the yield components were largely governed by additive gene effects.

Plant development characters. Broad-sense heritability for plant height (Table 4) varied from 7.9 to 96.9 per cent and that of branches per plant ranged from 31.0 to 85.2 per cent. Significant GCA and SCA were observed for plant height by Swindell and Poehlman (1976). Tiwari and Ramanujam (1974) found a large proportion of variance due to GCA for plant habit, number of leaf axils per plant, number of effective leaf axils per plant, and

Table 4
Broad-sense heritability estimates for seed yield and some yield related characters in mungbean

Days to flowering	Days to maturity	Plant height	Branches per plant	Pod number	Pod length	Seeds per pod	100-seed weight	Seed yield	Reference[a]
—	—	96.8	66.0	87.9	66.0	41.3	85.3	69.9	1
94.9	—	87.3	34.5	53.8	88.6	39.0	98.6	56.9	2
62.0	71.2	27.0	—	24.6	—	10.0	51.2	8.6	3
99.0	—	96.9	—	77.0	89.7	70.2	97.0	79.1	4
39.5	—	—	50.9	47.8	89.9	58.7	92.9	51.6	5
88.8	81.5	7.9	—	90.4	—	12.3	—	84.2	6
69.0	—	75.0	57.0	78.0	74.0	78.0	—	75.0	7
—	—	96.3	85.2	21.7	95.0	68.3	95.3	59.0	8
65.1	61.4	—	—	30.9	33.9	30.5	88.9	27.5	9
—	—	83.0	31.0	43.0	31.0	26.0	97.0	38.0	10

[a]References: (1) Bhargava *et al.*, 1966; (2) Chowdhury *et al.*, 1971; (3) Empig *et al.*, 1970; (4) Giriraj, 1973; (5) Gupta and Singh, 1969; (6) Joshi and Kabaria, 1973; (7) Murty *et al.*, 1976; (8) Pokle and Nomulwar, 1975; (9) Singh and Malhotra, 1970b; (10) Tomar *et al.*, 1972.

number of branches per plant. Similarly, GCA was preponderant in Yohe and Pohelman's (1975) study on the variability in plant height and branch length. For plant height and number of branches, dominance × dominance gene action was demonstrated (Murty *et al.*, 1976), although Godhani *et al.* (1978) implicated additive gene action for branch number, and additive, dominance, and epistatic gene actions for plant height. Since plant height has high heritability and high genetic advance, it was suggested that additive gene action is predominant (Bhargava *et al.*, 1966; Chowdhury *et al.*, 1971; Giriraj, 1973; Veeraswamy *et al.*, 1973). Pokle and Nomulwar (1975) reported that genetic advance for plant height is low, even though it is highly heritable, and therefore is governed by non-addiive gene action. Reddy and Sreeramulu (1932) also reported non-additive gene action for height.

Flowering. Broad-sense heritability estimates for days to flowering ranged from 39.5 to 99 per cent. Based on the results of several studies, it was suggested that earliness is governed by additive gene action (Chowdhury *et al.*, 1971; Giriraj, 1973; Singh and Singh, 1974a; Tiwari and Ramanujam, 1974; Yohe and Poehlman, 1975; Swindell and Poehlman, 1976; Godhani *et al.*, 1978). Dominance × dominance gene action for days to flowering was reported by Murty *et al.* (1976) and Swindell and Poehlman (1978).

Pod length. Heritability of pod length was fairly high in most environments, although reports conflicted on type of gene action (Chowdhury *et al.*, 1971; Singh and Singh, 1971; Pokle and Nomulwar, 1975; Murty *et al.*, 1976; Godhani *et al.*, 1978). Such reports show the complexity of the inheritance and the environmental response of this character and, therefore, the selection for this trait will encounter difficulty.

Seeds per pod. The range in heritability for number of seeds per pod varied from 10.0 to 78.0 per cent. Additive and non-additive gene action was implicated for number of seeds per pod (Singh and Jain, 1971a; Tiwari and Ramanujam, 1974). Murty *et al.*, (1976) reported that dominance × dominance gene action was important.

Seed weight. Heritability of seed weight is high, ranging from 51.2 to 98.6 per cent. Several workers reported high heritability estimates and high genetic advance indicating that additive gene action is apparent in the inheritance of seed size.

Protein content. Singh and Singh (1973b) and Singh (1974) demonstrated both additive and non-additive gene action for the inheritance of protein content in mungbean. Tiwari and Ramanujam (1976a) observed predominant additive gene action.

Path Coefficient Analysis of Yield and Its Components

Because yield is a complex trait and its expression is governed by many factors, it is logical to expect that a number of plant traits jointly contribute to enhance yield. A knowledge then of the interrelationships and contributions of various yield components, such as pods per plant, seeds per pod, pod length, leaf size, or 100-seed weight, to yield would be useful if selection for simultaneous improvement of these traits is to be most effective. To establish such relationships, a path coefficient analysis of correlation in a system of related variables may be used. Path coefficient analyses in mungbean are reported by Singh and Malhotra (1970b), Singh and Singh (1973a), Chandel *et al.* (1973), Giriraj and Vijayakumar (1974), Malhotra *et al.* (1974b), Pokle and Patil (1975), Bhaumik and Jha (1976), Sandhu *et al.* (1980), Abilay and Lantican (1982), and Vidyadhar *et al.* (1984). The direct and indirect effects of specific yield components on yield were shown in the above reports. The character making major contributions directly to yield was pods per plant, with indirect contributions being made through seeds per pod and seed weight. Plant height and number of branches contributed indirectly to yield through pods per plant. Pods per plant also has a significant relationship with yield as measured by correlation analyses. These results point to the importance of selecting for higher number of pods per plant to improve yield. Uniform spacing of plants is required for valid comparisons of pods per plant among mungbean lines.

Genetic Divergence

Mahalanobis' D^2 analysis was used to study the yield and yield components of different strains in mungbean (Gupta and Singh, 1970; Thulasidass, 1984). Seed weight and pod length were found to be important characters contributing toward genetic divergence. Malhotra *et al.* (1974b) reported that geographic diversity had no direct association with genetic diversity. Thulasidass (1984), in a study of 30 genotypes of different origins, showed an association of genetic diversity with geographic diversity. In the latter study, days to maturity made the greatest contribution toward genetic divergence. Ramanujam *et al.* (1974) studied genetic divergence in 10 parents and 25 F_1 populations using D^2 and canonical analysis. He reported genetic divergence contributed by flowering time, maturity, seed density, and 100-seed weight. Canonical analysis supported the results of the D^2 analysis.

Heterosis

Heterosis in mungbean was reported for yield and the yield components days to flowering, plant height, leaf size, and branch number (Bhatnagar and Singh, 1964; Misra *et al.*, 1970; Singh and Jain, 1970; Ramanujam *et al.*, 1974; Singh and Singh, 1974a, b; Swindell and Poehlman, 1976). Overall, yield increase of the F_1 over mid-parent yield varied from 17 to 208 per cent and over the high parent from 1 to 188 per cent. The high yield increases reported may not be realistic in experiments where heterosis was measured from spaced plants, as spaced plants are notoriously variable in growth. In 58 crosses, the F_1 was significantly superior to the midparent in only 23 crosses, while out of 33 crosses, the F_1 was superior to the high parent in only 6 crosses.

Heterosis for yield and pods per plant was attributed to overdominance and non-fixable genic interactions (Singh and Singh, 1974a, b). Evidence of significant inbreeding depression was observed for yield in F_2 and F_3. Presence of genes with oppositional dominance was reported to be responsible for lack of heterosis for protein content (Singh and Singh, 1973b; Tiwari and Ramanujam, 1976a). In 25 crosses, three showed significant heterosis over midparent for methionine content (Tiwari and Ramanujam, 1976a). Due to the low frequency of crosses with significant heterosis and the closed pollination system in mungbean, the exploitation of hybrid vigour *per se* as a breeding procedure is not feasible.

Induced Mutations

Many studies have been conducted on induced mutations in mungbean, especially in India, Pakistan, and Bangladesh. The mutagenic agent used most frequently has been ^{60}Co gamma radiation, but X-rays, neutrons, and chemical mutagens such as EMS and DMSO have also been used. The mutagenic agents cause varying kinds of undesirable mutations in addition to the desirable ones. More point mutations and fewer chromosomal aberrations are obtained with EMS or DMSO than with radiations. With gamma radiation, a dose of 30 to 40 KR ^{60}Co radiation was found to be most effective in producing desirable mutants (Singh *et al.,* 1979), although the sensitivity dosage of the mutagenic agent may vary with the genotype used (Murray and Newcombe, 1970; Rajput, 1973; Rangaswamy *et al.,* 1973; Yahya *et al.,* 1975; Khan and Hashim, 1978; Singh and Chaturvedi, 1981).

Mutants commonly reported have been characterized by the following traits: shorter plants, earlier and synchronous maturity, more branches per plant, more pod clusters, greater pod set, MYMV resistance, higher 100-seed weight, and higher protein content (Santos, 1969; Bhatt *et al.,* 1972; Dahiya, 1973, 1978; Rajput, 1974; Tikoo and Jain, 1974; Prasad, 1976; Subramanian, 1980; Singh and Chaturvedi, 1981; Shaikh *et al.,* 1982). Four mungbean cultivars have been released from mutation breeding programmes, 'Pant Mung 2', 'TAP-7', and 'Co4' from India and 'NIAB Mung 2B' from Pakistan (Shanmugasundaram, 1988). The cultivar 'Pant Moong 2', released in 1982, is reported to be moderately resistant to MYMV, and 'TAP-7', also released in 1982, is tolerant to powdery mildew and a leaf spot disease.

Cytology and Cytogenetics

Cytological and cytogenetic studies with mungbean have been relatively few owing to the small size of the mungbean chromosomes and the difficulty of working with them. The diploid chromosome number of mungbean is $2n = 2x = 22$ (Karpechenko, 1925; Kumar, 1945;' Krishnan and De, 1965). A pachytene chromosome analysis disclosed that the chromosomes varied in length from 28.1 to 73.3 μ with two bivalents associated with the nucleolus at pachytene (Krishnan and De, 1965). The nuclear chromosomes varied in length and the position of secondary constrictions at pachytene and somatic metaphase stages also varied. The karyotypes of three wild types had shorter somatic chromosomes than two cultivated varieties (Shrivastava *et al.,* 1973). By comparing the chromosome complements of several related species, Bhatnagar *et al.* (1974) proposed the following karyotype formula for mungbean: $4L^{sm}$ +

$4M^{sm} + 3M^{m}$, where L = long (2.7 to 3.5 μ), M = medium (1.9 to 2.6 μ), sm = submedian, and m = median centromere. An idiogram of prometaphase chromosomes was presented by Joseph and Boukamp (1978). Since mungbean has long metacentric and submetacentric chromosomes, it is considered to be more primitive than blackgram (De and Krishnan, 1966a, b; Sarbhoy, 1980).

Polyploids

Autotetraploids have been induced, following colchicine treatment by Kumar and Abraham (1942a, b), Kumar (1945), Sen and Ghosh (1960a), Mitra (1967), and Kabi and Bhaduri (1978). Generally, the autotetraploids had *gigas* floral parts, slow gemination, late maturity, high sterility, poor seed set, and larger seed stomata. Autotetraploidy *per se* is not expected to make a significant impact in the development of new mungbean cultivars. No natural amphidiploids have been identified with a genome homologous to *V. radiata*.

Interspecific Crosses

Several closely related species (see Table 2) have the same chromosome number as mungbean, 2n = 2x = 22. Based on cytogenetic studies, Dana (1966a, b) proposed the genome designation of AA for *V. radiata* and *V. mungo* and A_1A_1 for *A. umbellata.* Blackgram, rice bean, and adzuki bean have genes for disease resistance and nutritional quality, factors that would be useful in mungbean. So far, not much has been accomplished in transferring genes for useful characters from related species to mungbean. Reviews of interspecific crosses with mungbean have been made by Yadav *et al.* (1986) and Poehlman (1991).

Crosses of mungbean × blackgram were attempted by Sen and Ghosh (1960b), Dana (1966a), De and Krishnan (1966b), Chowdhury and Chowdhury (1974), Verma (1977), Ahn and Hartmann (1978a, b, c), AVRDC (1979), Chen *et al.* (1983), and Yadav *et al.* (1986). Generally crosses have been successful only when mungbean is used as the female parent. Success of the crosses, as measured by viability of the crossed seeds and F_1 plant survival, varied with the parent genotype in the cross (AVRDC, 1979). Ahn and Hartmann (1978c) found that compatibility was improved if a mungbean F_1 was used as the female in crosses with blackgram but germination of F_2 seeds was still erratic. Generally, the hybrids are sterile, although their amphidiploids have normal meiosis and up to 80 per cent fertility. The amphidiploids are vigourous, but shy in pod production and pod filling, are unstable, and tend to revert back to parental forms by gradual elimination of chromosomes (Yadav *et al.*, 1986). Selection for recombinant types can be made among the derivatives. Amphidiploid-derived lines appear to be superior in yield, tryptophan and methionine content, and protein digestibilty, which is attributed to gene introgression between mungbean and blackgram (Yadav *et al.*, 1986).

Attempts have been made to cross mungbean with ricebean (*V. umbellata*), which has resistance to Cercospora leaf spot and powdery mildew pathogens. The sterility of the F_1 was partly over come by doubling the chromosomes to produce an amphidiploid (Dana, 1966b; Sawa, 1974; Ahn and Hartmann, 1978c; AVRDC, 1979; Chen *et al.*, 1983).

The cross between mungbean and adzuki bean (*V. angularis*) produces a few seeds that abort prematurely (Ahn and Hartmann, 1978b, c). A few hybrid plants grown through embryo culture reached maturity without producing seeds. Chromosome pairing in the hybrid was low, suggesting that the two species are distantly related.

A cross of mungbean with *V. radiata* var. *sublobata,* the wild progenitor of mungbean, was successful when mungbean was used as the female (Ahuja and Singh, 1977). A wide range of segregation was observed up to the F_4 generation. Successful reciprocal crosses were obtained at AVRDC (AVRDC, 1979). The F_1 plants from a cross between mungbean and *V. trilobata* were sterile, but if the chromosomes were doubled the amphidiploids were fertile (Dana, 1966c, d).

Vigna glabrescens is a natural tetraploid ($2n = 44$) and its cross with *V. radiata* is not possible. Chen and Mok produced F_1 plants by the embryo rescue culture (Chen *et al.*, 1989). The F_1 plants were highly sterile upon selfing, but backcrossing to mungbean parent yielded viable seeds.

Genetic Engineering and Biotechnology

Cell cultures, protoplast fusion, and genetic transformation of cells through vectors form part of a new generation of genetic engineering and biotechnology research. The novel approach involves the potential transfer of genes between species and even genera at the cellular level. For biotechnology to be successful the ability to regenerate whole plants from isolated single cells in a consistent, reproducible manner is essential. Until recently most of the legume species have been considered recalcitrant (Bhojwani and Mukhopadyay, 1986). Plant regeneration varies among species and genotypes within species. In addition to choice of the right genotype, choice of the medium and its constituents is required. Before biotechnology becomes a useful tool for transfer of valuable genes by the mungbean breeder, stable regeneration and transfer procedures must be developed.

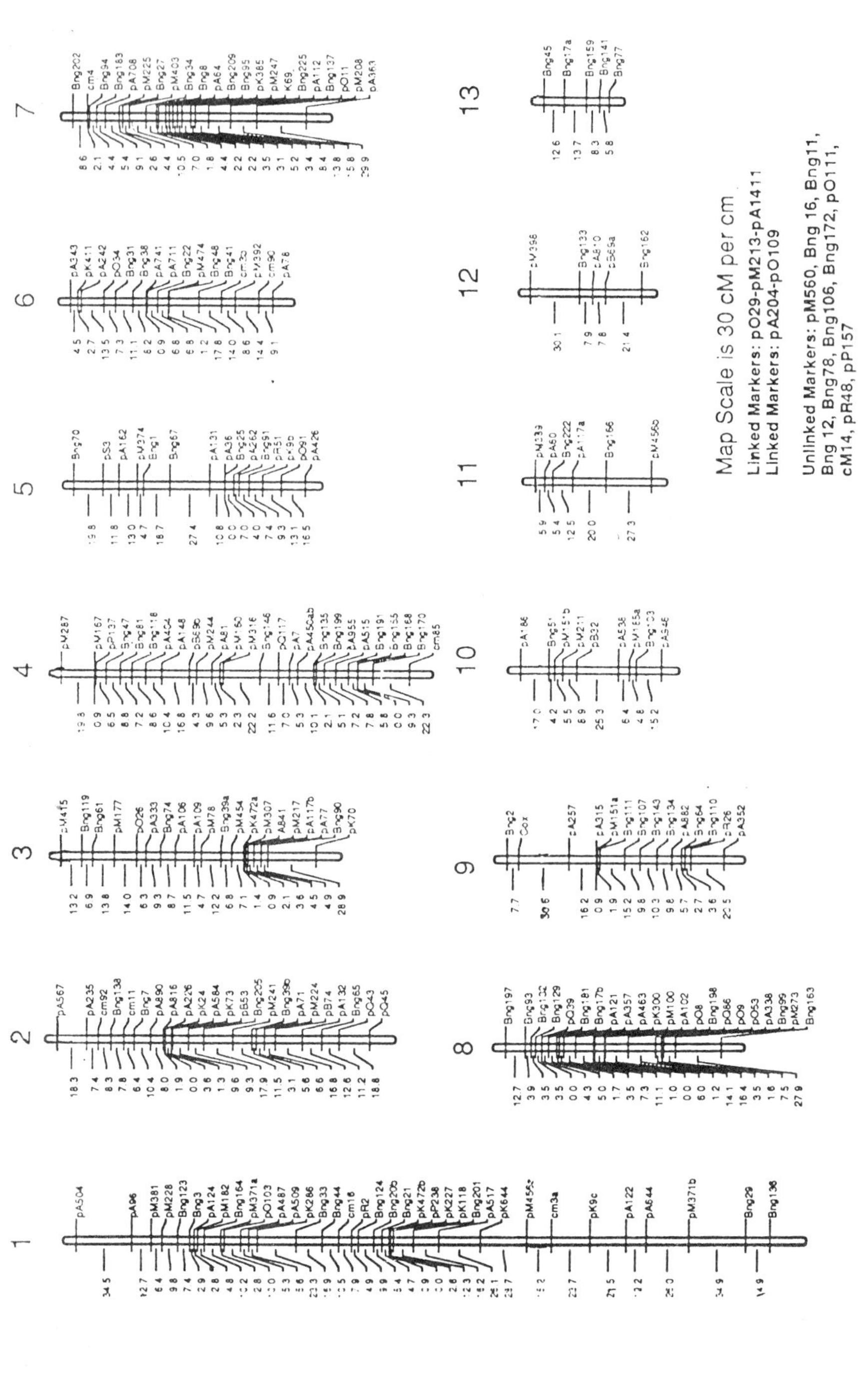

Fig. 1. RFLP Linkage Map of Mungbean *(Vigna radiata)*

Genetic materials evolved through these procedures must then be carefully integrated into traditional plant breeding programmes to develop improved cultivars.

Whole plants were regenerated from shoot tips in mungbean, and calli were initiated from mungbean root hypocotyl and shoot explants (Singh *et al.,* 1986). Root regeneration was obtained with low levels of auxin but shoot regeneration appears to be difficult. Recently, at AVRDC, explants from cotyledons developed multiple shoots on MS or modified medium with 5 mg of cytokinin. All regenerated shoots were able to develop root after subculturing in MS media, which lays the foundation for biotechnology research in mungbean. Useful genes for disease and stress resistance are reported to be present in *Vigna* species (Chen *et al.,* 1983) but appropriate technology for transfer to mungbean is necessary. Because biotechnology is an inordinately expensive area of research, major early developments will be in crops with higher economic value.

A new molecular tool, restriction fragment length polymorphism (RFLP), is used to construct the molecular genetic map for most major crop plants and these maps provide a more direct method for selecting desirable genes via their linkage to easily detectable RFLP markers. N.D. Young has constructed a saturated RFLP map of mungbean in collaboration with AVRDC (Young *et al.,* 1992). The latest RFLP map is given in Fig. 1 (Young *et al.,* unpublished).

Breeding

Until 1971, major mungbean breeding programmes operated primarily in India, the Philippines, and the United States. In 1972, an International Mungbean Nursery (IMN) was developed by the University of Missouri, Columbia, Missouri (USA), an activity that stimulated international cooperation in mungbean breeding (Poehlman *et al.,* 1976; Poehlman, 1978). With the establishment of the Asian Vegetable Research and Development Centre (AVRDC) in 1971/72, the mungbean was selected as a mandate crop and an integrated, interdisciplinary approach to mungbean improvement has since been developed. Since 1976, the IMN has been coordinated through AVRDC (Ahn *et al.,* 1985).

Problems of Mungbean Production and Breeding

Relatively minor research attention has been given to mungbean breeding compared to the cereals or soybeans. Yield potential of most present-day cultivars is low as the world average yield for mungbean is less than 500 kg/ha. Yield is unstable over locations and seasons due to the susceptibility of mungbean cultivars to environmental stresses and diseases. This has led to mungbean being grown on marginal lands with a minimum of management inputs. Natural selection under such situations has favoured types that can survive under unfavourable environments and culture. Mungbean is less responsive to management inputs than the cereal crops. Although resistance to certain diseases is present in varying degrees in some cultivars, diseases such as powdery mildew and Cercospora leaf spot are still major risks in growing mungbean in some production areas. Mungbean yellow mosaic virus (MYMV) is a serious threat for mungbean production in certain seasons in the Indian subcontinent. In addition, insect pests such as beanflies and pod borers can cause considerable loss. In countries where labour is becoming expensive, the non-synchronous pod maturity that necessitates multiple harvesting by hand is yet another problem.

Breeding Objectives

High and stable yield is the primary breeding objective. Since yields are commonly reduced by diseases and insect pests, resistance to Cercospora leaf spot, powdery mildew, MYMV, beanfly, pod broer, bruchids, and other regional pests is a specific objective where yield losses from these pests occur. Reduced sensitivity to photoperiod and temperature, early and synchronous flowering, tolerance to lodging and drought, reduced weather damage to seeds, and less pod shattering are common general objectives.

Germplasm Resources

A knowledge of sources of genetic variability for specific traits is available, and information on its inheritance, breeding behaviour and response in varying environments is highly valuable for use of germplasm resources in breeding programmes. Many accessions of *V. radiata* are being maintained in different countries, but there is a considerable amount of duplication of the accessions among the collections. The major collections of mungbean germplasm are being maintained at AVRDC, Shanhua, Taiwan; University of the Philippines, Los Baños; US Department of Agriculture, Experiment, Georgia; and Indian Agricultural Research Institute, New Delhi (Anishetty and Moss, 1988; Poehlman, 1991).

AVRDC has been designated by the International Board for Plant Genetic Resources (IBPGR) as the base collection centre for mungbean. A set of descriptors was perpared for mungbean so that a uniform system of classification could be used by all agencies maintaining mungbean germplasm collections (IBPGR, 1985). The AVRDC collection contains more than 5000 accessions and has been systematically catalogued for various morphological, agronomical, and biochemical traits and for disease and insect resistance (Shanmugasundaram and Tschanz, 1985). At AVRDC, the mungbean germplasm has been evaluated for 35 characters using the IBPGR descriptors and a few additional traits. The variability in 29 characters among 68 promising cultivars at AVRDC is reported in Table 5. These data suggest a wide range of variability for all characters examined. In a similar screening of the AVRDC mungbean collection at the Malang Institute for Food Crops (Malang, Indonesia), the variability among accessions was larger than at AVRDC. Thulasidass (1984), using multiple regression and

Table 5
Genetic variability in 29 characters of 68 promising mungbean cultivars[a]

Character	Mean ± SE	Range
Yield (kg/ha)	866 ± 19	339–1222
Pods/plant	26 ± 0.8	11–47
Seeds/pod	11 ± 0.1	9–14
1000-seed weight (g)	46 ± 1.2	23–89
Pod length (mm)	88 ± 1	65–139
Pod thickness (mm)	5 ± 0.1	4–7
No. of clusters	8 ± 0.2	4–16
No. of clusters on main stem	5 ± 0.1	3–9
No. of pods on main stem	20 ± 0.5	11–37
Av. no. of pods/cluster	3 ± 0.04	2–5
Position of 1st pod (node)	7 ± 0.06	5–9
Length of peduncle (cm)	16 ± 0.2	9–21
Plant height (cm)	52 ± 0.9	23–74
No. of nodes on main stem	12 ± 0.1	9–15
Average length of internode (cm)	4 ± 0.1	2.5–5.8
Protein (%)	21.8 ±	19–25
Days to flowering (DAP)	41 ± 0.3	34–52
Days to first pod maturity (DAP)	63 ± 0.4	55–76
Mean maturity (DAP)	89 ± 0.2	86–96
Photoperiod response[b]	19 ± 0.9	3–43
Hairiness (1–5)	0.0419 ± 0.9	0.8–3.1
Stem diameter (mm)	0.119 ± 0.9	3.6–9.3
Total no. of branches	0.0519 ± 0.9	1.8–4.6
Longest branch length (cm)	0.819 ± 0.9	4.4 –45.7
No. of branches with pods	0.0619± 0.9	0.2–2.9
Width of middle leaflet (cm)	0.1319 ± 0.9	6.4–13.3
Length of middle leaflet (cm)	0.1119 ± 0.9	8.1–13.5
Leaf shape (27/26)	0.0119 ± 0.9	0.98–1.27
Petiole length (cm)	0.219 ± 0.9	9.6–21.7

[a]Planted at AVRDC on Sept. 8, 1977, four rows 6 m in length, 50 cm apart, with three replications (H.G. Park, AVRDC, unpublished).
[b]Difference in days to flowering between September and December plantings (12-hour and 16-hour photoperiods).

classificatory analysis, grouped 30 accessions from AVRDC into seven clusters. Cultivars identified with desirable characters for crossing are: V-3388 (USA), for high yield and pods per plant; V-3404 (Thailand), for pod length and 100-seed weight; V-1789 (India) and V-2184 (Philippines), for short-duration; and V-1948 (Philippines) and V-3092 (India), for number of seeds per pod.

Crossing Methods

The mungbean plant flowers in phases with axillary or terminal racemes containing a cluster of 10 to 20 cleistogamous flowers. Pollination occurs from 9 to 10 pm until midnight (Bose, 1939). Flower shedding is common, averaging about 60 per cent. Outcrossing averages around 2 to 5 per cent (van Rheenen, 1964; Bhadra and Shill, 1986), but varies with the cultivar and the season. A crossing technique was outlined by Boling *et al.* (1961). For emasculation, the standard and wing petals are separated with a dissecting needle and the keel petals and anthers are carefully removed. For pollination, a stigma covered with fresh pollen from dehisced anthers is rubbed across the stigma of the emasculated flower and the standard and wing petals returned to their place. Emasculation in the evening and pollination the following morning gives maximum seed set. From 20 to 60 per cent pod set has been reported with this procedure (Singh and Malhotra, 1975; Park and Yang, 1978).

An improved technique was described by Cupka and Edwards (1986) in which the tip of the bud is opened with fine-tipped forceps to expose the style and the stigma, and the anthers are slowly dislodged. After the emasculated bud is pollinated, a cellophane tape is placed over the opening to seal the bud and reduce humidity and temperature fluctuations within the floret. Emasculation and pollination can be accomplished in one minute with 60 per cent success.

Breeding Methods

Mungbean is a self-pollinated crop and the breeding methods are those commonly used with self-pollinated species. These are:

1) Collection of local cultivars, or introduction of promising cultivars or breeding lines.

2) Pure-line selection: Isolation of a superior pure line from a mixed seed lot is a rapid and efficient procedure for obtaining a new cultivar.

3) Selection from hybrid progenies: Pedigree selection is generally practised in segregating generations following hybridization since individual plants are easily identified and evaluated.

4) Multiple-parent crosses: Multiple-crosses are generally designed to

combine two or more useful genes from undesirable parents simultaneously into an adapted germplasm.

5) Backcrosses: A desirable gene may be added to an adapted variety by a succession of crosses using the adapted parent as the recurrent parent, or by making, each backcross to a different recurrent parent, a procedure termed modified backcrossing.

6) Repetitive intercrossing: It involves intermating for one or two cycles before starting selection (Dahiya and Singh, 1985). As mungbean is a self-pollinating crop, this is a labour-intensive procedure.

Most early varieties were developed by pure-line selection from native germplasm. Currently, hybridization is the most important breeding procedure. (See discussion in Poehlman, 1991).

Varietal Improvement Programmes

India

Systematic breeding programme on different pulse crops in India started after the organization of the All India Coordinated Pulse Improvement Programmes (AICPIP) in 1967, though varietal improvement work on mungbean was earlier going on in many states of India resulting in the development of a number of cultivars, including Mung Type 1 in Uttar Pradesh (Mehta and Sahai, 1955) and Co 1 in Tamil Nadu (Premsekar and Srinivasan, 1961). Pusa Baisakhi, developed at the Indian Agricultural Research Institute, was the first variety released under AICPIP. Breeding programmes are now conducted in different state agricultural universities, the Indian Institute of Pulse Research, Kanpur, and the Indian Agricultural Research Institute, New Delhi. The breeding lines are regularly tested over a series of environments under AICPIP before identification and release of varieties for different regions of the country. A list of mungbean cultivars released in India has been reported by Anon (1988), Shanmugasundaram (1988), and Poehlman, (1991).

Varietal improvement in India has concentrated on increasing yield and disease resistance. To improve yield potential, a 'plant type' concept was proposed that focussed on short, compact plants, high harvest index, reduced photoperiod sensitivity, early maturity, and more determinate growth habit (Jain, 1971, 1975). Consideration has been given to suitability for planting in multiple cropping systems (Singh, 1975). The compact plant and early maturity concept is based on obtaining maximum yield per hectare per day, a desirable feature in a short-season, multiple cropping system. Ramanujam (1975) and Thulasidass (1984) have generated information suggesting that more crosses should be made to develop cultivars for specific seasons and locations.

A major mungbean improvement programme at the Punjab Agricultural University, Ludhiana, has led to the development of many disease and insect-resistant cultivars and breeding lines, beginning with release of the cultivar ML 1 in 1974. The new ML (Mung Ludhiana) and LM (Ludhiana Mung) cultivars have resistance, singly or combined, to disease pathogens inciting MYMV, Cercospora leaf spot (*Cercospora canescens* E. and M.), bacterial leaf spot (*Xanthomonas phaseoli* (Smith) Dowson), root-knot nematode (*Meloidogyne incognita* (Kofold and White) Chitwood, and insect pests whitefly (*Bemisia tabaci* Genn.), leafhoppers or jassids (*Empoasca* spp.), and pod borers (*Lampides boeticus* L., *Maruca testulalis* (Gey.). (See reviews by Singh *et al.*, 1988; Chhabra *et al.*, 1988; Shanmugasundaram, 1988; Poehlman, 1991.) In addition to this, very aggressive and broad-based mungbean improvement programmes at IARI, New Delhi and many other universities across the country have also developed and released a number of high yielding and disease resistant varieties for different agro-climatic zones of the country.

The Philippines

As early as 1916, San Miguel (1916) emphasized that mungbean should be improved in productivity, evenness in maturity, total growth, adaptability to seasons, and resistance to diseases. In 1956, a varietal improvement programme was initiated by the Bureau of Plant Industry (BPI) Economic Garden, Los Baños, with pure lines selected from local varieties (Ballon *et al.*, 1978). Hybridization and selection resulted in the release of MG-50-10A, MD15-2, Galbrous No. 3, and MG-50-10A-Y. The University of the Philippines, Los Baños, released CES-14, CES 55, and several PAGASA lines (Lantican and Catedral 1977; Catedral and Lantican, 1978). Since 1975, the AVRDC/Philippine Outreach Programme has been located at the BPI Economic Garden where AVRDC mungbean breeding lines are evaluated. Cooperatively, the BPI Economic Garden and AVRDC released BPI Mg 2 in 1984 and BPI Mg 4 in 1986 (Catipon *et al.*, 1988). BPI Mg 7 and 9 were released in 1989 and the latter has been promoted nationwide as 'Taiwan Green. Punto and Lantican (1982), in a study of 10 mungbean genotypes over nine locations for two years, found 'PAGASA 1' to be the most stable genotype studied.

Indonesia

Varietal improvement of mungbean was started in 1935 by collection of local varieties (Somaatmadja and Sutarman, 1978). In 1965, variety 'Jala', introduced from Sri Lanka, was distributed to farmers. During recent years, the Sukamandi Research Institute for Food Crops (SURIF), Bogor Research Institute for Food Crops (BORIF), and Malang Research Institute for Food Crops (MARIF) cooperated with AVRDC in evaluation of AVRDC breeding lines. 'Manyar' and 'Nuri' were released in 1983, 'Walet' and 'Gelatik' in 1986 (Shanmugasundaram, 1988), 'Merpati' in 1991 and 'Sriti' in 1992.

Thailand

Native cultivars with varied seed characters such as dull or shiny seed coats and golden or black seed colour were originally grown by farmers (Bhumiratna, 1978). An introduced 'Philippine-type' was released as 'Uthong 1' (Nalampang, 1978). Chainat Field Crop Research Center of the Thailand Department of Agriculture and its associated Phitsanulok Field Crop Experiment Station, Kasetsart University, and Prince of Songkhla University cooperate with AVRDC in evaluating breeding lines and segregating populations. The AVRDC-Asian Regional Center is located at Kasetsart University, Bangkok.

All the recommended cultivars in Thailand except 'Uthong 1' were introduced from AVRDC, such as Kamphaeng Saen 1 (KPS 1), Kamphaeng Saen (KPS 2), Chainat 60 (CN 60), Chainat 36 (CN 36), and Prince of Songkhla University 1 (PSU 1). These cultivars account for over 90 per cent of the total planted area (0.4 million ha) in 1992 (Jansen *et al.*, 1993).

The United States

Mungbean breeding is conducted at the Oklahoma Agricultural Experiment Station, Stillwater. A large-seeded, pure-line selection was developed from 'Jumbo'; an introduction from China was named 'Berken'; and a medium-sized cultivar, 'Kiloga', and a small-seeded cultivar, 'Oklahoma 12', were selected from introductions from China (Matlock and Oswalt, 1963). Mungbean genotype/biological nitrogen-fixation studies are conducted at the Texas A & M University, College Station. These universities evaluate the AVRDC breeding lines almost every year and Texas A & M University released 'Texsprout' in 1988.

Australia

A multidisciplinary research programme is sponsored by the CSIRO Division of Tropical Crops and Pastures (Lawn and Ahn, 1985). Three cultivars, 'Berken' from Oklahoma, 'Celera' (Kingston, 1975), and 'King', an AVRDC breeding line that was released in 1982, account for almost the entire area planted to mungbean. Recently, MYMV-resistant lines have been introduced from AVRDC and evaluated.

Bangladesh

Bangladesh Agricultural Research Institute (BARI) has been conducting research on mungbean since 1976. A MYMV-tolerant cultivar 'Mubarik' was developed and released by BARI in 1982 for March/April sowing and 'Kanti' in 1987 for August/September sowing. Bangladesh Institute for Nuclear Agriculture (BINA) has been working on mutation breeding using gamma radiation and released BINA Mung-I.

Since 1993, many lines have been introduced from AVRDC and evaluated by BARI and the Institute of Postgraduate Studies in Agriculture (IPSA). Among them, NM-92, NM 94, VC 6144, and VC 6146 have been selected for high yield, MYMV resistance, and early (60–65 days) and uniform maturity (Malik *et al.,* in press; Sarker *et al.,* in press).

Pakistan

Mungbean breeding improvement in Pakistan was mainly confined to the collection and evaluation of local and exotic germplasm until the 1970s. The only approved mungbean variety was '6601' until 1983 (Bashir *et al.,* 1988). Mungbean cultivars, M-28, NM 20 and 21, NM 121 to 25, NM 13-1, and NM 19-19, were developed through mutation breeding by the Nuclear Institute of Agriculture and Biology (NIAB) and released in the 1980s. AVRDC breeding materials were introduced and the National Agricultural Research Centre (NARC) and NIAB have started the hybridization programme, from which NIAB has developed and released NM-51 and NM-54 (Malik, 1991). Recently, NIAB has intensified the collaboration with AVRDC by shuttle breeding and developed several lines with MYMV resistance, high yield, large seed size, and early and uniform maturity.

Sri Lanka

Mungbean breeding is conducted at the Mahaillupallama Agricultural Experiment Station in the dry zone. Cultivars 'ML 1', 'ML 2', and 'ML 3' were introduced from India. A large-seeded cultivar 'M 4' that matures in 75 days and 'Type 51', a small-seeded cultivar that matures in 60 days, are also grown (Vignarajah, 1978). In cooperation with AVRDC, 'Type 77' was released to farmers in 1982.

Asian Vegetable Research and Development Center

AVRDC is the only international agricultural research centre with a mandate on mungbean and has played an important role in mungbean improvement since 1971–72. It was designated as the world mungbean collection centre by IBPGR. Its germplasm bank has more than 6000 accessions and most of them have been characterized. The objective of mungbean improvement at AVRDC is to develop early and uniformly maturing genotypes with high and stable yield, reduced sensitivity to photoperiod and temperature, and resistance or tolerance to major biotic and abiotic stresses. The initial step was to identify germplasm desirable in these characteristics. Accessions from the Philippines had high yield, earliness, and uniform maturity. From the germplasm collection, accessions were identified, mostly from India, with insensitivity to photoperiod and temperature, resistance to Cercospora leaf spot and powdery mildew, and resistance or tolerance for beanflies, bruchids, and pod borers (AVRDC, 1979; 1981a, b). Crosses between the Philippine and Indian germplasm have generated a new group of breeding lines. A combination of bulk, modified bulk, pedigree, single-seed-descent, and backcross methods of breeding were used in their development.

AVRDC-improved lines carry moderate resistance to Cercospora leaf spot and powdery mildew diseases, reduced photothermal sensitivity, increased 1000-seed weight (70 g), and improved plant type with pods above the canopy. Yield as high as 3 t/ha can be obtained in experimental plots. Each year the selected AVRDC's elite lines and superior accessions from different countries are evaluated in the IMN for adaptability in many countries. Fifty-four cultivars have been officially released from AVRDC breeding lines in 19 countries by national partners. AVRDC's elite lines have been extensively grown in all the countries except in South Asia due to MYMV.

At AVRDC, powdery mildew is prevalent in spring and autumn season crops when the weather is cool and dry, and Cercospora leaf spot is prevalent during summer when the weather is hot and humid. The segregating generations are subjected alternatively to these diseases in order to select for combined resistance (Shanmugasundaram and Tschanz, 1985). Photoperiod and temperature variations during these seasons permit selection for these environmental stresses. The IMN provides an opportunity to evaluate superior genotypes over a range of environments.

Imrie *et al.* (1981) used the cluster analysis to determine the genotypic and environmental responses and stability of genotypes, and Imrie and Shanmugasundaram (1987) analysed the 7th and 8th, and the 9th and 10th IMN trials. They found 70–80 per cent of the variation was due to environmental causes with 17–27 per cent due to genotype × environmental interaction, and less than 3 per cent due to variation among genotypes. A segmented regression analysis was used by Fernandez and Shanmugasundaram (1988) to detect genotypes that are high yielding and less sensitive to environmental fluctuation in unfavourable environments, but have the capacity to yield well in favourable environments.

The AVRDC ideotype for mungbean is a monocrop with yield of 3 to

4 t/ha at maturity of 70 days (57.1 kg/ha/day) under optimum environmental and cultural conditions. This assumes having 20 pods / plant, 10 seeds / pod, and seed weight of 6 g /100 seeds in a plant population of 250,000 to 400,000 plants/ha. To provide sufficient source, a rapid and vigorous vegetative growth is needed. The characters visualized to attain this goal are rapid seedling growth, dark green lanceolate leaflets, no branching, a strong erect stem, vigourous root system, abundant nodulation, extended photosynthesis, a vegetative/reproductive ratio of 4 : 3, high and efficient photosynthesis, and efficient translocation of the photosynthates. Better distribution of the sink would require a high harvest index with a large number of nodes and short internodes, multiple peduncles at each node with abundant pods per cluster, synchronized flowering and ripening, non-shattering pods, with pods above the seed canopy, large number of seeds per pod, and large seeds. For yield stability, resistance to diseases and insects, tolerance to temperature and moisture stresses, and resistance to seed weathering is needed.

Employing a multidisciplinary approach, mungbean yield potential has been more than doubled in breeding lines at AVRDC. Plant architecture is being altered to provide for more pods exposed above the canopy and more synchronous maturity. Progress is being made in combining less sensitivity to photoperiod and temperature and resistance to multiple disease.

Breeding lines from AVRDC have been released in many countries (Costa Rica, China, Ecuador, Fiji, India, Indonesia, Kenya, Republic of Korea, Pakistan, the Philippines, Sri Lanka, Taiwan, Tanzania, Thailand, the United States, and Vietnam (see Shanmugasundaram, 1988). Combining the disease and pest resistance and high yield of AVRDC breeding lines with resistance to local stress factors from native cultivars poses a major challenge to country breeding programmes.

AVRDC's objectives for the 1990s are to increase and stabilise yield, improve early and uniform maturity, maintain Cercospora leaf spot resistance and photothermal insensitivity, increase resistance to powdery mildew, and incorporate resistance genes for MYMV, bruchid, beanfly, and pod borer into the advanced breeding lines.

Bruchid is a very serious pest of mungbean during storage. A wild mungbean accession, *V. radiata* ssp. *sublobata* (TC 1966), and two mungbean accessions (V 2709 and V 2802) are higly resistant to bruchid *Callosobruchus chinensis* (L.), and only TC 1966 is resistant to *C. maculatus* (F.) The bruchid-resistant elite lines developed at AVRDC will be distributed to the national partners.

AVRDC is committed to strengthening its upstream research activities, particularly in biotechnology. RFLP mapping is a new and powerful tool that can assist breeding programmes, improve management of germplasm collections by DNA fingerprinting, and provide the basis for cloning genes. AVRDC has constructed a saturated genetic map of mungbean using RFLP markers in collaboration with the University of Minnesota. About 200 RFLP markers using a cross between mungbean and its wild progenitor, *V. radiata* ssp. *sublobata,* have been identified (Young *et al.,* 1992). RFLP markers linked to the genes for powdery mildew and bruchid resistance, seed weight, pod length, and days to maturity have already been identified. Recombinant inbred lines are being developed to provide a permanent mapping source that can be distributed worldwide to map markers in a single genetic population. The genes controlling Cercospora leaf spot and MYMV were mapped using different mungbean populations. These genes will be the first targets of RFLP marker-assisted breeding.

RFLP markers were used to accelerate backcross breeeding for bruchid resistance. With only two backcross generations, RFLP-based selection provided the bruchid-resistant lines, while more than six backcross generations were necessary with the conventional backcross breeding. MYMV is the most serious constraint in mungbean production in South Asia. The first international MYMV nursery, including 17 resistant lines originating from India, Pakistan, and Sri Lanka, were sent to more than 20 collaborators in South Asia and neighbouring countries.

MYMV ratings in New Delhi, India, and Faisalabad, Pakistan, were very similar, indicating probably no difference in race flora between the two regions. However, only four Pakistani and one Sri Lankan line were resistant in Bangladesh. It may be due to high pressure of virus or different races in Bangladesh. A DNA probe specific for MYMV in India has been developed in collaboration with the University of Wisconsin.

AVRDC and NIAB in Pakistan have been collaborating through the shuttle breeding. After AVRDC breeding lines were crossed with MYMV resistant lines, F_2 and F_3 seeds were sent to NIAB in Pakistan and selected for MYMV resistance since MYMV is not present either in Taiwan or in Thailand. The selected materials were sent back to AVRDC in Thailand selected for CLS resistance and other traits. After one more year of shuttle breeding, the seeds were multiplied in Thailand for yield trials. These selected lines have resistance to MYMV and CLS besides high yield, large seed size, and early and uniform maturity.

With the establishment of the South Asian Vegetable Research Network (SAVERNET), the MYMV network was informally organized after the international MYMV Expert Consultation Workshop in 1991 in Bangkok. The AVRDC/USAID/BARI Project in Bangladesh is ensuring in-country and AVRDC-Bangladesh collaboration. Furthermore, the mungbean session during the International Symposium on Pulses Research in 1994 in New Delhi has reinforced the collaborative spirit for mungbean research in this region.

Problems and Future Prospects

Traditionally, mungbean is grown in a low-yield environment, with little attention to the yield inputs routinely given to other crops. Breeding efforts need to be continued and augmented to further improve yield potential and yield stability. To achieve tolerance to heat and drought and resistance to Cercospora, powdery mildew, and virus diseases still pose a challenge to breeders and pathologists. While MYMV constitutes a serious hazard in the Indian subcontinent, good progress has been made in India on breeding for MYMV resistance combined with resistance to Cercospora leaf spot and various pests. Other viruses attack mungbean and reduce yields in local areas. In addition to traditional breeding methods, innovative new approaches such as genotype stability analysis (Dahiya and Singh, 1985; Verma *et al.,* 1978), would help to identify superior parent materials.

Future strategies should be directed toward sustaining high yields in farmers' fields. Use of the mungbean's short maturity as an intercrop, or as a short-season crop in multiple cropping system, would increase profitability of growing mungbean. Varietal research needs to be directed toward cultivars with this special utility and resistance to the disease and insect hazards associated with specific cropping patterns. Quality of seeds would be improved by enhancing the sulphur-containing amino acids. Mungbean breeding and evaluation of cultivars must be integrated into the total production system, which involves fertility maintenance, and disease, insect, and weed control, and efficient labour management.

Traditionally, mungbean used to be planted in the marginal lands. Recently, however, the newly developed cultivars with high yield, early and uniform maturity, and disease and insect resistance have been planted in extensive areas with appropriate inputs and management to enhance sustainability and farmer's income in Asia.

REFERENCES

Abilay, R.M., and Lantican, R.M. 1982. Characters associated with yield performance of grain legume crops under partial shade condition. 1. Mungbean (*Vigna radiata* (L.) Wilczek. *Phil. J. Crop Sci.* 7(2): 88–93.

Ahn, C.S., Chen, J.H. and Chen, H.K. 1985. Performance of the ninth (1981) and tenth (1983) International Mungbean Nursery. AVRDC Publication No. 85–242. The Asian Vegetable Research and Development Center, Shanhua, Taiwan, 86 pp.

Ahn, C.S., and Hartmann, R.W. 1978a. Interspecific hybridization among four species of the genus, *Vigna* Savi. *Proc. 1st Int. Mungbean Symp.* Asian Vegetable Research and Development Center, Shanhua, Taiwan, pp. 240–246.

Ahn. C.S., and Hartmann, R.W. 1978b. Interspecific hybridization between mungbean (*Vigna radiata* (L.) Wilczek) and adzuki bean (*V. angularis* (Willd). Ohwi and Ohashi). *J. Amer. Soc. Hor. Sci.* 103: 3–6.

Ahn, C.S., and Hartmann, R.W. 1978c. Interspecific hybridizaion between rice bean (*Vigna umbellata* (Thunb.) Ohwi (Ohashi) and adzuki bean (*V. angularis* (Willd.) Ohwi and Ohashi). *J. Amer. Soc. Hort. Sci.* 103: 435–436.

Ahuja, M.R., and Singh, B. 1977. Induced genetic variability in mungbean through interspecific hybridization. *Indian J. Genet.* 37: 133–136.

Ahuja, S.L., and Chowdhury, R.K. 1984. Graphical and component analysis of mungbean. *Indian J. Hered.* 16: 4–10.

Anishetty, N.M., and Moss, H. 1988. Vigna genetic resources: Current status and future plans. *Mungbean: Proc. 2nd. Int. Symp. Asian* Vegetable Research and Development Center, Shanhua, Taiwan.

Anonymous. 1988. *Technology for Increasing Pulse Production in India.* Directorate of Pulses Research, Kanpur.

Appa Rao, S., and Jana, M.K. 1973. Inheritance of anthocyanin colouration in *Phaseolus* mutants. *Indian Sci. Congr. Assoc. Proc.* 60: 302.

AVRDC. 1979. *Progress Report, 1978.* Asian Vegetable Research and Development Center, Shanhua, Taiwan, pp. 71–91.

AVRDC. 1981a *Progress Report, 1979.* Asian Vegetable Research and Development Center, Shanhua, Taiwan, pp. 42–55.

AVRDC. 1981b. *Progress Report, 1980.* Asian Vegetable Research and Development Center, Shanhua, Taiwan, pp. 40–50.

Bailey, L.H. 1970. *Manual of Cultivated Plants* (rev.). Macmillan, New York, 1116 pp.

Ballon, F.B., Legaspi, B.M., and Catipon, E.M. 1978. Mungbean varietal improvement programme of the Bureau of Plant Industry (Philippines). *Proc. 1st Int. Mungbean Symp.*, Los Baños. Asian Vegetable Research and Development Center, Shanhua, Taiwan, pp. 217–219.

Banks, D.J. 1958. A classification of certain characteristics of mungbean strains as an aid to improvement. M.S. thesis (unpublished), Oklahoma State Univ. Agric. and App. Sci., Stillwater, Oklahoma.

Bashir, M., Zubair, M., and Malik. 1988. Disease resistance sources and utilization in breeding improved mungbean in Pakistan. In *Mungbean: Proc. 2nd. Int. Symp.* 16–20 Nov. 1987. Bangkok, Thailand. S. Shanmugasundaram, ed. AVRDC, Shanhua, Taiwan, pp. 297–307.

Bhadra, S.K., and Shill, M. 1986. Natural outcrossing in mungbean (*Vigna radiata* (L.) Wilczek). *Bangladesh J. Bot.* 15: 217–220.

Bhargava, P.D., Johri, J.N., Sharma, S.K., and Bhatt, B.N. 1966. Morphological and genetic variability in green gram. *Indian J. Genet.* 26: 370–373.

Bhatnagar. C.P., Chandola, R.P., Saxena, D.K., and Sethi, S. 1974. Cytotaxonomic studies on the genus *Phaseolus. Indian J. Genet.* 34A: 800–804.

Bhatnagar, P.S., and Singh. B. 1964. Heterosis in mungbean. *Indian J. Genet.* 24: 89–91.

Bhatt, B.N., Kotwal, S.V., and Chandola, R.P. 1972. A giant variant in greengram (*Phaseolus aureus* Roxb.). *Curr. Sci.* 41: 854–855.

Bhaumik, P.K., and Jha, A.R. 1976. Estimation of physiological relationship through path coefficient analysis in mungbean (*Phaselus aureus* Roxb.). *Indian Agric.* 20: 1–10.

Bhojwani, S.S., and Mukhopadyay, A. 1986. Some aspects of plant regeneration in tissue cultures of legumes. In *Genetics and Crop Improvement.* P.K. Gupta and J.R. Bahl, eds. Rastogi and Co., Meerut, India.

Bhumiratana, A. 1978. Mungbean and its utilization in Thailand. *Proc. 1st Int. Mungbean Symp.* Asian Vegetable Research and Development Center, Shanhua, Taiwan, pp. 46–48.

Boling, M., Snader, D.A., and Matlock, R.S. 1961. Mungbean hybridization technique. *Agron. J.* 53: 54–55.

Bose, B.D. 1939. Studies in Indian pulses. IX. Contributions to the genetics of mung (*Phaseolus radiatus* Linn., *Ph. aureus* Roxb.)

Catedral, I.G., and Lantican, R.M. 1978. Mungbean breeding programme of UPLB, Philippines. *Proc. 1st Int. Mungbean Symp.* Asian Vegetable Research and Development Center, Shanhua, Taiwan, pp. 225–227.

Catipon, E.M., Legaspi, B.M., and Jarilla, F.A. 1988. Development of mungbean varieties from AVRDC lines for the Philippines. *Mungbean: Proc. 2 Int. Symp.* Asian Vegetable Research and Development Center, Shanhua, Taiwan, pp. 88–97.

Chandel, K.P.S., Joshi, B.S. and Pant, K.C. 1973. Yield in mungbean and its components. *Indian J. Genet.* 33: 271–276.

Chen, H.K., Mok, M.C., Shanmugasundaram, S., and Mok, D.W.S. 1989. Interspecific hybridization between *Vigna radiata* (L.) Wilczek and *V. glabrescens. Theor. Appl. Genet.* 78: 641–647.

Chen, N.C., Baker, L.R., and Honma, S. 1983. Interspecific crossability among four species of *Vigna* food legumes. *Euphytica* 32: 925–937.

Chhabra, K.S., Kooner, B.S., Sharma, A.K., and Sareen, A.K. 1988. Sources of resistance in mungbean to insect pests and mungbean yellow mosaic virus. *Mungbean: Proc. 2nd Symp.* Asian Vegetable Research and Development Center, Shanhua, Taiwan, pp. 308–314.

Chowdhury, J.B., Chowdhury, R.B., and Kakar, S.N. 1971. Studies on the genetic variability in moong (*Phaseolus aureus*). *J. Res. PAU,* 8: 169–172.

Chowdhury, R.K., and Chowdhury, J.B. 1974. Induced amphidiploidy in *Phaseolus aureus* Roxb. × *P. mungo* L. hybrids. *Crop Improv.* 1: 46–52.

Cupka, T.B., and Edwards, L.H. 1986. A new technique for crossing mungbeans. *Crop Sci.* 26: 830–831.

Dahiya, B.N., and Singh, V.P. 1985. Comparative efficacy of pedigree selection and selective intermating in greengram (*Vigna radiata* (l.) Wilczek). *Theor. Appl. Genet.* 71: 129–132.

Dahiya, B.S: 1973. Improvement of mungbean through induced mutations. *Indian J. Genet.* 33: 460–468.

Dahiya, B.S. 1978. Mutation breeding in mungbean. *Proc. 1st Int. Mungbean Symp.* Asian Vegetable Research and Development Center, Shanhua, Taiwan, pp. 253–258.

Dana, S. 1966a. Cross between *Phaseolus aureus* Roxb. and *P. mungo* L. *Genetica* 37: 259–274.

Dana, S. 1966b. Species cross between *Phaseolus aureus* Roxb. and *P. trilobus* Ait. *Cytologia* 31: 176–187.

Dana, S. 1966c. Spontaneous amphidiploidy in F_1 *Phaseolus aureus* Roxb. × *P. trilobus* Ait. *Curr. Sci.* 35: 629–630.

Dana, S. 1966d. Cross between *Phaseolus aureus* Roxb. and *P. ricciardinaus* Ten. *Genetica Iberica* 18: 141–156.

De, D.N., and Krishnan, R. 1966a. Studies on pachytene and somatic chromosomes of *Phaseolus mungo* L. *Genetica* 37: 581–587.

De, D.N., and Krishnan, R. 1966b. Cytological studies of hybrid, *Phaseolus aureus* x *P. mungo. Genetica* 37: 588–600.

De Candolle, A. 1886. *Origin of Cultivated Plants.* Hafner Publ. Co., New York (reprint of 2nd edn. 1959).

Empig, L.T., Lantican, R.M., and Escuro, P.B. 1970. Heritability estimates of quantitative characters in mungbean (*Phaseolus aureus* Roxb.). *Crop Sci.* 10: 240–241.

Fernandez, G.C.F., and Shanmugasundaram, S. 1988. The AVRDC mungbean improvement programme: The past, present and future. In *Mungbean. Proc. 2nd Int. Symp.* S. Shanmugasundaram and N.B.T. McLean, eds. AVRDC, Shanhua, Taiwan. pp. 58–70.

Fery, R.L. 1980. Genetics of *Vigna.* In *Horticultural Reviews,* vol. 2. J. Janick, ed. AVI Publishing Co., Inc., Wesport, Connecticut, pp. 311–394.

Ganguli, P.K. 1972. Cytogenetics study of an induced flower-colour mutation in greengram (*Phaseolus aureus* Roxb.). *Indian Sci. Congr. Assoc. Proc.* 59: 553.

Ganguli, P.K. 1973. Radiation induced sterile mutant in greengram (*Phaseolus aureus* Roxb.). *Indian Sci. Congr. Assoc. Proc.* 60: 640–641.

Giriraj, K. 1973. Natural variability in greengram (*Phaseolus aureus* Roxb.). *Mysore J. Agric. Sci.* 7: 184–187.

Giriraj, K., and Vijayakumar, S. 1974, Path coefficient analysis for yield attributes in mungbean. *Indian J. Genet.* 34: 27–30.

Godhani, P.R., Jaisani, B.G., and Patel, G.J. 1978. Epistatic and other genetic variances in greengram varieties. *Gujarat Res. J.* 4: 1–6.

Gupta, M.P., and Singh, R.B. 1969. Variability and correlation studies in greengram (*Phaseolus aureus* Roxb.). *Indian J. Agric. Sci.* 39: 482–493.

Gupta, M.P., and Singh, R.B. 1970. Genetic divergence for yield and its components in greengram. *Indian J. Genet.* 30: 212–221.

Gupta, S.N., Lal, S., Rai, L., and Tomer, Y.S. 1982. Correlations and path coefficient analysis in mungbean (*Vigna radiata* (L.) Wilczek). *J. Res. Haryana Agric. Univ.* 12: 287–291.

He, X., He, T., Xion, Y., and Jiao, C. 1988. Research and use of mungbean germplasm resources in Hubei, China. *Mungbean: Proc. 2nd Int. Symp.* Asian Vegetable Research and Development Center, Shanhua, Taiwan, pp. 35–41.

Imrie, B.C., Drake, D.W. de Lacy, I.H., and Byth, D.E. 1981. Analysis of genotypic and environmental variation in international mungbean trials. *Euphytica* 30: 301–311.

Imrie, B.C., and Shanmugasundaram, S. 1987. Source of variation for yield in international mungbean trials. *Field Crops Res.* 16: 197–208.

International Board for Plant Genetic Resources. 1985. *Descriptors for* Vigna mungo and V. radiata (rev.). IBPGR Secretariat, Rome.

Jain, H.K. 1971. New plant type in pulses. *Indian Fmg.* 21(8): 9–10, 15.

Jain, H.K. 1975. Breeding for yield and other attributes in grain legumes. *Indian J. Genet.* 35: 169–187.

Jain, H.K., and Mehra, K.L. 1980. Evolution, adaptation, relationships and uses of species of *Vigna* cultivated in India. In *Advances in Legume Science.* R.J. Summerfield and A.H. Bunting, eds. HMSO, London, pp. 459–468.

Jansen, Hans G.P., Charnmarongkul Somchai and Kim, Doo-Hwan. 1993. Economic analysis of mungbean production in Thailand. AVRDC Working Paper No. 6, 96 pp.

Joseph, L.S., and Bouwkamp, J.C. 1978. Karyomorphology of several species of *Phaseolus* and *Vigna*. *Cytologia* 43: 595–600.

Joshi, S.N., and Kabaria, M.M. 1973. Interrelationship between yield and yield components in *Phaseolus aureus* Roxb. *Madras Agric. J.* 60: 1331–1334.

Kabi, M.C., and Bhaduri, P.N. 1978. Nodulating behaviour of colchicine induced polyploids of *Phaseolus aureus* Roxb. *Cytologia* 43: 467–475.

Karpechenko, G.D. 1925. Chromosomes of Phaseolinae. *Bull. Appl. Bot. Leningrad* 14: 143–148.

Khan, I.A., and Hashim, M. 1978. Radiation induced variability in quantitative traits of mungbean (*Phaseolus aureus* Roxb.). *J. Cytol. Genet.* 13: 12–15.

Kingston, R.W. 1975. Berken—a new mungbean variety. *Queensland Agric. J.* 101: 659–661.

Kitamura, K., Ishimoto and Sawa, M. 1988. Inheritance of resistance with adzuki been weevil in *Vigna radiata* ssp. *sublobata* and successful incorporation to *Vigna radiata*. *Jap. J. Breed.* 38: 459–464.

Krishnan, R., and De, D.N. 1965. Studies on pachytene and somatic chromosomes of *Phaseolus aureus*. *Nucleus* 8: 7–16.

Kumar, L.S.S. 1945. A comparative study of autotetraploid and diploid types in mung (*Phaseolus radiatus* Linn.). *Proc. Indian Acad. Sci.* 21B: 266–268.

Kumar, L.S.S., and Abraham, A. 1942a. Induction of polyploidy in crop plants. *Curr. Sci.* 11: 112–113.

Kumar, L.S.S., and Abraham, A. 1942b. A study of colchicine induced polyploidy in *Phaseolus radiatus* L. *J. Univ. Bombay* 11: 30–37.

Lantican, R.M., and Catedral, I.G. 1977. Evaluation of legumes for adaptation to intensive cropping systems. I. Mungbean, *Vigna radiata* (L.) Wilczek. *Philippine J. Crop Sci.* 2(2): 62–66, 87.

Lawn, R.J., and Ahn, C.S. 1985. Mungbean (*Vigna radiata* (L.) Wilczek/*Vigna mungo* (L.) Hepper). In *Grain Legume Crops,* R.J. Summerfield and E.H. Roberts, eds. Collins, London, pp. 584–623.

Lukoki, L., Marechal and Otoul, E. 1980. Les ancétres sauvages des haricots cultivés: *Vigna radiata* (L.) Wilczek et *V. mungo* (L.) Hepper. *Bull. Jardin Bot. Nat. de Belique* 50: 385–391

Malhotra, V.V., Singh, S., and Singh, K.B. 1974a. Yield components in greengram (*Phaseolus aureus* Roxb.). *Indian J. Agric. Sci.* 44: 136–141.

Malhotra, V.V., Singh, S., and Singh, K.B. 1974b. Relation between geographic diversity and genetic divergence and the relative role of each character toward maximizing divergence in greengram. *Indian J. Agric. Sci.* 44: 811–815.

Malik, I.A. 1991. Two improved varieties of mungbean in Pakistan. *IAEA Mutation Breeding Newslett.* 37: 4.

Malik, I.A., Ali, Y., and Saleem, M. 1988. Incorporation of tolerance to mungbean yellow mosaic virus from local germplasm into exotic large-seeded mungbeen. In *Mungbean: Proc. 2nd Int. Symp.* S. Shanmugasundaram, ed. 16–20 Nov., 1987. Bangkok, Thailand. AVRDC, Shanhua, Taiwan, pp. 297–307.

Malik, B.P.S., Singh, V.P. and Singh, M. 1983. Correlation, correlated response, and relative selection efficiency in greengram. *Indian J. Agric. Sci.* 53: 101–105.

Malik, .A., Rahman, M.M., Sarker, and Siddique. 1994. Research and development of mungbeen in Bangladesh. Paper Presented at *Int. Symp. Pulses Res.* 2–6 April. New Delhi. (in Press)

Matloc, R.S., and Oswalt, R.M. 1963. Mungbean varieties for Oklahoma. *Oklahoma Agric. Exp. Stn. Bull.* B-612, pp. 1–15.

Mehta, T.R., and Sahai, J.M. 1955. A review of breeding work on mung (*Phaseolus aureus* Roxb). in Uttar Pradesh. *Agric. and Animal Husb.* 6 (2/3): 56–57.

Menancio, D.I., and Ramirez, D.A. 1977. Genetic polymorphism and ontogenetic isozyme patterns of Cercospora leaf spot resistant and susceptible varieties of mungbean. *Philippine J. Crop Sci.* 2: 197–202.

Misra, R.C., Sahu, R.C., and Tripathy, D. 1970. A note on heterosis in greengram (*Phaseolus aureus* Roxb.). *Curr. Sci.* 39: 190–191.

Mitra, R.K. 1967. Effect of colchicine on 'mung' (*Phaseolus aureus* Roxb.). *Kanpur Agr. Coll. J.* 27: 13–21.

Murray, S.A., and Newcombe, C.L. 1970. Synergistic inhibition of growth of mungbean seedlings after a combination of low-level X-irradiation and low-level shock treatment. *Radiation Bot.* 10: 563–567.

Murty, B.K., and Patel, G.J. 1972. Inheritance of seed coat colour in greengram. *Indian J. Genet.* 32: 373–378.

Murty, B.K., and Patel, G.J. 1973. Inheritance of some morphological characters in mungbean. *Bansilal Amrital Coll. Agric. Mag.* 25: 1–9.

Murty, B.K., Patel, G.J., and Jaisani, B.G. 1976. Gene action and heritability estimates of some quantitative traits in mung. *Gujarat Res. J.* 2: 1–4.

Nalampang, A. 1978. Mungbean production in Thailand. *Proc. 1st Int. Mungbean Symp.* Asian Vegetable Research and Development Center, Shanhua, Taiwan, pp. 12–14.

Park, H.G., and Yang, C.Y. 1978. The mungbean breeding programme at the Asian Vegetable Research and Development Center. *Proc. 1st Int. Mungbean Symp.* Asian Vegetable Research and Development Center, Shanhua, Taiwan, pp. 214–216.

Paroda, R.S., and Thomas, T.A. 1988. Genetic resources of mungbean (*Vigna radiata* (L.) Wilczek) in India. *Mungbean: Proc. 2nd Int. Symp.* Asian Vegetable Research and Development Center, Shanhua, Taiwan, pp. 19–28.

Pathak, G.N., and Singh, B. 1963. Inheritance studies in greengram. *Indian J. Genet.* 23: 215–218.

Poehlman, J.M. 1978. What we learned from the International Mungbean Nurseries. *Proc. 1st Int. Mungbean Symp.* Asian Vegetable Research and Development Center, Shanhua, Taiwan, pp. 97–100.

Poehlman, J.M. 1991. *The Mungbean.* Oxford and IBH Publishing Co., New Delhi, 375 pp.

Poehlman, J.M., Sechler, D.T., Swindell, R.E., and Sittiyos. P. 1976. Performance of the Fourth International Mungbean Nursery. Univ. Missouri, Columbia, Agric. Exp. St., Special Report 191.

Pokle, Y.S. 1972. Spontaneous mutation in mung (*Phaseolus aureus* Roxb.). *Sci. Cult.* 38: 142.

Pokle, Y.S., and Nomulwar, M.T. 1975. Variability studies in some quantitative characters of mung (*Phaseolus aureus* Roxb.). *Punjabrao Krishi Vidyapeeth Res. J.* 3: 129–132.

Pokle, V.S., and Patil, V.N. 1975. Path coefficient analysis in greengram (*Phaseolus aureus* Roxb.). *Sci. Cult.* 41: 265–266.

Polhill, R.M., and van der Maesen, L.J.G. 1985. Taxonomy of grain legumes. In *Grain Legume Crops.* R.J. Summerfield and E.H. Roberts, eds. Collins, London, pp. 3–36.

Prasad, M.V.R. 1976. Induced mutants in greengram. *Indian J. Genet.* 36: 218–222.

Premsekar, S., and Srinivasan, V. 1961. Two high yielding selections of greengram (*Phaseolus aureus* Roxb.) for Madras. *Madras Agric. J.* 48: 259–260.

Punto, N.G., and Lantican, R.M. 1982. Genotype × environmental interaction for yield in field legumes. II. Mungbean (*Vigna radiata* (L.) Wilczek.) *Philippine J. Crop. Sci.* 7: 137–140.

Rachie, K.O., and Roberts, L.M. 1974. Grain legumes of the lowland tropics. *Adv. Agron.* 26: 1–132.

Rajput, M.A. 1973. Gamma-irradiation M_1 studies in mungbean (*Phaseolus aureus* Roxb.). *Pakistan J. Sci. Res.* 25: 167–172.

Rajput, M.A. 1974. Increased variability in the M_2 of gamma- irradiated mungbeen (*Phaseolus aurius* Roxb.). *Radiat. Bot.* 14: 85–89.

Ramanujam, S. 1975. Genetic diversity, stability and plant type in pulse crops. *Proc. Int. Symp. Grain Legumes.* ICRISAT, Hyderabad, India, pp. 167–176.

Ramanujam, S. 1978. Biometrical basis for yield improvement in mungbean. *Proc. 1st Int. Mungbean Symp.* Asian Vegetable Research and Development Center, Shanhua, Taiwan, pp. 210–213.

Ramanujam, S., Tiwari, A.S. and Mehra, R.B. 1974. Genetic divergence and hybrid performance in mungbean. *Theor. Appl. Genet.* 45: 211–214.

Rangaswamy, S.R.S., Oblisami, G., and Krishnaswami, S. 1973. Nodulation and productivity in the induced mutants of greengram by gamma-rays. *Madras Agric. J.* 60: 359–361.

Reddy, P.R., and Sreeramulu, C. 1982. Heterosis and combining ability for yield and its components in greengram (*Vigna radiata* (L.) Wilczek). *Genetica Agraria* 36: 297–308.

San Miguel, L.A. 1916. Tests and selections of mungo beans. *The Philippine Agriculturalist* 5: 164–179.

Sandhu, T.S., Bhullar, B.S., Cheema, H.S., and Brar, J.S. 1980. Path-coefficient analysis for grain yield and its attributes in greengram. *Indian J. Agric. Sci.* 50: 541–544.

Sandhu, T.S., Brar, J.S., Sandhu, S.S., and Verma, M.M. 1985. Inheritance of resistance to mungbean yellow mosaic virus in greengram. *J. Res. PAU,* 22: 607–611.

Santos, I.S. 1969. Induction of mutations in mungbean (*Phaseolus aureus* Roxb.) and genetic studies of some of the mutants. In *Induced Mutations in Plants.* E. Doyle, ed. International Atomic Energy Agency, Vienna. pp. 169–179.

Sarbhoy, K. 1980. Karyological studies in the genus *Phaseolus* Linn. *Cytologia* 45: 363–373.

Sareen, P.K. 1985. Further genetic analysis of the trilobate leaf mutants in mungbean (*Vigna radiata,* var. *aureus* (L.) Wilczek). *Curr. Sci.* 54: 930–931.

Sarker, A., Rahman, M.M., Malik, A., Miah, A. and Siddique A.R. 1994. Performance of exotic mungbean lines in Bangladesh. Paper presented *at Int. Symp. Pulses Res.* 3–6 April, New Delhi (in Press).

Sawa, M. 1974. On the interspecific hybridization between the adzuki bean, *Phaseolus aureus* (Willd.) W.F. Wight, and the greengram, *Phaseolus radiatus* L. II. On the characteristics of amphidiploid in C1 generation from the cross greengram × rice bean, *Phaseolus calcaratus* Roxb. *Jap. J. Breed.* 24: 282–286.

Sen, N.K., and Ghosh, A.K. 1959. Genetic studies in greengram. *Indian J. Genet.* 19: 210–227.

Sen, N.K., and Ghosh, A.K. 1960a. Studies on the tetraploids of six varieties of greengram. *Proc. Nat. Inst. Sci. India* 26B: 291–299.

Sen, N.K., and Ghosh, A.K. 1960b. Interspecific hybridization between *Phaseolus aureus* Roxb. (greengram) and *Ph. mungo* L. (blackgram). *Bull. Bot. Soc. Bengal* 14: 1–4.

Sen, N.K., and Ghosh, A.K. 1960c. Inheritance of a chlorophyll deficient mutant in greengram (*Phaseolus aureus* Roxb.). *Bull. Bot. Soc. Bengal* 14: 80–81.

Sen, N.K., and Murty, A.S.N. 1960. Inheritance of seed weight in greengram (*Phaseolus aureus* Roxb.). *Genetics* 45: 1559–1562.

Shaikh, M.A.Q., Majid, M.A., Ahmed, Z.U., and Shansuzzaman, K.M. 1982. Induced mutations for new plant types and disease resistance in mungbean and blackgram. In *Induced Mutations for Improvement of Grain Legume Production. II.* Proc. IAEA, Vienna, pp. 7–17.

Shanmugasundaram, S. 1988. *A Catalogue of Mungbean Cultivars Released around the World.* Asian Vegetable Research and Development Center, Shanhua, Taiwan.

Shanmugasundaram, S., and Tschanz, A.T. 1985. Breeding for mungbean and soybean disease resistance, In *Varietal Improvement of Upland Crops for Rice-based Farming Systems.* International Rice Research Institute, Los Baños, Philippines.

Shrivastava, M.P., Singh, L., and Sharma, D. 1973. Karyomorphology of different ecotypes of greengram (*Phaseolus aureus* Roxb.). *Jawaharlal Nehru Krishi Vishwa Vidyalaya Res. J.* 7: 86–90.

Shukla, G.P., and Pandya, B.P. 1985. Resistance to yellow mosaic in greengram. *SABRAO J.* 17: 165–171.

Singh. B.V., and Ahuja, M.R. 1977. *Phaseolus sublobatus* Roxb.: A source of resistance to yellow mosaic virus for cultivated mung. *Indian J. Genet.* 37: 130–132.

Singh, D., and Mehta, T.R. 1953. Inheritance of lobed leaf margin in mung (*Phaseolus aureus* L.). *Curr. Sci.* 22: 348.

Singh, D., and Patel, P.N. 1977. Studies on resistance in crops to bacterial diseases in India, Part VIII. Investigations on inheritance of reactions to bacterial leaf spot and yellow mosaic diseases and linkage, if any, with other characters in mungbean. *Indian Phytopathol.* 30: 202–206.

Singh, D., and Saxena, J.K. 1959. A semi-dominant lethal leaf mutation in *Phaseolus aureus. Indian J. Genet.* 19: 83–89.

Singh, D.P. 1980. Note on the inheritance of lobed leaf in greengram. *Indian J. Genet.* 40: 608–609.

Singh, D.P. 1981. Breeding for resistance to diseases in greengram and blackgram. *Theor. Appl. Genet.* 59: 1–10.

Singh, D.P. 1982. Genetics and breeding of blackgram and greengram. Res. Bull. 109, G.B. Pant University of Agriculture and Technology, 68 pp.

Singh, D.P. 1986. Genetics and breeding for disease resistance in mungbean and urdbean. In *Genetics and Crop Improvement.* P.K. Gupta and J.R. Bahl, eds. Rastogi and Co., Meerut, India, pp. 125–131.

Singh, D.P., Vaidya, K.R., and Bhatt, D.D. 1979. Gamma induced variability for flowering and chlorophyll mutations in greengram. *Indian J. Agric. Sci.* 49: 835–838.

Singh, G., Kapoor, S., and Singh, K. 1988. Multiple disease resistance in mungbean with special emphasis on mungbean yellow mosaic virus. In *Mungbean: Proc. 2nd Int. Symp.* S. Shanmugasundaram and B.T. McLean, eds. AVRDC, Shanhua, Taiwan. pp. 290–296.

Singh, H.B., Joshi, B.S., and Thomas, T.A. 1970. The *Phaseolus* group. In *Pulse Crops of India.* P. Kachroo and M. Arif, eds. Indian Council Agric. Res., New Delhi, pp. 136–164.

Singh, K.B., and Jain, R.P. 1970. Heterosis in mungbean. *Indian J. Genet.* 30: 251–260.

Singh, K.B., and Jain, R.P. 1971a. Combining ability for pod length and seed size in mungbean. *Indian J. Genet.* 31: 145–148.

Singh, K.B., and Jain, R.P. 1971b. Analysis of diallel cross in *Phaseolus aureus* Roxb. *Theor. Appl. Genet.* 41: 279–281.

Singh, K.B., and Malhotra, R.S. 1970a. Estimates of genetic and environmental variability in mung (*Phaseolus aureus* Roxb.). *Madras Agric.* 57: 155–159.

Singh, K.B., and Malhotra, R..S. 1970b. interrelationships between yield and yield components in mungbean. *Indian J. Genet.* 30: 244–250.

Singh, K.B., and Singh, J.K. 1970. Inheritance of black spot on seed coat and shiny seed surface in mungbean (*Phaseolus aureus* Roxb.). *Indian J. Hered.* 2: 61–62.

Singh, L. 1975. Breeding pulse crop varieties for inter and multiple cropping. *Indian J. Genet.* 35: 221–228.

Singh, M.K. 1973. Inheritance of seed coat colour in *Phaseolus radiatus* L. (syn. *P. aureus* Roxb.). Part III. *Indian Sci. Congr. Assoc. Proc. 60th.* Jan. 3–9, 1973, Punjab Univ., Chandigarh, India, Indian Sci. Congress Assoc., Calcutta, pp. 321–322.

Singh, R.P., Singh, B.D., and Singh, R.M. 1986. Organ regeneration in mungbean callus cultures. In *Genetics and Crop Improvement.* P.K. Gupta and J.R. Bahl, eds. Rastogi and Co., Meerut, India, pp. 387–393.

Singh, T.P. 1974. Epistatic bias and gene action for protein content in greengram (*Phaseolus aureus* Roxb.). *Euphytica* 23: 459–465.

Singh, T.P., and Malhotra, R.S. 1975. Crossing technique in mungbean (*Phaseolus aureus* Roxb.). *Curr. Sci.* 44–64–65.

Singh, T.P., and Singh, K.B. 1970. Inheritance of clusters per node in mungbean (*Phaseolus aureus* Roxb.). *Curr. Sci.* 39: 265.

Singh, T.P., and Singh, K.B. 1971. Mode of inheritance and gene action for yield and its components in *Phaseolus aureus. Canad. J. Genet. Cytol.* 14: 517–525.

Singh, T.P., and Singh, K.B. 1972. Combining ability in mungbean. *Indian J. Genet.* 32: 67–72.

Singh, T.P., and Singh, K.B. 1973a. Association of grain yield and its components in segregating population of greengram. *Indian J. Genet.* 33: 112–117.

Singh, T.P., and Singh, K.B. 1973b. Combining ability for protein content in mungbean. *Indian J. Genet.* 33: 430–435.

Singh, T.P., and Singh, K.B. 1974a. Components of genetic variance and dominance pattern for some quantitative traits in mungbean (*Phaseolus aureus* Roxb.). *Z. Pflanzenzüchtg.* 71: 233–242.

Singh, T.P., and Singh, K.B. 1974b, Heterosis and combining ability in *Phaseolus aureus* Roxb.).*Theor. Appl. Genet.* 44: 12–16.

Singh, V.P., and Chaturvedi, S.N. 1981. Gamma-ray induced male sterile mutants in mungbean. *Pulse Crops Newslett.* 1: 41–42.

Sittiyos, P., Poehlman, J.M., and Sehgal, O.P. 1979. Inheritance of resistance to cucumber mosaic virus infection in mungbean. *Crop Sci.* 19: 51–53.

Smartt, J. 1985. Evolution of grain legumes. III. Pulses in the genus *Vigna. Exp. Agric.* 21: 87–100.

Somaatmadja, A.S., and Sutarman, T. 1978. Present status of mungbean breeding in Indonesia. *Proc. 1st Int. Mungbean Symp.* Asian Vegetable Research and Development Center, Shanhua, Taiwan, pp. 230–232.

Subramanian, D. 1980. Effect of gamma-radiation in *Vigna. Indian J. Genet.* 40: 187–194.

Swindell, R.E., and Poehlman, J.M. 1976. Heterosis in the mungbean (*Vigna radiata* (L.) Wilczek). *Trop. Agr.* 53: 25–30.

Swindell, R.E., and Poehlman, J.M. 1978. Inheritance of photoperiod response in mungbean (*Vigna radiata* (L.) Wilczek). *Euphytica* 27: 325–333.

Thakur, R.P., Patel, P.N., and Verma, J.P. 1977a. Studies on resistance in crops to bacterial diseases in India. XI. Genetic make-up of mungbean differentials of the races of bacterial leaf spot pathogen. *Xanthomonas phaseoli. Indian Phytopathol.* 30: 217–221.

Thakur, R.P., Patel, P.N., and Verma, J.P. 1977b. Genetic relationship between reactions to bacterial leaf spot, yellow mosaic and Cercospora leaf spot diseases in mungbean (*Vigna radiata*), *Euphytica* 26: 765–774.

Thakur, R.P., Patel, P.N., and Verma, J.P. 1977c. Independent assortment of pigmentation and resistance to Cercospora leaf spot in mungbean. *Indian Phytopathol.* 30: 264–265.

Thulasidass, G. 1984. Multiple regression and classificatory analysis in *Vigna radiata* (L.). *Field Crops Res.* 9: 183–191.

Tikoo, J.L., and Jain, H.K. 1974. Mutation studies in mungbean. *Mutation Breed. Newslett.* 3: 10–11.

Tiwari, A.S., and Ramanujam, S. 1974. Partial diallel analysis of combining ability in mungbean. *Z. Pflanzenzüchtg.* 73: 103–111.

Tiwari, A.S., and Ramanujam, S. 1976a. Combining ability and heterosis for protein and methionine contents in mungbean. *Indian J. Genet.* 36: 353–357.

Tiwari, A.S., and Ramanujam, S. 1976b. Genetics of flowering response in mungbean. *Indian J. Genet.* 36: 418–419.

Tomar, G.S., Singh, L., and Sharma, D. 1972. Effects of environment on character correlation and heritability in greengram. *SABRAO Newslett.* 4: 49–52.

van Rheenen, H.A. 1964. Preliminary study of natural cross-fertilization in mungbean, *Phaseolus aureus* Roxb. *Netherlands J. Agric. Sci.* 12: 260–262.

van Rheenen, H.A. 1965. The inheritance of some characters in the mungbean, *Phaseolus aureus* Roxb. *Genetica* 36: 412–419.

Vavilov, N.I. 1951. The origin, variation, immunity, and breeding of cultivated plants (translation by K.S. Chester). *Chron. Bot.* 13: 1–364.

Veeraswamy, R., and Kunjamma, V.K. 1958. A note on foliar variation in greengram or mung (*Phaseolus aureus* L.). *Madras Agric. J.* 45: 151–152.

Veeraswamy, R., Rathnaswamy, R., and Palanisamy, G.A. 1973. Genetic variability in some quantitative characters of *Phaseolus aureus* Roxb. *Madras Agric. J.* 60: 1320–1322.

Verdcourt, B. 1970. Studies in the Leguminosae-Papilionoideae for the Flora of Tropical East Africa: IV *Kew Bull.* 24: 507–569.

Verma, M.M., Chahal, G.S., and Murty, B.R. 1978. Limitations of conventional regression analysis, a proposed modification. *Theor. Appl. Genet.* 53: 89–91.

Verma, S.N.P. 1971. Inheritance of photosensitivity in mungbean (*Phaseolus aureus* Roxb.). *Mysore J. Agric. Sci.* 5: 477–480.

Verma, S.N.P. 1977. Recombined progenies in interspecific cross between greengram (*Phaseolus aureus* Roxb. syn. *Vigna radiatus*) and blackgram (*Phaseolus mungo* L. syn. *Vigna mungo*). *Mysore J. Agric. Sci.* 11: 431–434.

Verma, S.N.P., and Krishi, J.N. 1969. Inheritance of some qualitative characters in greengram (*Phaseolus aureus* Roxb.). *Indian J. Hered.* 1: 105–106.

Vidyadhar, G.S., Sharma, and Gupta, S.C. 1984. Path analysis in greengram. *Indian J. Agric. Sci.* 54: 144–145.

Vignarajah, N. 1978. Mungbean research and production in Sri Lanka. *Proc. Ist Int. Mungbean*

Symp. Asian Vegetable Research and Development Center, Shanhua, Taiwan, pp. 9–11.

Virmani, S.S., Singh, K.B., Singh, K., and Malhotra, R.S. 1983. Evaluation of mungbean germplasm. *Indian J. Genet.* 43: 54–58.

Watt, E.E., Poehlman, J.M., and Cumbie, B.G. 1977. Origin and composition of a texture layer on seeds of mungbean *Crop Sci.* 17: 121–125.

Wilczek, R. 1954. *Vigna. Flore du Congo Belge.* 6: 343–393.

Yadav, H.C., Chand, J.N., and Saharan, G.S. 1981. Inheritance of resistance in mungbean to bacterial leaf spot. In *Proc. 5th Int. Conf. Plant Pathogenic Bacteria.* J.C. Lazano, ed. CIAT, Cali., Colombia, pp. 580–583.

Yadav, R.C., Singh, R.K., and Singh, K.P. 1986. Amphidiploids of greengram × blackgram. In *Genetics and Crop Improvement.* P.K. Gupta and J.R. Bahl, eds. Rastogi and Co., Meerut, India, pp. 349–358.

Yahya, A., Alam, K., and Yousouf, M. 1975. The effect of ionizing radiations in mung (*Phaseolus aureus* Roxb.). *Pakistan J. Agric. Sci.* 12: 259–266.

Yohe, J.M., and Poehlman, J.M. 1972. Genetic variability in the mungbean, *Vigna radiata* (L.). Wilczek. *Crop Sci.* 12: 461–464.

Yohe, J.M., and Poehlman, J.M. 1975. Regression, correlations and combining ability in mungbean (*Vigna radiata* (L.). Wilczek). *Trop. Agric.* (Trinidad) 52: 343–352.

Young, N.D., Kumar, L., Menancio-Hautea, D., Danesh, D., Talekar, N.S., Shanmugasundaram, S., and Kim, D.H. 1992. RFLP mapping of major bruchid resistance gene in mungbean (*Vigna radiata*). *Theor. Appl. Genet.* 84: 839–844.

Young, N.D., Danesh, D., Menancio-Hautea, D., and Kumar, L. 1994. Mapping polygenic resistance to powdery mildew in mungbean with RFLPs. *Phytopathology* (in press).

Zhukovsky, P.M. 1950. *Cultivated Plants and Their Wild Relatives* (translation by P.S. Hudson, 1962). Commonwealth Agricultural Bureau, Farnham Royal, England, 107 pp.

5. SOYBEAN

P.S. Bhatnagar
Director, National Research Centre for Soybean (ICAR), Indore 452 001, M.P., India

S.P. Tiwari
Assistant Director General (Seeds), Indian Council of Agricultural Research, New Delhi 110001, India

Introduction

Soybean (*Glycine max* (L.) Merrill) tops in the world production of both oilseeds and edible oil. World harvest of soybean in 1991–92 was around 105 million t, which is about 50 per cent of the total world oilseed production. The United States, Brazil, China, Argentina, and India are the top five countries in the world with respect to area and production of soybean (Table 1). Besides oil (19–20 per cent), soybean seeds contain as much as 42–45 per cent protein. Since early times, soybean cultivation in Southeast Asia became popular mainly because of its high protein content.

Soybean has emerged as the third-most important oilseed crop, next only to groundnut and rapeseed-mustard in India. During 1993–94, the estimated area under soybean in the country was above 4 million ha.

Table 1
Annual area, production and yield in major soybean growing countries of the world (1990–91)

Country	Area ('000 ha)	Production ('000 t)	Yield (kg/ha)
United States	22,870	52,416	2,292
Brazil	9,650	15,750	1,632
China	7,560	11,000	1,455
Argentina	4,800	11,500	2,396
India	2,564	2,601	1,015
Indonesia	1,215	1,315	1,082
Paraguay	890	1,300	1,461
Thailand	375	530	1,413
North Korea	340	440	1,294
South Korea	152	233	1,533
Vietnam	140	105	750

Source: '92 Soya Blue Book (Soyatech Inc., USA Publication), p. 188.

Madhya Pradesh, a state in central India, is called the 'soybean state' as it alone accounts for more than 3 million ha, which is about 80 per cent of the total area under soybean in the country.

India is among the top three countries in the world to have registered a phenomenal growth rate in area expansion under the soybean crop. The other two are Argentina and Brazil. From 1981–82 to 1990–91, India has experienced a tremendous growth in area and production of soybean (Bhatnagar and Tiwari, 1991a), which showed increases of 312 and 458 per cent, respectively (Table 2). Notwithstanding the spectacular increase in area and production, the productivity of the crop has remained low. The current level of soybean yield, being around 1 t/ha, is about half of the world average productivity.

Table 2
Area, production and yield of soybean in India, 1979–80 to 1993–94

Year	Area ('000 ha)	Production ('000 t)	Yield (kg/ha)
1979–80	496	282	562
1980–81	608	442	728
1981–82	622	466	749
1982–83	770	491	638
1983–84	836	614	735
1984–85	1,243	955	768
1985–86	1,340	1,024	764
1986–87	1,527	891	584
1987–88	1,543	898	582
1988–89	1,655	1,501	907
1989–90	2,253	1,806	801
1990–91	2,564	2,601	1,014
1991–92	2,821	2,275	806
1992–93	3,545	3,109	877
1993–94	4,320	4,365	1,010

India has been able to earn valuable foreign exchange from export of soymeal. During the year 1992–93, a total of 0.45 million t of soybean oil was produced and export value of Rs 11,144 million was earned by exporting 1.83 million t of soymeal.

Presently, the bulk of the Indian soybean production is being used for oil and meal production in more than 100 solvent extraction plants with an annual capacity to crush about 7 million t of soybean. The edible oil extracted from soybean is consumed in the form of refined or hydrogenated oil. Concerted efforts are needed to diversify the food uses of soybean in India.

Origin and Early History

Although the origin and early history of soybean is not accurately known, it is believed that the origin of soybean dates back to 2800 BC in China. The evidences for the antiquity of soybean from the pictographical analysis of the archaic Chinese word for soybean 'Shu', the Book of Odes, and bronze inscriptions point to its first domestication in the eastern half of northern China around the 11th century BC (Hymowitz, 1970). The lower and middle Yellow River valley is believed to be the primary centre and Manchuria the secondary centre of origin of soybean.

Chang Ruzhen (1989) discussed different hypotheses suggesting a centre of origin in any of four regions: (1) the northeast of China, (2) Yangtze River valley and southern part of China, (3) lower and middle Yellow River valley, and (4) several regions such as northeast of China, the Yellow River valley, and the Yangtze River valley. It is now widely accepted that the lower and middle Yellow River valley is the main centre of origin of soybean. *Glycine soja* Sieb & Zucc. is the probable progenitor of cultivated soybean.

The available evidences suggest dissemination of cultivated form of soybean from northern China to Korea and then to Japan, sometime between 200 B.C. and the 3rd century AD (Nagata, 1959, 1960). Another route of dissemination could have been from northern China to southern Japan (Morse, 1950a).

In Europe, soybean was known in the 17th century as an exotic plant from the Orient. However, it was not until 1712 that the Western world finally recognized soybean as a food plant (Brooks, 1966). Its cultivation in Europe started in 1875. In the United States, mention of soybean in literature dates back to 1804 but its commercial production picked up only in the 1930s.

Cytotaxonomy and Chromosomes

The cultivated soybean, *Glycine max* (L.) Merrill, is a member of the family Leguminosae and subfamily Papilionaceae. The genus *Glycine* consists of two subgenera, *Glycine* and *Soja* (Moench) F.J. Herm. The subgenus *Soja* comprises the cultivated soybean *G. max* (L.) Merrill and its annual wild progenitor, *G. sojà* Sieb & Zucc (also known as *G. ussuriensis).* These two species have diploid chromosome number, 2n = 40. Prior to 1983, only seven species were recognized in the subgenus *Glycine*. Tindale and Craven added eight new species of *Glycine* from northwestern Australia (Tindale, 1984; Tindale and Craven, 1988; and their subsequent work). The subgenus *Glycine* is, thus, composed of 15 wild perennial species

(Table 3). Out of these, two species, namely, *G. tabacina* (2n = 40, 80) and *G. tomentella* (2n = 38, 40, 78, 80), are polyploid forms, whereas the other 13 species are diploids (2n = 40). Singh *et al.* (1988) and Hymowitz and Singh (1992) have assigned genome symbols to several of the species of both the subgenera (Table 3).

Table 3

Species of *Glycine*, their chromosome number and genome(s)

Species	2n	Genome
Subgenus *Glycine* (x = 10)		
G. albicans Tind. and Craven	40	—
G. arenaria Tind	40	—
G. argyrea Tind	40	$A_2 A_2$
G. canescens F.J. Herm	40	AA
G. clandestina Wendl.	40	$A_1 A_1$
G. curvata Tind.	40	$C_1 C_1$
G. cyrtoloba Tind.	40	CC
G. falcata Benth.	40	FF
G. hirticaulis Tind. and Craven	80	—
G. lactovirens Tind. and Craven	40	—
G. latifolia (Benth.) Newell and Hymowitz	40	$B_1 B_1$
G. latrobeana (Meissn.) Benth.	40	$A_3 A_3$
G. microphylla (Benth.) Tind.	40	BB
G. tabacina (Labill.) Benth.	40	$B_2 B_2$
	80	Complex
G. tomentella Hayata	38	EE
	40	DD
	78	Complex
	80	Complex
Subgenus *Soja* (Moench) F.J. Herm. (x = 10)		
G. soja Sieb. and Zucc.	40	GG
G. max (L.) Merr.	40	GG

Induced polyploids in the cultigen and many wild species have been reported (Biswas and Bhattacharyya, 1972; Sadanaga and Grindeland, 1981; Cheng and Hadley, 1983). Haploids and polyploids occurred among progeny of sterile 'ms_1 ms_1' plants especially from polyembryonic seed (Kenworthy *et al.*, 1973; Beversdorf and Bingham, 1977; Chen *et al.*, 1985). Natural occurrence of haploids (2n = 20) as reported by Pillai (1976) has not been verified. Details of polyploids, aneuploids, interchanges, and inversions have been given by Palmer and Kilen (1987).

Mitotic chromosomes in soybean have been studied by several workers. Twenty chromosomes of the haploid complement (2n = 40) comprise 2 large, 14 intermediate, and 4 small chromosomes (Fukuda, 1933; Sen and

Vidyabhusan, 1960). Ahmad *et al.* (1983) identified two metacentric, six sub-metacentric and one sub-telocentric (satellited) chromosomes; the remaining 11 chromosomes were characterized into classes based on chromosome length and arm ratio. Four satellites, consisting of two each on a pair of long and short chromosomes, have been reported in *G. max* (Pillai, 1976) and also in a 40-chromosome *G. soja* (Zheng *et al.*, 1984). Halvankar and Patil (1990) studied karyotypes of six species of the genus *Glycine* and found that chromosomes were median to submedian. Diploid species had one pair of satellited chromosomes, whereas tetraploid species, namely *G. tabacina* and *G. tomentella,* had two pairs of satellited chromosomes.

To study chromosomes of soybean, techniques used by Palmer and Heer (1973) and Zheng *et al.* (1984) for root-tip cells and by Sapra and Stewart (1981) for mitotic cells from very young leaves have been found suitable. The pretreatment of cells with 8-hydroxyquinoline gives better clarity of chromosome constrictions than that obtained by para-dichlorobenzene. Although Giemsa staining appears to have limited scope in soybean chromosome identification, four groups of soybean chromosomes have been distinguished using this technique (Ladizinsky *et al.*, 1979). The estimated DNA content per haploid nucleus varies from 1.84 to 2.61 pg (Doerschug *et al.*, 1978; Goldberg, 1978). Yamamoto and Nagato (1984) concluded that cellular DNA content in soybean species has been reduced during the course of soybean domestication.

Intra-subgeneric crosses among the subgenus *Soja* are possible except in some combinations where a degree of sterility is obtained in the hybrid. Members of the subgenus *Glycine* possess many desirable agronomic characters like resistance to insect-pests and diseases, drought and salt tolerance, tolerance to several herbicides and alternative photoperiodism response, which are not readily found in the subgenus *Soja.* Inter-subgeneric crosses involving the cultigen are, however, difficult to realize. Some success has been achieved in crossing *G. tomentella* Hayata with the cultigen (Singh *et al.*, 1990). Selective elimination of *G. tomentella* chromosome complement, subsequent to introgression of some genes in the cultigen genome, has been indicated in such hybrids (Shoemaker *et al.*, 1990).

Morphology and Chemical Composition

The details of vegetative and reproductive morphology of soybean have been discussed by Lersten and Carlson (1987) and Carlson and Lersten (1987).

Soybean is an annual, bushy, erect, and leafy plant. It has a taproot with a large number of secondary and higher order smaller roots. The lower portion of the hypocotyl gives rise to multibranched adventitious roots. Bacterial root nodules are present as conspicuous spheroidal swellings of the root cortex inhabited by *Bradyrhizobium japonicum.* Stem termination in soybean may be of three types leading to indeterminate, determinate, and semi-determinate types. There are four kinds of soybean leaves: (1) first pair of simple cotyledons or seed leaves, (2) second pair of simple primary leaves, (3) trifoliate foliage leaves, and (4) prophylls, which are a tiny pair of simple leaves without petioles occurring at the base of each lateral branch. Leaflets of trifoliate leaves have entire margins and may be oblong, ovate, or lanceolate in shape.

The inflorescence is axillary raceme and terminal raceme in determinate and axillary raceme in indeterminate types. About 20–80 per cent of the flowers produced on a plant drop in various soybean cultivars. The soybean flower is typically papillionaceous. Colour of flowers is generally white or purple or sometimes of an intermediate shade. The flowers contain two bracteoles and are perfect and hermaphrodite. The calyx is tubular, consisting of five unequal sepals united together. The corolla has free petals consisting of a standard, two wings, and two keels. The androecium consists of 10 stamens arranged in a diadelphous pattern. Just before anthesis, nine stamens are pushed forward and one is left behind. The stigma is capitate.

Fruits of soybean are classified as simple, dry legumes or pods. There may be 2 to more than 20 pods in a single inflorescence. The number of pods per plant may be as high as 400. The pod colour on maturity may be light yellow, yellow, grey, brown, or black. The pubescence on pods as well as on the stem and branches can be dense or sparse and is usually tawny or grey in colour. The calyx, being persistent, remains attached to the pods permanently. The two halves of a single carpel are joined by dorsal and ventral sutures in the fruit.

There are usually two to three and sometimes even four seeds in a single pod (Gour *et al.,* 1992). Seeds are usually oval in shape but may also be spherical, flattened, and elongate. Seed coat may be yellow, green, brown, or black and may be of one colour, bicoloured, or variegated. The seed is devoid of endosperm and consists of a large embryo surrounded by seed coat. The embryo includes two large fleshy cotyledons, a plumule, and a hypocotyl-radicle axis. The cotyledons are usually yellow but sometimes may be green or chalky yellow. Colour of hilum varies from dark black to light grey or brown. Hilum bleeding of different intensity, characterized by spreading and diffusion of hilum colour, is occasionally observed. Germination of seed is epigeal. Flowers are self-pollinated with natural cross-pollination occurring at a very low frequency of about 0.3 to 3 per cent (Sadanaga and Grindeland, 1981). Pollination takes place a little before opening of flower or sometimes at the time of flower opening.

The microstructure of soybean seed consists of palisade cells of epidermis, hourglass cells, parenchyma, aleurone cells, and compressed cells of the endosperm. The cotyledon, separated by the epidermis from the seed coat, consists of cotyledon cells containing protein bodies, called 'aleurones'. and oil in spherosomes or lipid bodies.

Fehr *et al.* (1971) proposed a system that is widely used for determining developmental stages in soybean plant. These are grouped into vegetative and reproductive stages. Vegetative stage extends from VE, denoting emergence, to V (n), where (n) denotes the number of nodes on the main stem with fully developed leaves. The reproductive stage commences from R-1 (beginning to bloom) and goes up to R-8, i.e., full maturity (Fehr and Caviness, 1977).

Soybeans are highly photosensitive and are known as short-day plants. Many varieties require 10 hours' darkness to flower. All varieties tend to flower with 16 hours' exposure to artificial night. There are 13 maturity classes recognized in the United States, i.e., maturity group 000, 00, 0, and I through X according to the need of cultivars for specific photoperiods. Varieties adapted to the northernmost area of the United States and southern Canada are designated as '000', '00' and '0' and are the earliest maturing, with a shorter night requirement for flower initiation. From north to south, the maturity groups move from 0 to X. The varieties adapted to the southernmost region of the United States are categorized in the VIII maturity group (Howell, 1960). Varieties adapted to northern latitudes will normally be shorter and mature earlier when planted in southern latitudes.

The economic importance of soybean is mainly due to its chemical constituents, primarily oil and protein. In addition, it contains carbohydrate and sugars, cellulose, inorganic matter, vitamins, and enzymes. The chemical properties of soybean have been summarized by Watanabe and Kishi (1984). The chemical composition of soybean and various foods that can be made from it is given in Table 4. Both variety and climate, which can be influenced by location and season, are known to affect chemical constitution of soybean. Generally, Japanese summer types of soybean have high protein content, whereas American varieties are high in oil. Japanese and Chinese varieties have high carbohydrate content. Cellulose, hemicellulose, and other polysaccharides are present in the cell wall. Sucrose is present in the range of 6 to 7 per cent and other oligosaccharides and stachyose vary from 1 to 2 per cent. The range of raffinose varies from 5 to 6 per cent. It is interesting that starch, although present in the bean prior to its ripening, is completely absent in the seed on maturity.

Protein, which is in the range of 30 to 42 per cent in soybean seed, is mostly salt-soluble globulin and partly albumins. From the viewpoint of essential amino acid composition, soybean protein is fairly well balanced

Table 4
Proximate composition and selected nutrient content of various soy foods in 100 g edible portions

Soy food	Water (g)	Kcal	Protein (g)	Fat (g)	Carbo-hydrate (g)	Crude fibre (g)	Cal-cium (mg)	Iron (mg)	Zinc (mg)	Thiamin (mg)	Ribo-flavin (mg)	Niacin (mg)	Vitamin B6 (mg)	Folacin (mcg)
Miso	41.5	206	11.8	6.1	28.0	2.5	66	2.74	3.32	0.10	0.25	0.86	0.22	33.0
Natto	55.0	212	17.7	11.0	14.4	1.6	217	8.60	3.03	0.16	0.19	0.00	n/a	n/a
Okara, fibre by-product of soy milk/tofu	81.6	77	3.2	1.7	12.5	4.1	80	1.30	n/a	0.02	0.02	0.10	n/a	n/a
Soy flour, defatted	7.3	329	47.0	1.2	38.4	4.3	241	9.24	2.46	0.70	0.25	2.61	0.57	305.4
Soy flour, full-fat raw	5.2	436	34.5	20.6	35.2	4.7	206	6.37	3.92	0.58	1.16	4.32	0.46	345.0
Soy flour, full-fat roasted	3.8	441	34.8	21.9	33.7	2.2	188	5.82	3.58	0.41	0.94	3.29	0.35	227.4
Soy flour, low-fat	2.7	326	46.5	6.7	38.0	4.2	188	5.99	1.18	0.38	0.29	2.16	0.52	410.0
Soy meal, defatted, raw	6.9	339	45.0	2.4	40.1	5.8	244	13.70	5.06	0.69	0.25	2.59	0.57	302.6
Soy protein, concentrate	5.8	332	58.1	0.5	31.2	3.8	363	10.78	4.40	0.32	0.14	0.72	0.13	340.0
Soy protein, isolate	5.0	338	80.7	3.4	7.4	0.3	178	14.50	4.03	0.18	0.10	1.44	n/a	176.0
Soy sauce, from HVP	75.7	41	2.4	0.1	7.7	0.0	5	1.49	0.31	0.04	0.11	2.83	0.14	13.0
Soy sauce, from soy (tamari)	66.0	60	10.5	0.1	5.6	0.0	20	2.38	0.43	0.06	0.15	3.95	0.20	18.2
Soy sauce, from soy and wheat (shoyu)	71.1	53	5.2	0.1	8.5	0.0	17	2.02	0.37	0.05	0.13	3.36	0.17	15.5
Soybeans, cooked, boiled	62.6	173	16.6	9.0	9.9	2.0	102	5.14	1.15	0.16	0.29	0.40	0.23	53.8
Soybeans, dry-roasted	0.8	450	39.6	21.6	32.7	5.4	270	3.95	4.77	0.43	0.76	1.06	0.23	204.6
Soybeans, raw	8.5	416	36.5	19.9	30.2	5.0	277	15.70	4.89	0.87	0.87	1.62	0.38	375.1
Soybeans, roasted	2.0	474	35.2	25.4	33.6	4.6	138	3.90	3.14	0.10	0.15	1.41	0.21	211.0
Soymilk, fluid	93.3	33	2.8	1.9	1.8	1.1	4	0.58	0.23	0.16	0.07	0.15	0.04	1.5
Tempeh	55.0	199	19.0	7.7	17.0	3.0	93	2.26	1.81	0.13	0.11	4.63	0.30	52.0
Tofu, raw, firm	69.8	145	15.8	8.7	4.3	0.2	205	10.47	1.57	0.16	0.10	0.38	0.09	29.3
Tofu, raw, regular	84.6	76	8.1	4.8	1.9	0.1	105	5.36	0.80	0.08	0.05	0.20	0.05	15.0

Source: 92 Soya Blue Book (Soyatech Inc., USA Publication), p. 223.
(Originally from *Legume and Legume Products,* US Department of Agriculture, Human Nutrition Information Service, Agriculture Handbook N. 8-16. Revised December 1986).
The proximate composition of soy foods shown here may differ signficantly from accepted industry standards.
n/a = data not listed or available.

(Table 5). Lysine is abundant and there are less sulphur-containing amino acids, such as methionine and cystine, than in animal as well as cereal proteins.

Table 5
Amino acid composition of soy protein

Amino acid	g/16 g N
Arginine	6.8
Histidine	3.2
Lysine	7.4
Tryptophan	1.3
Tyrosine	3.1
Phenylalanine	4.9
Methionine	1.4
Cystine	1.3
Threonine	4.0
Serine	5.8
Leucine	7.1
Isoleucine	4.5
Valine	5.0
Glutamic acid	19.9

Source: Singhal, S.C. 1992. Food applications of soy flour. Souvenir, 30th All India Convention of Oilseeds, Oil Trade and Industry. SOPA, Indore, p. 21.

Soybean oil, present in the seed to the extent of 20 per cent, consists largely of unsaturated fatty acids such as oleic and linoleic. Hawkins *et al.* (1983) reported that the range of fatty acids among the soybean lines, averaged across seven environments, was 10.2–11.9 per cent for palmitic, 2.9–4.2 per cent for stearic, 18.6–27.9 per cent for oleic, 50.0–58.6 per cent for linoleic, and 6.8–9.3 per cent for linolenic. Fatty acid composition of soybean oil is presented in Table 6.

Table 6
Mean fatty acid composition of soybeans

Fatty acid	Type	Mean/range
Palmitic (C 16)	Saturated	11% ± 1
Stearic (C 18)	Saturated	4% ± 1
Oleic (C 18 : 1)	Monounsaturated	25% ± 3
Linoleic (C 18 : 2)	Polyunsaturated	50% ± 3
Linolenic (C 18 : 3)	Polyunsaturated	9% ± 3

Lecithin, sterols, and chlorophyll are also found in soybean oil. Lecithin and cephaline, belonging to phosphatide group, together are present up to about 2 per cent. Lecithin obtained from soybean oil is an excellent emulsifying agent and is used in production of chocolate and medicines. It is believed that lecithin prevents accumulation of cholesterol in blood. Sterols are compounds similar to vitamin D and hormones in chemical nature. Saponins present in soybean up to 0.6 per cent form soapy foam when mixed in water. Soybean saponins have recently been reported to prevent formation of peroxide oil, which causes senility. In addition, isoflavon and phytin are present in soybean. Isoflavon, a yellow substance accounting for the water solubility of soybean colour, has the same effect as saponins. Soybean is also known to contain goitrogenic substances that have an enlarging effect on goiter caused by iodine deficiency. Nevertheless, this substance has not been chemically identified so far.

Vitamins present in soybean are B_1 and other members of B group such as riboflavin, niacin, pantothenic acid, carotene, the precursor of vitamin A and vitamin E. Composition of vitamins varies from variety to variety. The ash, which is generally about 5 per cent, contains minerals such as potassium, phosphorus, sodium, calcium and sulphur. Unlike most animal foods, which are acidic, soybean foods are alkaline.

Soybean is reported to have some undesirable antinutritional substances. These are trypsin inhibitor, hemagglutinins, and oligosaccharides such as raffinose, stachyose and verbascose. However, most of these antinutritional substances are heat-labile and are inactivated almost completely when subjected to moist heat of about 110°C.

Genetics of Qualitative Characters

Pigment Colour

The studies on inheritance of pigment colour in soybean has been summarized by Johnson and Bernard (1962) and Bhatt and Torrie (1968). Black and brown seed coat and hilum colour are controlled by two gene pairs, Tr and Rr, their expression being modified in certain gene combinations by Oo and Wiwi. Brown seed coat (TrO) was dominant over reddish-brown (Tro) and green seed coat (G) over yellow (g). The 'I' allele for light hilum gave grey hilum with TR and tRW_1 genotypes and yellow hilum with Tr_1, tRw_1, and tr genotypes.

The inheritance of hypocotyl pigmentation was studied by Sengupta (1975), who reported a new gene conditioning this character. Purple (W1) and white (w1) colour of soybean flower is controlled monogenically, purple being dominant over white. Pleiotropic effect of the 'W1' allele gives purple pigmentation on the hypocotyl of seedlings. The hypocotyl colour of seedlings with the 'w1' allele is green. Magenta flower colour is controlled by mutant allele 'wm' in the presence of 'W1'. Magenta flower colour is not desirable, as the allele controlling it is associated with lower flavonol glycoside content of flower and leaves, lower photosynthetic rate, earlier leaf senescence, and lower seed yield (Buzzell *et al.,* 1977). Two distinctly different pubescence types, i.e., brown and grey, are controlled by one gene, brown (tawny) being dominant over grey.

Growth Habit

Growth habit or stem termination has been found to be controlled by Dt_1 and Dt_2 genes (Bernard, 1972). The alleles Dt_1 (indeterminate) and Dt_2 (semideterminate) are completely dominant to dt_1 (determinate) and dt_2 respectively, where dt_1 is epistatic to Dt_2 and dt_2 (Bernard, 1972).

Dwarfness

Six genes, df_1, df_2, df_3, df_4, df_5, and df_6, controlling dwarfness, mostly by independent recessive nature, have been reported by Palmer and Kilen (1987) and Werner *et al.* (1987).

Maturity Duration

Flowering and maturity in soybean are known to be governed by five major and independent genes designated E_1 and E_2 (Bernard, 1972), E_3 (Buzzell, 1971), E_4 (Buzzell and Voldeng, 1980), and E_5 (McBlain and Bernard, 1987). The dominant alleles E_1 and E_2 caused delayed flowering and maturity. The effect of the E_3 allele on flowering was much greater if E_1 and E_2 were present (McBlain *et al.,* 1987).

Pod Colour

Pod colour on maturity, which is black, brown or tan, is controlled by two genes. Presence of two dominant genes, L_1 L_2 and L_1 l_2, gives black pod colour, while l_1 L_2 is responsible for brown colour. The digenic recessive condition, l_1 l_2, gives tan colour (Bernard and Weiss, 1973).

Seed Longevity

Inheritance studies on seed deterioration during storage have indicated non-cytoplasmic maternal influence (Kueneman, 1983; Verma and Ram, 1987). Longevity was found to be controlled by two to four major genes, good longevity being recessive.

Resistance to Diseases

Soybean rust. Soybean rust is caused by *Phakopsora pachyrhizi* Syd. Four independently inherited dominant genes, namely Rpp_1, Rpp_2, Rpp_3 (Hartwig and Bromfield, 1983), and Rpp_4 (Hartwig, 1986), confer specific resistance to soybean rust isolates.

Phytophthora rot. The resistance is controlled by six dominant genes— Rps_1, Rps_2, Rps_3, Rps_4, Rps_5, and Rps_6 — where Rps_4, Rps_5, and Rps_6 are non-allelic to Rps_2 (Kilen, 1986).

Brown stem rot. Three dominant genes singly control resistance to brown stem rot. These genes exhibit duplicate epistasis (Willmot and Nickell, 1989).

Bacterial pustules. Resistance to bacterial pustule caused by *Xanthmonas phaseoli* var. *sojensis* (Hedges) Starr and Burkh is controlled by a single recessive allele, 'rxp' (Bernard and Weiss, 1973).

Bacterial blight. Mukherjee *et al.* (1966) reported a dominant allele (Rpg_1) controlling resistance to race 1 of *P. glycinea* Coerper, which causes bacterial blight.

Frog eye leaf spot. Two dominant alleles control resistance to *Cercospora sojina* Hara, one of which (RcS1) is effective against race 1 and the other (RcS2) against race 2 (Bernard and Weiss, 1973).

Soybean viruses. Resistance to yellow mosaic virus is governed by a recessive gene pair for which the gene symbols rym_1 and rym_2 were assigned (Singh and Malik, 1978). Resistance to soybean mosaic virus was governed by a single dominant allele 'rsv'. However, the resistance in variety Ogden is governed by another allele at the same locus, which is dominant to susceptibility but recessive to Rsv (Kiihl and Hartwing, 1979). Other dominant genes controlling resistance to soybean mosaic virus have also been reported.

The inheritance of improtant diseases of soybean have been succinctly presented in the *Compendium of Soybean Diseases,* edited by Sinclair and Backman (1989).

Cyst nematode. The inheritance of cyst nematode resistance is complex. At least five races of the nematode are known and several genes imparting resistance have been identified. Anand and Rao-Arelli (1989) found that resistance to race 5 of the nematode is conditioned by two recessive genes each in 'Peking' and PI 90763, whereas PI 438489B has only one recessive gene. They further reported that at least one resistant gene was different in the three host genotypes.

Resistance to Insect-Pests

Van Duyn *et al.* (1971) identified three soybean introductions as highly resistant to Mexican bean beetle. These were subsequently found to possess multiple resistance against several insect-pests. Kilen and Lambert (1986) found that resistance in each of these three introductions, namely PI 171451, PI 227687, and PI 229358, is governed by at least one gene. Gai *et al.* (1991) studied the inheritance of resistance to the beanfly (*Melanogromyza sojae* Zehntner) in soybean. Segregation for resistance to the beanfly in crosses involving highly resistant, moderately resistant, and highly susceptible varieties indicated that resistance was controlled by a single gene, Rms.

Pubescence

Pubescence is important for imparting resistance to insects. Lines with appressed pubescence, more easily noted on the upper leaf surface, are often heavily fed upon by insects like potato leafhopper (*Empoasca fabae*). It is common but not universal among *G. soja* accessions. Two alleles (pa1 and pa2) are necessary for imparting appressed pubescence (Bernard, 1975a). Several alleles controlling the density of pubescence are known. Alleles 'Pd1' and 'Pd2' increase the density of pubescence. The alleles 'Ps' and 'Ps-s' cause sparse and semi-sparse pubescence with fewer trichomes per unit area than normal (Bernard, 1975b).

Lipoxygenase Activity

Absence or low level of lipoxygenase (EC 1.13.11.12) isozymes is desirable to improve the quality of soybeans. Lack of L-1 (Hildebrand and Hymowitz, 1982), L-2 (Kitamura *et al.,* 1985), and L-3 (Kitamura *et al.,* 1983) lipoxygenase isozymes is governed by three independent recessive genes with a tight linkage between the lx_1 and lx_2 loci (Kikuchi and Kitamura, 1985).

Trypsin Inhibitor

Three different electrophoretic forms of Kunitz trypsin inhibitor are controlled by a co-dominant multiple allelic system (Ti-a, Ti-b, Ti-c) at a single locus (Hymowitz and Headley, 1972; Orf and Hymowitz, 1977). The fourth variant of the Kunitz trypsin inhibitor is a null and is inherited as the recessive allele (ti) (Orf and Hymowitz, 1979). The homozygous recessive condition results in the lack of the inhibitor.

Fatty Acid Content of Seed Oil

The linolenic acid content of seed oil in soybean is simply inherited and is more influenced by the genotype of the embryo than that of the maternal parent (Wilcox and Cavins, 1985). Studying the genotypes A 16 and A 17, Fehr *et al.* (1992) recently reported that besides major genes ('fan' and 'fan2'), minor genes and environmental effects also are important in the expression of linolenic acid content, and the low linolenic acid content in soybean can thus be considered a quantitative character for cultivar development programmes. Graef *et al.* (1985) reported that high stearic acid in each mutant was controlled by a different recessive allele at a single locus. Erickson *et al.* (1988) found that the inheritance of palmitic acid was controlled by two independent loci that were additive in gene action. Bubeck *et al.* (1989) found that the allele controlling low palmitic and low stearic acid in the variety Coles was partly dominant to the alleles controlling high palmitic or high stearic acid in the seven mutants studied.

The genetics of qualitative characters has been reviewed by Palmer and Kilen (1987). Halvankar *et al.* (1990) have recently reviewed genetics of morphological characters. Twenty linkage groups have been established in soybeans (Palmer *et al.,* 1992). The linkage groups contain more than 40 known loci (Yee and Palmer, 1984; Palmer and Kilen, 1987; Palmer and Shoemaker, 1987; Rennie and Tanner, 1989). The linkage groups have been established using the morphological and biochemical markers as well as the primary trisomics.

Genetics of Quantitative Characters

An early review of inheritance of quantitative characters in soybean was made by Brim and Cockerham (1961). Flowering time, period from flowering to maturity, plant height, resistance to lodging, seed yield, seed weight, and oil content were reported to be governed by additive genetic variance (Gates *et al.,* 1960). Kaw and Menon (1983), Talwar and Singh (1983), and Harer and Deshmukh (1991) reported that both additive and dominance gene action are important in the inheritance of yield and its components. Two specific components of seed protein in soybean are controlled by co-dominant genes of a single locus (Larsen and Caldwell, 1968). However, recent studies indicate that seed protein is mainly controlled by additive gene action (McKendry *et al.,* 1985).

Although the individual fatty acids have been shown to display qualitative inheritance, the oil content in seeds as such behaves as a quantitative character. Singh and Hadley (1968) studied F_1, F_2 , and backcrosses from crosses between high (21–22%) and low (13–16%) seed oil lines. The oil content of F_1 plants was intermediate between the oil contents of seeds of the two parents. About 60 per cent of the total variability among F_2 plants was due to additive genetic effects. No cytoplasmic effect was detected for oil content. It was concluded that the oil synthesis is determined by the genotype of the plant producing the seed.

Van Schaik and Probst (1958) reported quantitative inheritance with dominance and complementary gene effects for long raceme, high flower number, and flower shedding/abortion in soybean. Similar indications of quantitative inheritance were obtained by Sharma *et al.* (1990) for flower abortion, raceme length, flowers per raceme, and pod type.

Heterosis

In most of the studies, mean heterotic response for grain yield varies from 13 to 23 per cent, although Weber *et al.* (1970) reported up to 90 per cent heterosis. High positive heterosis for seed yield per plant was reported in crosses involving the varieties Improved Pelican, MACS 111, Gaurav, and Nimsoy 7 (Mehta *et al.*, 1984; Raut *et al.*, 1988; Taware *et al.*, 1990). High heterosis for seed yield (74.5%), compared to better parent, was reported by Taware *et al.* (1990) in crosses of Improved Pelican × Nimsoy 7, followed by Gaurav × Improved Pelican (69.2%) and MACS 111 × Nimsoy 7 (65.9%). With respect to other yield components, MACS 111 × Gaurav combination gave the highest significant positive heterosis over better parent for pods per plant (125.92%) and seeds per plant (81.17%). Heterosis for different characters has also been reported by Choudhary and Singh (1974), Singh *et al.* (1974), Chauhan and Singh (1982), and Nelson and Bernard (1984). Despite the numerous reports of significant heterosis for seed yield and proposition of a method using genetic male sterility for production of experimental hybrid soybean seeds (Burton and Carter, 1983), the hybrid soybean has not yet reached the level of commercial exploitation.

Correlations and Path-coefficient Analysis

Weatherspoon and Wentz (1934) noticed soybean yield to be positively correlated with plant height and number of pods. Similar results have been reported by Weber and Moorthy (1952). While studying the correlations among the characters, Johnson *et al.* (1955) and Kwon and Torrie (1964) reported positive correlation of plant height and maturity with seed yield. A close and positive association of grain yield with number of pods, seeds, and branches, and 1000-seed weight was observed by Bhatt *et al.* (1968), Gopani and Kabaria (1970), Manzoor *et al.* (1971), Khurana and Sandhu (1972), Kaw and Menon (1973), and Shettar *et al.* (1978). Raut (1972) observed that yield was positively correlated with plant height, number of branches, leaf length, leaf breadth, number of pods, pod weight, pod length and number of seeds. Singh and Ram (1983) and Pushpendra and Ram (1988) reported significant positive correlation of dry-matter yield, pods per plant, and harvest index with grain yield. Anand and Torrie (1963), Harer and Deshmukh (1992), and Prabhakar and Tiwari (1993) reported strong and positive association between pods per plant and seed yield. In general, it has been observed that the genotypic correlations are higher than the phenotypic correlations between grain yield and yield-determining characters such as number of branches, number of pods, number of seeds, and 1000-seed weight.

Path-coefficient analyses based on genotypic correlations suggest that number of pods per plant or per unit area, seeds per pod, seed weight, and number of branches are the major yield-determining characters. Malhotra *et al.* (1972) observed that yield is mostly dependent on pod number. Pandey and Torrie (1973) reported number of pods per unit area and number of seeds per pod as important factors in determining seed yield. Similar observations have been reported by Yap and Lee (1975), Srivastava *et al.* (1976), Patil and Pokle (1976), Gautam and Singh (1977), Sharma (1979), and Dixit and Patil (1984). Although many of these studies lead to the conclusion that yield is subscribed mainly by number of pods, number of seeds, 1000-seed weight, and number of branches, other studies have indicated pods per plant and seeds per pod to be more important than seed size because reduced seed number is only partly compensated by increased seed size (Pandey and Torrie, 1973; Herbert and Litchfield, 1982; Fehr *et al.*, 1985; Prabhakar and Tiwari, 1993). Quantitative genetics as relevant to soybean breeding has been reviewed by Burton (1987).

Genetic Improvement

Germplasm Utilization

Genetic diversity in the cultigen as well as the wild relatives is the basis of crop improvement as it helps in the development of superior cultivars with improved traits. The use of exotic soybean germplasm in breeding programmes often results in the increased genetic variability for desired selection. Vello *et al.* (1984) realized increased genetic variability in crosses involving exotic and indigenous soybean germplasm. Khalaf *et al.* (1984) reported that there was greater variability for yield in populations where the exotic genotype contributed 25 per cent germplasm to the cross, as in the case of a three-way cross such as (domestic × exotic) × domestic. A greater number of superior lines were identified from the more variable than from the less variable populations.

There are about 14,000 accessions in the world collection of soybean germplasm, out of which about 1000 accessions belong to *Glycine soja* and the rest to the cultigen. In India, the National Research Centre for Soybean, Indore, has a germplasm collection of about 2800 accessions of the cultigen. Pantnagar, Pune, and Akola are the other centres having sizeable collections of soybean germplasm. Besides the cultigen, several wild species are being maintained at Indore and Pune.

Several black-seeded land races of India have been collected under the names of Kalitur, Bhat, Bhatmash, and Kala Hulga from the northern hills of Kumaon region and cetral Indian plains. It is noteworthy that the first ever sizeable increase in soybean area during mid-1970s in Madhya Pradesh was due to the sweeping coverage by Kalitur. This variety is still cultivated in certain pockets and is the most commonly used donor for high seed longevity and hardiness in breeding programmes.

The systematic evaluation of soybean germplasm has resulted in the identification of donors for several specific characters (Table 7). Some of the accessions have been directly entered in the multilocation trials under the All India Coordinated Research Project on Soybean for evaluating their suitability as varieties to be released.

Soybean varieties earlier cultivated in India were largely introductions from the United States. The exotic soybean varieties released in India are Bragg, Clark 63, Davis, Hardee, Improved Pelican, KM 1 (introduced from AVRDC, Taiwan), and Monetta.

Selection

Selection in exotic and indigenous germplasm/land races has resulted in identification, development, and release of several soybean varieties. In India, about one-fourth of the total released soybean varieties owe their evolution to selection. Varieties thus evolved are Co 1, Gujarat Soy 1, Gujarat Soy 2, JS 2, JS 71-05, Punjab 1, Shilajeet, Shivalik, Type 49, and VL Soya 2.

Hybridization

Soybean is normally a self-pollinated crop with natural cross-pollination occurring at a very low frequency of 0.3 to 3 per cent. Hybridization for breeding purpose is generally done between carefully selected parents, which involves preparation of female flowers and pollination (Fehr, 1978). The

Table 7
Donors of desirable characters of soybean

Character	Donors/sources
High yield	JS 335, Bragg, PK 472, Carroll, Pusa 16, JS 80-21
Seed longevity	Type 49, Kalitur, Punjab 1, JS 80-21, NRC 1
Pod shattering resistance	Bragg, PK 416, NRC 7, Pusa 22, VLS 1, Himso 1520, JS 71-05, Nathan
Resistance to insect-pests	*Glycine soja* (for Bihar hairy caterpillar, girdle beetle etc.), PI 171451, PI 227687 and PI 229358, L 592 (for girdle beetle)
Resistance to diseases	
Yellow mosaic virus	*Glycine soja,* UPSM 534, PK 416
Bacterial pustules	Bragg, Hardee, PK 564, PK 472, Pusa 16
Aerial blight	PK 472, Jawa 16, Kalitur
Rust	PI 200492, PI 462312, PI 230970, PI 459025
Early maturity	Monetta, Shelby, Punjab 1, JS 71-05, PK 327, MACS 330
Oil quantity and quality	PK 416 (high oil), A16 and A17 (low linolenic acid)
Lodging resistance	PK 262, JS 71–05
Enhanced efficiency and promiscuity for nodulation	Carroll, JS 335
Higher insertion point for the first pod	PK 416, MACS 58, JS 80–21
Suitability for food uses	Hardee, Punjab 1, PK 472, PI 408251, PI 86023, 'Wase Natsu' (PI 417458), 'I-Higo-Wase' (PI 205085), 'Kunitz'

proper stage for emasculation of female flowers is generally indicated by the faint visibility of petals through the top of the calyx. The five lobes of the calyx and the entire petals are removed to expose the pistil and stamens. Anthers are collected from newly opened flowers and placed on the stigma of the flowers prepared for the purpose. Such hand cross-pollinated flowers are identified with tags.

The stigma is receptive to pollination at least one day before anthers shed pollen. Therefore, many breeders do not prefer to emasculate a variety if a marker gene like purple flower colour (dominant) versus white flower colour (recessive) is present in the parents (Paschal, 1976; Walker et al., 1979). Several such marker genes are available. The female parent with a recessive character should be pollinated by using the male parent with the marker gene. It facilitates identification of genuine hybrids in F_1. If the calyx has been removed while preparing the flowers for hybridization, the crossed pods are identifiable at maturity by their lack of sepals. To ensure the synchronized flowering in the late- and early-maturing types, staggered sowing of parental lines is suggested. To hasten the onset of flowering, the plants may be covered with a box made of black plastic to give a 12-hour dark period. The flowering in the early-maturing lines can be prolonged by supplemental exposure to artificial light.

Breeding Problems and Objectives

Many of the agronomic plant characteristics are directly or indirectly dependent on each other and are functions of ultimate objective, i.e., high yield in variable magnitude. Important breeding objectives are described below and sources of desirable characters for use in hybridization programme have been listed in Table 7.

Seed longevity. In soybean, poor seed germinability is the main cause of poor plant stand, limiting its productivity. With reference to relative storability index, soybean seed has been included in the least storable group (Justice and Bass, 1978). Seed germinability and vigour in soybean is ascribable to inherent genetic features of the variety and several factors such as soil type, date of planting and harvest, fertilizer application, use of pesticides, harmful allelopathic effects of weeds, and seed handling (Green and Pinnel, 1968; Yaklich *et al.,* 1979; Mugnisjah and Nakamura, 1984a, b; Tiwari *et al.,* 1985). Cultivars with high seed germinability, such as T 49, Punjab 1, NRC 1, and Kalitur, are available in India but these are beset with defects like low yield, high pod shattering, and late maturity. Recently developed cultivars such as JS 80-21, NRC 2, and JS 335 have high germinability combined with high yield and other desirable agronomic characters.

Effect of seed size on germinability, seed vigour, and yield performance in soybean is well documented (Edwards and Hartwig, 1971; Singh *et al.,* 1972; Paschal and Ellis, 1978; Tiwari and Joshi, 1989). Smaller seeds are known for better and faster germination. Tiwari and Joshi (1989) found that increased hypocotyl length and smaller seed size could be used as reliable parameters in selection for seed longevity and seed vigour. However, due to negative relationship of seed size with impermeable seed and ability to store well, rigid selection in favour of small seeds could lead to impermeable or hard seed (Hill *et al.,* 1986). Speed of germination has also been found to be a criterion of seed vigour in soybean (Pinthus and Kinnel, 1979). A combination of seed vigour tests has also been used to predict field emergence (Yaklich and Kulik, 1979). Screening methods for assessment of seed quality in soybean have been summarized by Kueneman (1982) and Dassou and Kueneman (1984).

Early maturity and photoinsensitivity. For adaption of soybean varieties to a particular region or latitude, proper maturity of the cultivar is an improtant factor. Photosensitivity and also to some extent thermosensitivity of a soybean variety influence its flowering and pod formation at a location. Highly photosensitive varieties will be incapable of producting flowers unless they receive 10 hours or more of darkness each day. Almost all varieties will flower in dark period of 14–16 hours. Period of darkness is influenced by latitude of the location. Varieties requiring longer period of darkness if grown in areas with shorter dark period may not flower or mature. Conversely, varieties suited to shorter period of darkness, if grown in areas with longer nights, will flower early without attaining the needed plant growth and will, thus, yield less.

From the viewpoint of photosensitivity and requirement of dark period, soybean varieties in the United States were classified into maturity groups ranging from 'OO' in the high latitudes to 'X' at low latitudes (Hartwig, 1973). The soybean varieties range from very early-maturing types, generally adapted to regions with long days, to very late-maturing types, adapted to areas with short days and long nights. The early-maturing types are designated 'OOO', 'OO', and 'O' and the late types requiring long nights and short days are progressively classified in higher maturity groups such as VIII, IX and X. Soybean cultivars of maturity group 'OOO' and 'OO' are adapted to the long days at high latitudes having a long critical photoperiod and are photoperiod-insensitive (Criswell and Hume, 1972). The critical photoperiod of cultivars decreases from the high to low latitudes. Thus, the photoperiod requirement limits the latitudes to which a soybean variety is best adapted.

One of the major aims of soybean improvement programme in India is to breed photo-insensitive and early-maturing genotypes. Such types, in addition to their wider adaptability, can be fitted into different cropping patterns and farming systems. Limited progress in this direction has been made so far. Varieties like Bragg and MACS 57 have a desirable degree of photoinsensitivity and are adapted to different regions during rainy and post-rainy or summer/spring season.

Determinate vs. indeterminate plant type. Plant habit of soybean varies from prostrate, viny wild types to the erect commercially grown varieties. Based on plant habit, cultivated varieties of soybean could be classified into indeterminate, determinate, and semi-indeterminate. The indeterminate varieties are generally tall and the growth of terminal bud continues till late, resulting in usnynchronized maturity and lesser number of pods per node but larger number of fruiting nodes and logner internodes. In contrast, determinate varieties are dwarf or semidwarf and stiff-stemmed. The growth of terminal bud ceases after a limited period of growth and flowering. Usually, there are a larger number of pods per node with shorter internodes. Semi-indeterminate plant types are in between the two extermes.

Cooper (1981) and others reported that lodging of indeterminate soybeans results in low yield due to adverse effects on several yield-contributing factors. The stiff-stemmed determinate semidwarf varieties are not only resistant to lodging but are also more responsive to higher fertility and intensive input conditions of cultivation. In the United States, remarkable progress has been made in evolving soybean varieties that can stand erect until harvest, which also helps in mechanized harvesting.

In India, determinate varieties such as PK 564, PK 472, and JS 71-05 and indeterminate or semi-indeterminate varieties such as MACS 58, PK 416, and JS 335 have been bred and released for cultivation in different agroclimatic regions. Semi-determinate habit of varieties such as MACS 58 and JS 335 helped to achieve yield stability and wider adaptability (Bhatnagar and Tiwari, 1989), whereas determinate semidwarf varieties have shown adaptability to a narrow, specific region, as in the case of JS 71-05, which is suitable for cultivation in the Malwa plateau of central India (Bhatnagar *et al.,* 1992a). Under specific and high input conditions, the determinate varieties give high yields.

Mechanical harvesting. During harvesting, even with hand, one or two pods are generally left on stubbles of soybean plant, particularly in varieties where podding starts too close from the ground. This leads to harvesting losses of at least 1 or 2 quintals /ha. Moreover, because of the pod-shattering nature of soybean, considerable yield losses are encountered due to delay in harvesting. Mechanized harvesting can help in reducing such losses. Besides, due to increasing shortage of farm labour, there is a need to develop varieties with enough gap between the first pod and the ground level to enable mechanical harvesting. Varieties such as MACS 58, JS 80-21, and Durga meet this requirement to a fair degree and are suitable for mechanical harvesting.

Suitability for food uses. For acceptance of any item as food, its taste and price are two basic criteria. The beany flavour of soybean is not liked in India. The presence of lipoxygenase is reported to be responsible for the beany flavour in soybean foods. Lipoxygenase catalyses the hydroperoxidation of certain unsaturated fatty acids and various polyunsaturated lipids containing cis cis-1, 4-pentadiene moieties. The action is so strong that oxidation can occur when the enzyme and oil are mixed in the process of crushing of dry seeds. The oxidation products further decompose to middle chain aldehydes and alcohols, which are the major constituents of the undesirable grassy-beany flavour and bitter taste in soybean products. Mutants lacking lipoxygenase-1 (PI 408251), lipoxygenase-2 (PI

86023) and Lipoxygenase-3 (Wase Natsu and I-Higo-Wase) isozymes have been reported (Hildebrand and Hymowitz, 1982; Kitamura *et al.*, 1983, 1985). Trypsin inhibitor proteins, especially the protein SBTI-A_2 (Kunitz trypsin inhibitor), are considered one of the major antinutritional factors present in raw mature soybeans. Mutants lacking in trypsin inhibitor are also available (Hymowitz and Hadley, 1972; Orf and Hymowitz, 1977, 1979), which could enable the extension of the use of soybean in raw and flour forms without the need for heat-inactivation of the inhibitor. With the use of such mutants, qualitative improvement of soybean for food uses could be expected. Bhatnagar and Tiwari (1991b) reported genotypic differences among soybean varieties being grown in India for organoleptic acceptance of soy *paneer* ('tofu'). Varieties Punjab 1 and PK 472 were found to be better suited for making soy *paneer* and other soy products.

Other breeding objectives for soybean, which are described in detail later in this chapter, are resistance to pod-shattering, high yield, resistance to diseases and pests, higher oil and protein content with better quality, as well as resistance to abiotic stresses such as drought, waterlogging, and frost.

Breeding Methods

The breeding methods generally employed for soybean improvement are pure-line selection, pedigree method, bulk or modified bulk method, recurrent selection, backcrossing and single-seed descent.

Pedigree selection. Pedigree selection has been successful in the development of many soybean varieties with improved yield and other desirable characters. Selection generally begins in the F_2 generation and is continued until the selected line breeds true. Pedigree record is maintained through the generations of selection. In the advanced generations, each uniform line can be bulked because selection of individual plants within a line would be of little value. Pedigree selection is effective for those characters where the traits under selection are easily discerned. Boerma and Cooper (1975) reported that pedigree selection procedure in soybean retained more yield variability among lines than either early generation testing or single-seed descent methods. Byron and Orf (1991) found that the pedigree selection procedure in soybean was most expensive because it involves phenotypic evaluation of single plants in the F_2 and of progeny rows in the F_3, F_4, and F_5 generations. It requires more time, land area, and labour than the single-seed descent method.

Bulk method. Bulk method of breeding depends primarily on natural selection to increase the frequency of superior genotypes. Natural selection in bulk populations is considered to be effective for improving quantitative characters. Reports on soybean, however, do not support this conclusion. Torrie (1958) found slight improvement in seed yield after following bulk method in soybean but no impact was observed for either plant height or lodging index. Reaber and Weber (1953) did not find any difference between the bulk and pedigree methods for these characters. Leudders *et al.* (1973) reported that lines selected from populations advanced by the bulk, pedigree, and early generation testing methods showed no difference in seed yield. Degago and Caviness (1987) concluded that bulk method of breeding for soybean is more effective when a constant year-to-year selection pressure for a disease or other stress factors is practised.

Recurrent selection. Recurrent selection has been applied to a limited extent in soybean because of the difficulty in obtaining a large number of crossed seeds. Piper and Fehr (1987) found that several generations of intermating did not enhance genetic gain. However, many workers applied recurrent selection successfully in soybean to increase seed yield (Kenworthy and Brim, 1979; Sumarno and Fehr, 1982), per cent protein and oil in the seed (Brim and Burton, 1979; Burton and Brim, 1981), resistance to iron-deficiency chlorosis in calcareous soils (Prohaska and Fehr, 1981), and manipulation of the proportion of various fatty acids in soybean oil (Carver *et al.*, 1986).

Although most of the hybrid (F_1) seed is produced by hand emasculation, genetic male sterility is sometimes used to produce hybrid seed for recurrent selection programmes. Tinius *et al.* (1991) increased seed size and seed yield through such recurrent mass selection by making use of genetic male sterility.

Backcross method. The backcross method of breeding is used to transfer one or a few genes from one variety to another. In soybean, varieties Chippewa 64, Amsoy 71, Cutler 71, and Hood 75 have been developed by this method by transferring genes controlling specific resistance to diseases and pests such as phytophthora rot, cyst nematode, and root-knot nematode from donor parents to susceptible cultivars.

Other methods. Single-seed descent method, proposed as a modified pedigree method of selection in soybean by Brim (1966), requires the least handling, besides resulting in rapid advance of early generation segregating populations. Empig and Fehr (1971) found that maximum genetic gain per unit time can best be achieved without loss in efficiency with the use of single-seed descent method by growing multiple generations using post-rainy or winter crops or artificial environments. Byron and Orf (1991) observed that the single-seed descent method was the most cost-effective procedure because it required the least resources.

Mixtures or blends of soybean cultivars have been reported to have some yield advantage over their constituent components owing to favourable intergenotypic competition (Schweitzer *et al.*, 1986). However, often the increase is small and difficult to detect or non-significant (Sumarno and Fehr, 1980; Sutton and Weaver, 1989). Beneficial effects of heterogeneity could, however, lead to better yield stability across environments.

Early generation testing has been found to be useful in soybean and several varieties have emanated using this procedure. The reason could be that a large proportion of genetic variance for yield in soybean is additive (Gates *et al.*, 1960; Brim and Cockerham, 1961).

Breeding for High Yield

Yield in soybean, as in any other crop plant, is the most complex polygenic character. Major components of yield in soybean are number of nodes per plant, number of pods per node, number of seeds per pod and seed size. One of the major aims of a breeding programme is to synthesize a variety with most desirable combination of these characters. It may not be possible to combine all the desirable components in the right combination in one genotype. In different varieties different components and their combinations could be responsible for high yield. For example, larger number of pods per plant could be responsible for high yield in some, and larger seed size in others.

During the last three to four decades, plant breeding and other related disciplines have made significant advances in upgrading the yield level of soybean. For further enhancement of the present yield level, plant breeders have to look for existing or induced genetic diversity for yield and related characters.

In India, soybean varietal improvement work, mainly on genetic enhancement of yield, started around 1965. This work got an impetus with the launching of the All India Coordinated Research Project on Soybean (AICRPS) by ICAR in 1967. Since then, a large number of high-yielding varieties have been evolved and released for different agroclimatic regions of the country (Table 8). Bhatnagar and Tiwari (1990) have described these varieties in detail.

Breeding for Pod-shattering Resistance

Under tropical conditions, harvesting losses of yield due to pod shattering on an average could be about 15–20 per cent. Under unfavourable weather conditions soybean yield can be devastatingly eroded due to pod shattering. Pod shattering is a primitive character probably inherited from *Glycine soja*

Table 8
Soyben varicties of India

Variety (1)	Area of adaptability (2)	Maturity duration (days) (3)	Yield (q/ha) (4)
ADT 1	Rice fallows of Tamil Nadu	85–90	20–22
Alankar	Northern plains	115–120	20–25
Ankur	Northern plains	115–120	20–25
Birsa Soy 1	Bihar plateau	106–110	20–25
Bragg	Throughout India	112–115	15–20
Clark 63	Northern and central zone	80–85	15–20
Co 1	Tamil Nadu	85–90	17–20
Davis	Southern zone	110–115	15–20
Durga	Central zone	102–105	20–22
Gaurav	Central zone	104–106	20–25
Guj Soy 1	Low rainfall areas of Gujarat	90–95	15–20
Guj Soy 2	Medium to high rainfall areas of Gujarat	105–110	22–25
Hardee	Karnataka	105–110	22–25
Improved Pelican	Southern zone	112–115	15–20
JS 2	Central zone	100–106	25–30
JS 76-205	Central zone	105–110	20–25
JS 80-21	Central zone	90–109	25–30
JS 71-05	Malwa plateau of MP	90–95	22–25
JS 335	Central zone	98–101	30–35
Kalitur	MP and Bundelkhand region	120–130	18–20
KHSb 2	Karnataka	115–120	20–25
Lee	Northern hill zone	105–115	21–25
MACS 13	Central zone	90–100	25–30
MACS 58	Central zone	90–100	25–35
MACS 124	Southern zone	90–105	25–35
Monetta	Central and southern zone	80–85	20–23
NRC 2	Malwa plateau of MP and northern hill zone	94–102	25–35
Pant Soya 564	Northern plain zone	95–100	30–35
PK 262	Northern hill and northern plain zones	120–125	25–30
PK 308	Northern hill and northern plain zones	110–115	20–25
PK 327	Northern hill and northern plain zones	100–105	20–25
PK 416	Northern hill and northern plain zones	115–120	30–35

Contd.

Table 8 Contd.

(1)	(2)	(3)	(4)
PK 472	Northern plain and central zones	100–105	30–35
PK 471	Southern zone and northern hill zones	90–95	25–30
Punjab 1	Central, northern plain and northern hill zones	95–100	30–35
Pusa 16	Northern plain and northern hill zones	105–115	25–30
Pusa 24	Northern plain and northern hill zones	110–115	25–30
Pusa 20	Northern hill zone	105–120	25–30
Pusa 22	Northern plain zone	105–110	25–30
Pusa 37	Northern plain zone	105–115	25–30
Pusa 40	Southern zone	110–115	25–30
Shilajeet	Northern hill zone and northern plain zone	100–105	20–25
SL 4	Northern plain zone	100–106	16–20
SL 96	Punjab	110–112	18–20
Type 49	Central zone	125–130	20–25
VL Soya 1	UP	110–113	20–25
VL Soya 2	Northern hill zone	104–116	25–30
Shivalik	Himachal Pradesh	120–125	25–30

Sieb & Zucc, the progenitor of *Glycine max* (L) Merrill, which could be of survival value through seed dispersal. The cultivars display a wide range of pod shattering (Tiwari and Bhatnagar, 1988, 1993).

In order to evaluate genotypes for pod shattering, a laboratory screening method based upon oven-drying of detached pods is being followed at the International Institute of Tropical Agriculture (IITA), Ibadan, Nigeria (Dashiell and Bello, 1988). Another laboratory method of Mundel and Mains (1979) relates to the use of measurement of physical force to open pods whereby a force of 1.72 kg was found necessary to open pods of non-shattering variety compared with 0.68 kg for a shattering genotype. A field screening method for quantifying pod shattering was earlier standardized at the National Research Centre for Soybean (NRCS), Indore. Later on, a simplified oven-drying method using detached pods for confirmation of the field-selected shattering-resistant genotypes has been found to be a fairly reliable indicator of pod shattering of varieties in the field.

A number of genotypes showing resistance to pod shattering have been identified under various soybean improvement programmes (Hartwig and Epps, 1982; Tiwari and Bhatnagar, 1993). These have been listed in Table 7. Caviness (1969) and Tiwari and Bhatnagar (1991a) have reported that pod shattering is not beset with any negative associations of significance that could encumber its combination with yield and other desirable economic characters. Susceptibility to pod shattering in soybean is dominant or partly dominant with a preponderance of additive gene action, although dominant gene action also played an important role in the expression of this character (Tsuchiya and Sunada, 1979; Tiwari and Bhatnagar, 1991b). Since the causes of pod shattering are multifactorial, comprehensive measures to minimize such harvesting losses are needed (Tiwari and Bhatnagar, 1989).

Breeding for Resistance to Insect-pests

Sources of genetic resistance to insect-pests of soybean, especially foliar-feeding insects, were identified in the early 1970s (Van Duyn *et al.,* 1971; Clark *et al.,* 1972). Subsequet efforts in identification and transfer of insect resistance resulted in the availability of a large number of donors. Some of these sources possessed multiple insect resistance (Hatchet *et al.,* 1979; Kilen and Lambert, 1986; Beach and Todd, 1988). The progress in developing insect-resistant cultivars with acceptable agronomical characters has, however, been very slow.

Mechanisms of insect resistance in plants have been reviewed by Maxwell and Jennings (1980). Pubescence, i.e., density, length, and orientation of hairs, in soybean appears to be one of the main resistance-imparting factors. It seems to interfere with oviposition, feeding, attachment mechanism, and normal development and behaviour of insects (Hallowell and Johnson, 1934; Turnipseed, 1977; Lee, 1983).

The insect-pest problem on soybean crop in India has reached serious proportions. This is probably due to area expansion as indicated in the 'species-area hypothesis' of Strong (1979). Important insect pests of soybean in India are listed in Table 9. Of all the insect-pests, green semi-looper (*Diachrysia orichalcea*), girdle beetle (*Obereopsis brevis*), stemfly (*Melanagromyza sojae*), and blue beetle (*Cneorane* sp.) are the major insect-pests in the predominant soybean-growing area of central India. Leaf miner (*Bilobata subsecivella*) in the Marathwada region and Bihar hairy caterpillar (*Spilosoma obliqua*) in parts of the northern plains and Terai region are also becoming serious. Breeding for insect-pest resistance is now receiving attention of soybean breeders in India. Two insect-resistant strains, UPIR-1 and UPIR- 2, found promising at Pantnagar in northern India are being used as resistance sources. At Parbhani (Marathwada region of the country), emphasis is being given to resistance breeding for leaf miner (*Bilobata subsecivella*). Resistant donors MACS 6613 and MACS 74 are being used in the breeding programme. At NRCS, Indore, in central India, *G. soja* as well as germplasm line L 592 were found to have no infestation of girdle beetle. Some mutant lines were identified to be resistant to jassids. Ram *et al.* (1989) reported *G. soja* to be resistant to Bihar hairy caterpillar. The resistance appeared to be due to feeding preference rather

Table 9
Major insect-pests of soybean in India

Common name	Scientific name
Seed and seedling feeders	
Blue beetle	*Cneorane* sp.
Field cricket	*Gryllus* sp.
Seed maggot	*Delia platura* (Meigen)
Cut worms	*Agrotis ipsilon* (Huf.)
Stem borers	
Girdle beetle	*Obereopsis brevis* (Swed.); *Nupserha bicolor* (Thomson); *N. nitidior* Brug.
Stem fly	*Melanagromyza sojae* (Zehnt.)
Foliage feeders	
Green semiloopers	*Diachrysia orichalcea* (Fab.) *Chrysodexis actua* (Walk.)
Brown-striped semilooper	*Mocis undata* (Fab.)
Leaf miner	*Bilobata subsecivella*
Tobacco caterpillar	*Spodoptera litura* (Fab.)
Linseed caterpillar	*Spodoptera exigual* (Hub.)
Bihar hairy caterpillar	*Spilosoma obligua* (Walk.)
Gram pod borer	*Heliothis armigera* (Hub.)/ *Helicoverpa armigera*
Grey weevil	*Myliocerus maculosus* (Desb.)
Sap suckers	
Whitefly	*Bemisia tabaci* Genn.
Green jassids	*Emposaca terminalis* Distant
Thrips	*Caliothrips indicus* Bag. *Thrips tabaci* Lin *Ayyaria chaetophora* Karny
Green stink bug	*Nezara viridula* (Linn.)

than antibiosis or avoidance. The species was used in hybridization with the cultigen and a highly tolerant line, PK 515, was developed.

During the last few years of research at AVRDC, Taiwan, four accessions—PI 171444, PI 171451, PI 227687, and PI 229358—were found to be resistant to various defoliators and stink bugs. All of them were not resistant to all insect species but together they covered practically all insect-pests prevalent in Taiwan. The resistance present in three multiple insect resistance donors, i.e., PI 171451, PI 227687, and PI 229358, was reported to be controlled by different genes (Kilen and Lambert, 1986). Breeding advances using the available sources of resistance are needed in order to develop agronomically acceptable insect-resistant varieties.

Breeding for Resistance to Diseases

With the spread of soybean to new regions in India, several diseases have

Table 10
Major diseases of soybean in India

Name of disease	Causal organism
Seedling and stem/root diseases	
Charcoal-rot, ashy or stem blight or dry root rot	*Macrophomina phaseolina* (Tassi) Goid. (*Rhizoctonia bataticola* (Taub.) Butler)
Collar rot or sclerotial blight	*Sclerotium rolfsii* Sacc.
Foliage/pod (seed) diseases	
Bacterial pustule	*Xanthomonas campestris* pv. *Glycines* (E.F. Smith) Dowson var. *sojensis* (Hedges) Starr and Burkh.
Frog eye leaf spot	*Cercospora sojina* Hara
Pod blight or anthracnose	*Colletotrichum dematirum* fsp. *truncatum* (Schw.) Andrus & Morre
Rust	*Phakopsora pachyrhizi* Syd.
Purple seed stain	*Cercospora kikuchii* Mat. and Tom.
Soybean mosaic	Soybean mosaic virus
Yellow mosaic	Mungbean yellow mosaic virus
Rhizoctonia aerial blight	*Rhizoctonia solani* Kuhn

been observed to infect the crop. Major diseases of soybean reported in India are listed in Table 10. The important ones from the viewpoint of occurrence and economic loss are bacterial pustule, pod blight, yellow mosaic virus, rhizoctonia aerial blight, and bud blight. Importance of different diseases varies from region to region.

Hartwig (1973) and Sinclair and Backman (1989) have summarized sources of disease resistance for use in breeding disease-resistant soybean varieties. Resistance sources against important diseases prevalent in India are listed in Table 7.

With rapid expansion of area under soybean crop in India, several diseases previously not known are now threatening its successful cultivation. However, the interdisciplinary approach being followed under AICRP on soybean has helped to evaluate a large number of segregating populations for resistance to different diseases. As a result of these efforts, soybean varieties highly resistant to yellow mosaic virus (PK 416), bacterial pustule (PK 416, PK 472, PK 564, Bragg), pod blight (Bragg, Hardee, PK 472), and rhizoctonia aerial blight (PK 472, PK 262, PK 416) have been developed.

Breeding for Oil and Protein Quantity and Quality

Oil and protein content in soybean are of economic importance. Besides genetic factors, the environment in which a variety is grown also influences oil and protein content in soybean (Morse, 1950b; Fehr *et al.*, 1992). However, genotypes with high oil content always contain relatively higher oil content under different growing environments. Soybean grown under lower temperature conditions contain relatively higher oil percentage in the seeds. Brim *et al.* (1968) and Singh and Hadley (1968) have reported maternal control on the inheritance of oil content in soybean. According to Singh and Hadley (1968) and McKendry *et al.* (1985), additive gene action is important for oil content in soybean seeds. Fehr *et al.* (1968) have suggested seed density and specific gravity along with the use of oil estimation methods such as nuclear magnetic resonance and solvent extraction as reliable parameters for breeding varieties with higher oil content.

Besides percentage of oil, its quality is also important in soybean breeding. There is need to bring down linolenic acid content and enhance desirable polyunsaturated fatty acids. Linolenic acid is responsible for poor keeping quality of oil. Variability for low linolenic acid content, up to about 3 per cent, has been reported in soybean (Hymowitz *et al.*, 1972; Hammond and Fehr, 1983).

Protein content in soybean varieties varies from 35 to 50 per cent (Hartwig, 1979; Bahrenfus and Fehr, 1980). The relationship between oil and protein content in soybean has been generally found to be inverse, although some indications of disrupting this negative association through mutagenesis have been obtained (Bhatnagar *et al.*, 1992b). From the viewpoint of essential amino acids, the quality of soy protein is considered good. While it is rich in lysine, which is markedly deficient in cereals, it is low in sulphur-containing amino acids such as methionine and cystine. Efforts have been made to increase methionine content in soybean protein (Howell *et al.*, 1972). Creation of variability by hybridization or mutagenesis followed by either direct selection for methionine content or identification of genotypes superior for the character on the basis of selection indices such as N : S ratio and 7S/11S ratio in protein could yield rewarding results.

Various breeding methods have been used to evolve soybean varieties with higher protein content. Selection for high protein content should be done judiciously so that it does not lead to low yield and oil content due to interrelationship of the characters under reference. Selection index, stratified selection, restricted index selection, selection based on total protein yield and family selection have been suggested for improving protein content.

Breeding for Drought Resistance

When non-availability of soil water adversely affects plant growth and crop yield, the condition is biologically called drought (Quizenberry, 1982). There could be morphological, physiological, and biochemical effects of drought on plant growth and development (Pendelton and Hartwig, 1973) that ultimately reduce the yield. Characters such as root density, ability to maintain turgor in the tissues, frequent closure of stomata, stability in the activity of nitrate reductase, protein synthesis, proline accumulation, and lesser increase in abscisic acid can be used to identify drought-resistant or tolerant plants (Singh *et al.*, 1973; Sarkar *et al.*, 1991). Genetic variability for some of these characters that directly or indirectly contribute towards drought resistance has been reported in soybean (Sivkumar and Shaw, 1978; Brown *et al.*, 1985; Sarkar *et al.*, 1991). Sivkumar *et al.* (1978) and Sloane *et al.* (1990) reported that the seed yield of PI 416937 was less sensitive to drought than currently grown cultivars. This genotype had larger and thicker leaves and was superior in the ability to maintain leaf turgor, transpiration, and net C-exchange rates under severe drought stress. Sarkar *et al.* (1991) observed that Indian strains VHC 3068, VLS 4, and Himso 107 showed greater stability of grain yield under both stress and non-stress conditions.

Mutation Breeding

Several reviews covering the early mutation work in soybean are available. Irradiation of soybean seeds with X-rays and thermal neutrons and resultant induced variability for maturity, seed weight, and oil and protein content is well documented (Rawlings *et al.*, 1958; Papa *et al.*, 1961). Constantin *et al.* (1976) reported radiosensitivity with reference to the dose-response for efficiency and effectiveness of mutagens in a soybean variety following seed treatment with gamma rays, fission spectrum neutrons, ethyl methane sulphonate, and diethyl sulphate. Conger *et al.* (1976) studied M_3 populations for yield and its components and observed that variability for individual components of yield was enhanced in M_3 populations without a significant change in mean yield. Of special interest are the mutants lacking in lipoxygenase-1 (Hildebrand and Hymowitz, 1982), lipoxygenase-2 (Kitamura *et al.*, 1985), and lipoxygenase-3 (Kitamura *et al.*, 1983) and those lacking in Kunitz trypsin inhibitor (Hymowitz and Hadley, 1972; Orf and Hymowitz, 1977, 1979), which could improve taste and utility of soybean foods. Similarly, increased soybean nodulation and yields are of special significance as reported in supernodulating mutants of 'Bragg' (Carroll *et al.*, 1985) and 'Williams' (Gremaus and Harper, 1989). The mutant of 'Bragg' displayed a dramatic increase of nodule numbers (up to 1000 per plant) when compared with about 50 in the parent. The supernodulating variety Carroll has been developed at the Australian National University. As per the FAO/IAEA database, 41 mutant varieties of soybean have been released, out of which three have emanated from crosses of mutants (Maluszynski *et al.*, 1991).

Induced mutations in soybean varieties in India have been reported for seed coat colour (Raut *et al.*, 1982; Bhatnagar *et al.*, 1990), shattering habit

(Misra *et al.*, 1981), leaf and floral modifications (Singh and Jha, 1978; Singh *et al.*, 1974a), oil and protein content (Bhatnagar *et al.*, 1992b), male sterility (Singh *et al.*, 1974b), improved nodulation and seed yield (Khurana and Lakshminarayana, 1991), and certain other quantitive traits (Upadhyay and Singh, 1979; Dwivedi and Pandey, 1981). Mutation breeding was successful in rectifying the defect of low germinability in the variety 'Bragg' (Bhatnagar *et al.*, 1989) as well as in changing the seed coat colour of high seed longevity cultivar Kalitur from black to yellow, which fetches a higher price (Raut *et al.*, 1982; Patil *et al.*, 1982; Bhatnagar *et al.*, 1990). Some soybean mutants promise commercial acceptability as cultivars (Bhatnagar and Tiwari, 1991c; Bhatnagar *et al.* 1992c).

Future Problems and Prospects

Genetic improvement in soybean has resulted in development of several high-yielding and disease-resistant cultivars. However, there is still ample scope not only to enhance the productivity and quality of the crop but also to insulate it against biotic and abiotic stresses for ensuring stability and high yield.

Genetic incorporation of promiscuity for nodulation as reported from IITA (Pulver *et al.*, 1985; Dashiell *et al.*, 1986) as well as the use of supernodulating mutants of 'Bragg' (Carroll *et al.*, 1985) and 'Williams' (Gremaud and Harper, 1989) could lead to evolution of improved soybeans that could yield up to 25 per cent more than the present levels and yet require less fertilizers.

Introgression of desirable genes from wild relatives, at both intra-subgeneric and inter-subgeneric levels, could provide novel combinations of genes and stretch the range of present variability to ultimately help to evolve efficient soybean varieties suitable for diverse growing conditions.

There is an urgent need in India to develop moderately priced products and processes based on defatted soy flour and to diversfy food uses of soybean. Concerted efforts are required to establish a domestic market for defatted soy flour and other soy products by creating consumer awareness. This will enhance the availability of much needed low-cost, high-quality protein.

REFERENCES

Anand, S.C., and Torrie, J.H. 1963. Heritability of yield and inter-relationships among traits in the F_3 and F_4 generations of three soybean crosses. *Crop Sci.* 3: 508–511.

Anand, S.C., and Rao-Arelli, A.P. 1989. Genetic analyses of soybean genotypes resistant to soybean cyst nematode Race 5. *Crop Sci.* 29: 1181–1184.

Ahmad, Q.M., Britten, E.J., and Byth, D.E. 1983. A quantitative method of karyotype analysis applied to the soybean, *Glycine max* (L.) Merr. *Cytologia* 48: 879–892.

Bahrenfus, J.B., and Fehr, W.R. 1980. Registration of Vinton soybean. *Crop Sci.* 20: 673–674.

Beach, M.R., and Todd, J.W. 1988. Foliage consumption and development parameters of the soybean looper and the velvet bean caterpillar (Lepidoptera: Noctuidae) reared on susceptible and resistant soybean genotypes. *J. Econ. Entomol.* 81: 310–316.

Bernard, R.L. 1972. Two genes effecting stem termination in soybean. *Crop Sci.* 12: 235–239.

Bernard, R.L. 1975a. The inheritance of appressed pubescence. *Soybean Genet. Newslett.* 2: 34–36.

Bernard, R.L. 1975b. The inheritance of semi-sparse pubescence. *Soybean Genet. Newslett. 2: 33–34.*

Bernard, R.L., and Weiss, M.G. 1973. In *Soybeans: Improvement, Production and Uses.* B.E. Caldwell, ed. Agron. Ser. No. 16. Amer. Soc. Agron., Madison, Wisconsin pp. 117–154.

Beversdorf, W.D., and Bingham, E.T. 1977. Male-sterility as a source of haploids and polyploids of *Glycine max. Can. J. Genet. Cytol.* 19: 283–287.

Bhatnagar, P.S., and Tiwari, S.P. 1989. Phenotypic stability analysis of yield in soybean (*Glycine max* (L) Merrill). *Biovigyanam* 15: 90–93.

Bhatnagar, P.S., and Tiwari, S.P. 1990. Soybean varieties of India. Tech Bull. No. 2. NRCS (ICAR), Indore (India), pp. 1–12.

Bhatnagar, P.S., and Tiwari, S.P. 1991a. Soybean in edible oil economy of India. *Agricultural Situation in India.* XLVI(5): 295–301.

Bhatnagar, P.S., and Tiwari, S.P. 1991b. Genotypic differences for organoleptic acceptance of soypaneer ('tofu'). *Biovigyanam* 17(2): 90–93.

Bhatnagar, P.S., and Tiwari, S.P. 1991c. Soybean improvement through mutation breeding in India. In: *Plant Mutation Breeding for Crop Improvement,* Vol. I, Proceedings of an International Symposium. IAEA Publ., Vienna, pp. 381–384.

Bhatnagar, P.S., Prabhakar, Tiwari, S.P., and Sandhu, J.S. 1989. Improvement of soybean variety 'Bragg' through mutagenesis, *Mutation Breeding Newslett.* 33: 15.

Bhatnagar, P.S., Tiwari, S.P., and Prabhakar. 1990. Application of mutagenesis for improvement of indigenous soybean variety of India. *Mutation Breeding Newslett.* 36: 8.

Bhatnagar, P.S., Tiwari, S.P., and Kamra, S.K. 1992a. Stability for grain yield of soybean varieties in the predominant soybean belt of India. *Trop. Sci.* 34(3): 48–50.

Bhatnagar, P.S., Tiwari, S.P., and Singh. C. 1992b. Disrupting the negative association between oil and protein content in soybean seeds through mutagenesis. *Mutation Breeding Newslett.* 39: 7.

Bhatnagar, P.S., Tiwari, S.P., and Prabhakar. 1992c. NRC 2—A promising soybean mutant. *J. Oilseeds Res.* 9(2): 312–313.

Bhatt, G.M., and Torrie, J.H. 1968. Inheritance of pigment colour in soybean. *Crop Sci.* 8: 617–619.

Bhatt, G.M., Sandhu, R.S., and Singh, A. 1968. Study of genetic variability and correlations in soybean. *J. Res. Punjab Agric. Univ.* 5: 1–5.

Biswas, A.K., and Bhattacharyya, H.K. 1972. Induced polyploidy in legumes II. *Glycine max* (L.) Merr. *Cytologia* 37: 605–617.

Boerma, H.R., and Cooper, R.L. 1975. Comparison of three selection procedures for yield in soybean. *Crop Sci.* 15: 225–229.

Brim, C.A. 1966. A modified pedigree method of selection in soybean. *Crop Sci.* 6: 220.

Brim, C.A., and Burton, J.W. 1979. Recurrent selection in soybeans II. Selection for increased per cent protein in seeds. *Crop Sci.* 19: 494–498.

Brim, C.A., and Cockerham, C.C. 1961. Inheritance of quantitative characters in soybean. *Crop Sci.* 1: 187–190.

Brim, C.A., Schutz, W.M., and Collins, F.I. 1968. Maternal effect on fatty acid composition and oil content of soybean, *Glycine max* (L.) Merrill. *Crop Sci.* 8: 517–518.

Brooks, H. 1966. Applied and economic botany. *National Academy of Science, The Plant Sciences Now and in the Coming Decade.* Publication No. 1405. NRC, Washington, DC. pp. 100–114.

Brown, E.A., Caviness, C.E., and Brown, D.A. 1985. Response of selected soybean cultivars to soil moisture deficit. *Agron. J.* 77: 274–278.

Bubeck, D.M., Fehr, W.R., and Hammond, E.G. 1989. Inheritance of palmitic, stearic acid mutants of soybean. *Crop Sci.* 29: 652–656.

Burton, J.W. 1987. Quantitative genetics: Results relevant to soybean breeding. In *Soybeans: Improvement, Production and Uses.* J.R. Wilcox, ed. Agronomy Series No. 16, Madison, Wisconsin, U.S.A. pp. 211–247.

Burton, J.W., and Brim, C.A. 1981. Recurrent selection in soybeans III. Selection for increased per cent oil in seeds. *Crop Sci.* 21: 31–34.

Burton, J.W., and Carter, T.E., Jr. 1983. A method for production of experimental quantities of hybrid soybean seed. *Crop Sci.* 23: 388–390.

Buzell, R.I. 1971. Inheritance of a soybean flowering response to fluorescent day length conditions. *Can. J. Genet. Cytol.* 13: 703–707.

Buzell, R.I., Buttery, B.R., and Bernard, R.L. 1977. Inheritance and linkage of a magenta flower gene in soybeans. *Can J. Genet. Cytol.* 19: 749–751.

Buzell, R.I., and Voldeng, H.D. 1980. Inheritance of sensitivity to day length. *Soybean Genet. Newslett.* 7: 26–29. *Crop Sci.* 29: 652–656.

Byron, D.F., and Orf, J.H. 1991. Comparison of three selection procedures for development of early-maturing soybean lines. *Crop Sci.* 31: 656–660.

Carlson, J.B., and Lersten, N.R. 1987. Reproductive morphology. In *Soybeans: Improvement, Production and Uses.* J.R. Wilcox, ed. Agronomy Series No. 1b. Madison, Wisconsin, U.S.A., 16: 95–134.

Carroll, B.J., McNeil, D.L., and Gresshoff, P.M. 1985. A supernodulaion and nitrate-tolerant symbiotic (nts) soybean mutant. *Plant Physiol.* 78: 34–40.

Carver, B.F., Burton, J.W., Wilson, R.F., and Carter, J.E., Jr. 1986. Cumulative response to various recurrent selection schemes in soybean: Oil quality and correlated agronomic traits. *Crop Sci.* 26: 853–858.

Caviness, C.E. 1969. Heritability of pod dehiscence and its association with some agronomic characters in soybeans. *Crop Sci.* 9: 207–209.

Chang Ruzhen. 1989. Studies on the origin of cultivated soybean (*Glycine max* (L.) Merr.). *Oil Crops of China.* 1: 1–6 (in Chinese with English summary).

Chaudhary, A.N., and Singh, B.B. 1974. Heterosis in soybean. *Indian J. Genet.* 34(1): 69–74.

Chauhan, V.S., and Singh, B.B. 1982. Heterosis and genetic variability in relation to genetic divergence in soybean. *Indian J. Genet.* 42: 324–328.

Chen, L.F.O., Heer, H.E., and Palmer, R.G. 1985. The frequency of polyembryonic seedlings and polyploids from 'msl' soybean. *Theor. Appl. Genet.* 69: 271–277.

Cheng, S.H., and Hadley, H.H. 1983. Studies in polyploidy in soybeans: A simple and effective colchicine technique of chromosome doubling for soybean (*Glycine max* (L.) Merr.) and its wild relatives. *Soybean Genet. Newslett.* 10: 23–24.

Clark, W.J., Harris, F.A., Maxwell, F.G.. and Hartwig, E.E. 1972. Resistance of certain soybean cultivars to bean leaf beetle, striped blister beetle and bollworm. *J. Econ. Entomol.* 65: 1669–1672.

Conger, B.V., Skinner, L.W., and Skold, L.N. 1976. Variability for components of yield induced in soybeans by seed treatment with gamma radiation fission neutrons and ethylmethane sulfonate. *Crop Sci.* 16: 233–236.

Constantin, M.J., Klobe, W.D., and Skold, L.N. 1976. Effects of physical and chemical mutagenesis on survival, growth and seed yield of soybeans. *Crop Sci.* 16: 49–52.

Cocper, R.L. 1981. Development of short-saturated soybean cultivars. *Crop Sci.* 21: 127–131.

Criswell, J.G., and Hume, D.J. 1972. Variation in sensitivity to photoperiod among early maturing soybean strains. *Crop Sci.* 12: 657–660.

Dashiell, K.E., and Bello, L.L. 1988. Screening for resistance to pod-shattering. IITA, Grain Legume Improvement Programme, Annual Report for 1986. Ibadan, Nigeria, p. 120.

Dashiell, K.E., Kueneman, E.A., Root, W.R., and Singh, S.R. 1986. Breeding tropical soybean for superior seed longevity and for nodulation with indigenous rhizobia. In *Soybean in Tropical and Sub-tropical Cropping Systems.* Proc. of a Symp. Tsukuba, Japan, pp. 133–140.

Dassou, S., and Kueneman, E.A. 1984. Screening methodology for identification of soybean varieties resistant to field weathering of seed. *Crop Sci.* 24: 744–779.

Degago, Y., and Caviness, C.E. 1987. Seed yield of soybean bulk populations grown for 10 to 18 years in two environments. *Crop Sci.* 27: 207–210.

Dixit, R.N., and Patil, V.P. 1984. Path analysis studies in soybean. *J. Maharashtra Agric. Univ.* 9(3): 267–269.

Doerschug, E.B., Miksche, J.P., and Palmer, R.G. 1978. DNA content, ribosomal-RNA gene number, and protein content in soybeans. *Can. J. Genet. Cytol.* 20: 531–538.

Dwivedi, A.K., and Pandey, M.P. 1981. Evaluation of induced quantitative mutants in soybean. *Indian J. Agric. Sci.* 51: 715.

Edwards, C.T., and Hartwig, E.E. 1971. Effect of seed size upon rate of germination in soybeans. *Agron. J.* 63: 429–430.

Empig, I.T., and Fehr, W.R. 1971. Evaluation of methods for generation advance in bulk hybrid soybean populations. *Crop Sci.* 11: 51–54.

Erickson, E.A., Wilcox, J.R., and Cavins, J.F. 1988. Inheritance of altered palmitic acid percentage in two soybean mutants. *J. Hered.* 79: 465–468.

Fehr, W.R. 1978. Breeding. In *Soybean Physiology, Agronomy and Utilization.* N.G. Norman, ed. A.P. Inc., New York, pp. 119–155.

Fehr, W.R., Caviness, C.E., Burmood, D.T., and Pennington, J.S. 1971. Stages of development descriptions for soybeans. *Glycine max* (L.) Merr. *Crop Sci.* 11: 929–931.

Fehr, W.R., and Caviness, C.E. 1977. Stages of soybean development. Cooperative Extension Service, Iowa State University, spec. rep. 80, pp. 1–11.

Fehr, W.R., Collins, F.I., and Weber, C.R. 1968. Evaluation of methods for protein and oil determination in soybean seed. *Crop Sci.* 8: 47–49.

Fehr, W.R., Lynk, B.D., and Carlson, G.E. 1985. Performance of semi-determinate and indeterminate soybean genotypes subjected to defoliation. *Crop Sci.* 25: 24–26.

Fehr, W.R., Welke, G.A, Hammond, E.G., Duvick, D.N., and Cianzio, S.R. 1992. Inheritance of reduced linolenic acid content in soybean genotypes A 16 and A 17. *Crop Sci.* 32: 903–906.

Fukuda, U. 1933. Cytogenetical studies on the wild and cultivated Manchurian soybeans (*Glycine* L.). *Jap. J. Bot.* 6: 489–506.

Gai, J.Y., Liu, F.Y., Cui, Z.L., Xia, J.K., and Ma, Y.H. 1991. Inheritance of resistance to the beanfly (*Melanagromyza sojae* Zehntner) in soybean. *Soybean Genet. Newslett.* 18: 104–106.

Gates, C.E., Weber, C.R., and Horner, T.W. 1960. A linkage study of quantitative characters in a soybean cross. *Agron. J.* 52: 45–49.

Gautam, P.L., and Singh, H.B. 1977. Analysis of character association in soybean. *Crop Improv.* 4: 184–190.

Goldberg, R.B. 1978. DNA sequence organization in the soybean plant. *Biochem. Genet.* 16: 45–68.

Gour, V.K., Mehta, A.K., and Mehta, S.K. 1992. A line with four-seeded pods derived from an interspecific hybrid in soybean. *Plant Breeding* 108: 260–262.

Gopani, D.D., and Kabaria, M.M. 1970. Correlation of yield with agronomic characters and their heritability in soybean. *Indian J. Agric. Sci.* 40: 847–853.

Graef, G.L., Fehr, W.R., and Hammond, E.G. 1985. Inheritance of three stearic acid mutants of soybean. *Crop Sci.* 25: 1076–1079.

Green, D.E. and Pinnell, E.F. 1968. Inheritance of soybean seed quality, I. Heritability of laboratory germination and field emergence. *Crop Sci.* 8: 5–11.

Gremaud, M.F., and Harper, J.E. 1989. Selection and initial characterization of partially nitrate tolerant nodulation mutants of soybean. *Plant Physiol.* 89: 169–173.

Halvankar, G.B., and Patil, V.P. 1990. Inter-specific variation in the karyotypes of the genus *Glycine* Wild. *Cytologia* 55: 273–279.

Halvankar, G.B., Raut, V.M., and Patil, V.P. 1990. Genetic constitutions of morphological characters of soybean—a review. *Biovigyanam* 16(1): 38–50.

Hammond, E.G., and Fehr, W.R. 1983. Registration of A_5 germplasm line of soybean. *Crop Sci.* 23: 192.

Harer, P.N., and Deshmukh, R.B. 1991. Components of genetic variation in soybean (*Glycine max* (L.) Merrill). *J. Oilseeds Res.* 8(2): 220–225.

Harer, P.N., and Deshmukh, R.B. 1992. Genetic variability, correlation and path coefficient analysis in soybean. *J. Oilseeds Res.* 9(1): 65–71.

Hartwig, E.E. 1973. Varietal development. In *Soybeans: Improvement, Production and Uses.* B.E. Caldwell, ed. Agron. Ser. No. 16. Amer. Soc. Agron. Madison, Wisconsin, pp. 187–210.

Hartwig, E.E. 1979. Breeding productive soybeans with a higher percentage of protein. In *Seed Protein Improvement in Cereals and Grain Legumes,* Vol. 2. IAEA, Vienna, pp. 59–65.

Hartwig, E.E. 1986. Identification of 4th major gene conferring resistant to soybean rust. *Crop Sci.* 26: 1735–1736.

Hartwig, E.E. and Epps. J.M. 1982. Registration of Nathan soybeans. *Crop Sci.* 22: 1264.

Hartwig, E.E., and Bromfield, K.R. 1983. Relationship among three genes conferring specific resistance to rust in soybean. *Crop Sci.* 23: 237–239.

Hatchet, J.H., Beland, G.H., and Kilen, T.C. 1979. Identification of multiple insect resistance in soybean lines. *Crop Sci.* 19: 557–559.

Hawkins, S.E., Fehr, W.R., and Hammond, E.G. 1983. Resource allocation in breeding for fatty acid composition of soybean oil. *Crop Sci.* 23: 900–904.

Herbert, S.T., and Litchfield, C.V. 1982. Partitioning soybean seed yield components. *Crop Sci.* 22: 1074–1079.

Hildebrand, D.F., and Hymowitz, T. 1982. Inheritance of lipoxygenase-1 activity in soybean seeds. *Crop Sci.* 22: 851–853.

Hill, H.J., West, S.H., and Hinson, K. 1986. Soybean seed size influences expression of the impermeable seed coat trait. *Crop Sci.* 26: 634–637.

Hollowell, E.A., and Johnson, H.W. 1934. Correlation between rough-hairy pubescence in soybean and freedom from injury by *Empoasca fabae. Phytopathology* 24: 12.

Howell, R.W., Brim, C.A., and Rinne, R.W. 1972. The plant geneticists contribution toward changing lipid and amino acid composition of soybean. *J. Amer. Oil Chem. Soc.* 49: 30–32.

Howell, R.W. 1960. Physiology of the soybean. *Adv. Agron.* 12: 265–316.

Hymowitz, T. 1970. On the domestication of the soybean. *Econ. Bot.* 24(4): 408–421.

Hymowitz, T., Palmer, R.G., and Hadley, H.H. 1972. Seed weight, protein, oil and fatty acid relationship within the genus *Glycine. Trop. Agric.* 49: 245–250.

Hymowitz, T., and Hadley, H.H. 1972. Inheritance of a trypsin inhibitor variant in seed protein of soybean. *Crop Sci.* 12: 107–108.

Hymowitz, T. and Singh, R.J. 1992. Biosystématics of the genus *Glycine*. *Soybean Genet. Newslett.* 19: 184–185.

Johnson, H.W., Robinson, H.F. and Comstock. 1955. Genotypic and phenotypic correlations in soybean and their implications in selections. *Agron. J.* 47: 477–483.

Johnson, H.W., and Bernard, R.L. 1962. Soybean genetics and breeding. *Adv. Agron.* 14: 149–227.

Justice, O.L., and Bass, L.N. 1978. *Principles and Practices of Seed Storage.* US Dept. of Agric. Handbook 506.

Kaw, R.N., and Madhav Menon, P. 1973. Heritability estimates of agronomic characters in *Glycine max* (L.) Merrill. *Madras Agric. J.* 60(9/12): 1512–1516.

Kaw, R.N., and Madhav Menon, P. 1983. Diallel analysis in soybean. *Indian J. Genet.* 53(12): 991–997.

Kenworthy, W.J., Brim, C.A., and Wernsman. 1973. Polyembryony in soybeans. *Crop Sci.* 13: 637–639.

Kenworthy, W.J., and Brim, C.A. 1979. Recurrent selection in soybeans I. Seed yield. *Crop Sci.* 19: 315–318.

Khalaf, A.G.M., Brossman, G.D., and Wilcox, J.R. 1984. Use of diverse populations in soybean breeding. *Crop Sci.* 24: 358–360.

Khurana, S.R., and Lakshminarayana, K. 1991. Use of irradiations to improve seed yield and nodulation in soybean variety 'Cobb'. *Indian J. Genet.* 51 (2): 231–234.

Khurana, S.R., and Sandhu, R.S. 1972. Genetic variability and inter-relationship among traits of two soybean populations. *Crop Sci.* 4: 196–198.

Kiihl, R.A.S., and Hartwig, E.E. 1979. Inheritance of reaction to soybean mosaic virus in soybeans. *Crop Sci.* 19: 372–375.

Kikuchi, A., and Kitamura, K. 1985. Genetic elimination of soybean seed lipoxygenase 6. Genetic linkage relationship between L-1 and L-2 loci. *Japan J. Breed.* 35(Suppl. 1): 300–301.

Kilen, T.C. 1986. Relationships between Rps_2 and other genes controlling phytophthora rot in soybean. *Crop Sci.* 26: 711–712.

Kilen, T.C., and Lambert, L. 1986. Evidence for different genes controlling insect resistance in 3 soybean genotypes. *Crop Sci.* 26: 869–871.

Kitamura, K., Davis, C.S., Kaizuma, N., and Nielsen, N.C. 1983. Genetic analysis of a null-allele for lipoxygenase-3 in soybean seeds. *Crop Sci.* 23: 924–927.

Kitamura, K., Kumagai, T., and Kikuchi, A. 1985. Inheritance of lipoxygenase-2 and genetic relationship among for lipoxygenase-1, -2 and -3 isozymes in soybean seeds. *Jap. J. Breed.* 35: 413–420.

Kueneman, E.A. 1982. Genetic differences in soybean seed quality screening methods for cultivar improvements. *Soybean Seed Quality and Stand Establishment.* J.B. Sinclair and J.A. Jackobs, eds. INTSOY Ser. No. 22, Univ. Illinois, Urbana, Illinois, pp. 31–41.

Kueneman, E.A. 1983. Genetic control of seed longevity in soybeans. *Crop Sci.* 23: 5–8.

Kwon, S.H., and Torrie, J.H. 1964. Heritability and interrelationship among traits of two soybean populations. *Crop Sci.* 4: 196–198.

Ladizinksy, G., Newell, C.A., and Hymowitz, T. 1979. Giemsa staining of soybean chromosomes. *J. Hered.* 70: 415–416.

Larsen, A.L., and Caldwell, B.E. 1968. Inheritance of certain proteins in soybean seed. *Crop Sci.* 8: 474–476.

Lee, Y.I. 1983. The potato leafhopper, *Empoasca fabea,* soybean pubescence and hopper-burn resistance. Ph.D. dissertation, Univ. of Illinois, Urbana (Diss. Abstr. Int. 36B: 2057).

Lersten, N.R., and Carlson, J.B. 1987. Vegetative morphology. In *Soybeans: Improvement, Production and Uses.* J.R. Wlicox, ed. Agronomy Series No. 16:, Modison, Wisconsin, U.S.A. pp. 49–94.

Leudders, V.D., Duclos, L.A., and Watson, A.L. 1973. Bulk pedigree and early generation testing breeding methods compared in soybeans. *Crop Sci.* 13: 363–364.

Malhotra, R.S., Singh, K.B., and Dhaliwal, H.S. 1972. Correlation and path-coefficient analysis in soybean (*Glycine max* (L.) Merr.). *Indian J. Agric. Sci.* 42(1): 26–29.

Maluszynski, M., Sigurbjornsson, B., Amano, E., Stich, L., and Kamra, 0. 1991. Mutant varieties-data bank. *Mutation Breeding Newslett.* 38: 16–21.

Manzoor Ahmed, Baluch, A., and Karam Khan Kaleri. 1971. Correlation studies in soybean. *Agric. Pakistan* 22: 155–163.

Maxwell, F.G., and Jennings, P.R. 1980. Breeding Plants Resistant to Insects. John Wiley and Sons, New York.

McBlain, B.A., and Bernard, R.L. 1987. A new gene affecting the time of flowering and maturity in soybean. *J. Hered.* 78: 100–102.

McBlain, B.A., Hesketh, J.D., and Bernard, R.L. 1987. Genetic effects on reproductive phenology in soybean isolines differing in maturity genes. *Can. J. Plant Sci.* 67: 105–116.

McKendry, A.L., McVetty, P.B.E., and Voldeng, H.D. 1985. Inheritance of seed protein and seed oil content in early maturing soybean. *Can. J. Genet. Cytol.* 37: 603–607.

Mehta, S.K., Lal, M.S., and Beohar, A.B.L. 1984. Heterosis in soybean crosses. *Indian J. Agric. Sci.* 54: 682–684.

Misra, R.K., Singh, C.B., Sharma, S.M., and Mehta, S.K. 1981. Note on induced variation in shattering habit of soybean. *Indian J. Agric. Sci.* 51: 678.

Morse, W.J. 1950a. History of soybean production. In: *Soybeans and Soybean Products,* vol. 1. K.S. Markley, ed. Interscience Publ. Inc., New York.

Morse, W.J. 1950b. Chemical composition of soybean seed. Pages 135–156 In *Soybeans and Soybean Products, Vol 1, K.S. Markley, ed. Interscience Publ., New York.*

Mugnisjah, W.Q., and Nakamura, S. 1984a. Vigour of soybean seed produced from different nitrogen and phosphorous fertilizer application. *Seed Sci. Technol.* 12: 475–482.

Mugnisjah, W.Q., and Nakamura, S. 1984b. Vigour of soybean seed produced from different harvest data and phosphorous fertilizer application. *Seed Sci. Technol.* 12: 483–491.

Mukherjee, D., Lambert, J.W., Cooper, R.L., and Kennedy, B.W. 1966. Inheritance of resistance to bacterial blight in soybeans. *Crop Sci.* 6: 324–326.

Mundel, H.H., and Mains, W.H. 1979. A laboratory method for screening varieties for resistance to shattering. Research Highlights, Letwbridge Research Station, Pages Albertson, Canada, pp. 39–40.

Nagata, T. 1959. Studies on the differentiation of soybeans in the world with special regard to that of South-east Asia. *Proc. Crop Sci. Soc.,* Japan 28: 79–82.

Nagata, T. 1960. Studies on the differentiation of soybeans in Japan and the world. *Men. Hyogo Univ. Agric.* 3(2) Ser. 4: 63–102.

Nelson, R.L., and Bernard, R.L. 1984. Production and performance of hybrid soybeans. *Crop Sci.* 24: 549–553.

Orf, J.H., and Hymowitz, T. 1977. Inheritance of a second trypsin inhibitor variant in seed protein of soybeans. *Crop Sci.* 17: 811–813.

Orf, J.H., and Hymowitz, T. 1979. Inheritance of the absence of the Kunitz trypsin inhibitor in seed protein of soybeans. *Crop Sci.* 19: 107–109.

Palmer, R.G., and Heer, H. 1973. A root tip squash technique for soybean chromosomes. *Crop Sci.* 13: 389–391.

Palmer, R.G., and Kilen, T.C. 1987. Qualitative genetics and cytogenetics. In *Soybeans: Improvement, Production and Uses.* J.R. Wilcox, ed. Agronomy Series No. 16, Madison, Wisconsin, U.S.A., pp. 135–209.

Palmer, R.G., and Shoemaker, R.C. 1987. Linkage map of soybean *Glycine max* (L.) Merr In *Genetics Maps.* S.J.O. Brien, ed. Coldspring Harbor, New York, pp. 732–741.

Palmer, R.G., Lim, S.M., and Hedges, B.R. 1992. Testing for linkage between the Rxp locus and nine isozyme loci in soybean. *Crop Sci.* 32: 681–683.

Pandey, J.P., and Torrie, J.H. 1973. Path coefficient analysis of seed yield components in soybeans (*Glycine max* (L.) Merrill.). *Crop Sci.* 13(5): 505–507.

Papa, K.E., Williams, J.H., and Hanway, D.G. 1961. Effectiveness of selection for quantitative characters in third generation following irradiation of soybean seeds with X-rays and thermal neutrons. *Crop Sci.* I: 87–90.

Paschall, E.H. 1976. Crossing soybeans. In *World Soybean Research.* Proc. World Soybean Research Conf., L.D. Hill, ed. The Interstate Printers & Publishers, Inc., Danville, Illinois, pp. 266–267.

Paschall, E.H., and Ellis, M.A. 1978. Variation in seed quality characteristics of tropically grown soybeans. *Crop Sci.* 18: 837–840.

Patil, V.N., and Pokle, Y.S. 1976. Path-coefficient analysis in soybean. *Indian J. Agric. Sci.* 44(8): 544–546.

Patil, V.P., Raut, V.M., and Halvankar, G.B. 1982. Induced mutans for yellow seed and earliness in local soybean variety Kalitur. *J. Maharashtra Agric. Univ.* 1: 97–98.

Pendleton, J.W., and Hartwig, E.E. 1973. Management. In *Soybeans: Improvement, Production and Uses.* B.E. Caldwell, ed. ASA Publ. 16, Madison, Wisconsin, pp. 211–237.

Pillai, R.V.R. 1976. Diploids among the cultivars of soybean (*Glycine max* Linn.) in Manipur. *Sci. Cult.* 42: 519–521.

Pinthus, M.J. and Kinnel, V. 1979. Speed of germination as a criterion of seed vigour in soybean. *Crop Sci.* 19: 291–292.

Piper, T.E., and Fehr, W.R. 1987. Yield improvement in a soybean population by utilising alternative strategies of recurrent selection. *Crop Sci.* 27: 172–178.

Prabhakar and Tiwari, S.P. 1993. Multiple regression analysis for seed yield in soybean. *Soybean Genet. Newslett.* 20: 70–72.

Prohaska, K.R., and Fehr, W.R. 1981. Recurrent selection for resistance to iron deficiency chlorosis in soybean. *Crop Sci.* 21: 524–526.

Pulver, E.L., Kueneman, E.A., and Ranga Rao, V. 1985. Identification of promiscuous nodulating soybean efficient in N_2 fixation. *Crop Sci.* 25: 660–663.

Pushpendra, and Ram, H.H. 1988. Internationalships of quantitative characters with seed yield in F_2, F_3, F_4 and F_5 generations of soybean (*Glycine max*). *Indian J. Agric. Sci.* 58(4): 304–306.

Quizenberry, J.E. 1982. Breeding for drought resistance and plant water use efficiency. In *Breeding Plants for Less Favourable Environment.* M.N. Christiansen and C.F. Lewis, eds. Wiley, New York, pp. 193–212.

Raeber, J.G., and Weber, C.R. 1953. Effectiveness of selection for yield in soybean crosses by bulk and pedigree systems of breeding. *Agron. J.* 45: 362–366.

Ram, H.H., Pushpendra, Kamendra Singh and Ranjit. 1989. *Glycine soja*: a source of resistance for Bihar hairy caterpillar, *Spilosoma* (= *Diacrisia*) *obliqua* Wallace in soybean. *Soybean Genet. Newslett.* 16: 52–53.

Raut, V.M. 1972. Genetic and variability studies in soybean and tetraploid wheats. M.Sc. (Agric.) thesis, Mahatma Phule Agril. Univ., Rahuri (India), pp. 265.

Raut, V.M., Halvankar, G.B., and Patil, V.P. 1982. Induced variability in black seeded soybean variety Kalitur. *Indian J. Genet.* 42: 250–256.

Raut, V.M., Halvankar, G.B., and Patil, V.P. 1988. Heterosis in soybean (*Glycine max* (L.) Merr.) *Soybean Genet. Newslett.* 15: 57–60.

Rawlings, J.O., Hanway, G.B. and Gardener, C.O. 1958. Variation in quantitative characters of soybean after seed irradiation. *Agron. J.* 50: 524–528.

Rennie, B.D., and Tanner, J.W. 1989. Mapping of a second fatty acid locus to soybean linkage group 17. *Crop Sci.* 20: 1081–1083.

Sadanaga, K., and Grindeland, R. 1981. Natural cross-pollination in diploid and autotetraploid soybeans. *Crop Sci.* 21: 503–506.

Sapra, V.T., and Stewart, M.D. 1981. Leaves as a source of somatic chromosomes for cytogenetic studies of soybean. *Soybean Genet. Newslett.* 8: 22–23.

Sarkar, R.K., Saini, J.P., and Dube, S.D. 1991. Testing of soybean genotypes for drought tolerance. *Indian J. Agric. Sci.* 61(6): 369–373.

Schwetzer, L.E., Nyquist, W.E., Santini, J.B., and Kimer, T.M. 1986. Soybean cultivar mixtures in a narrow-row, noncultivable production system. *Crop Sci.* 26: 1043–1046.

Sen, N.K., and Vidyabhushan, R.V. 1960. Tetraploid soybeans *Euphytica* 9: 317–322.

Sengupta, K. 1975. A new gene in soybean (*Glycine max* (L.) Merr.) conditioning hypocotyl pigmentation. *Curr. Sci.* 44: 200–201.

Sharma, S.K. 1979. Note on path-coefficient analysis in F_2 populations on soybean grown at two locations. *Indian J. Agric. Sci.* 49(10): 2820–2821.

Sharma, K.P., Dybing, C.D., and Lay, C. 1990. Soybean flower abortion: genetics and impact of selection on seed yield. *Crop Sci.* 30: 1017–1072.

Shettar, B.I., Viswanatha, S.R., and Shivashankar, G. 1978. Correlation studies in soybean *Glycine max* (L.) Merrill. *Curr. Res. Univ. Agri. Sci.* Bangalore 7(10): 170–172.

Shoemaker, R.C., Health, M.S., Skorupksa, H., Delannay, X., Edge, M., and Newell, C.A. 1990. Fertile progeny of a hybridization between soybean (*Glycine max* (L.) Merr.) and *G. tomentella* Hayata. *Theor. Appl. Genet.* 80(1): 17–23.

Sinclair, J.B., and Backman, P.A. 1989. *Compendium of Soybean Diseases,* 3 edn. APS Press, The American Phytopathological Society, Minnesota, pp. 1–106.

Singh, B.B., and Hadely, H.H. 1968. Maternal control of oil synthesis in soybeans *Glycine max* (L.) Merr. *Crop Sci.* 8: 622–625.

Singh, B.B., and Jha, A.N. 1978. Abnormal differentiation of floral parts in a mutant strain of soybean. *J. Hered.* 69: 143–144.

Singh, B.B., and Malik, A.S. 1978. Inheritance of resistance to yellow mosaic virus in soybean. *Indian J. Genet.* 38: 258–261.

Singh, B.B., Gupta, S.C., and Singh, B.D. 1974a. An induced crinked leaf mutant in soybean. *Soybean Genet. Newslett.* 1: 16–17.

Singh, B.B., Gupta, S.C., and Singh, B.D. 1974b. Induced male sterility mutant in soybean. *Soybean Genet. Newslett.* 1: 19–20.

Singh, J.N., Tripathi, S.K., and Negi, P.S. 1972. Note on the effect of seed size on germination growth and yield of soybean. *Indian J. Agric. Sci.* 42: 83–86.

Singh, K., and Ram, H.H. 1983. Dry matter yield and branching as selection parameters in soybean. *Soybean Genet. Newslett.* 10: 13–15.

Singh, T.N., Aspinall, D., Paleg, L.G., and Bergess, S.F. 1973. Stress metabolism II. Changes in proline concentration in excised plant tissue. *Aust. J. Biol. Sci.* 26: 45–46.

Singh, R.J., Kollipara, K.P., and Hymowitz, T. 1988. Further data on the genomic relationships among wild perennial species (2n = 40) of the genus *Glycine* Willd. *Genome* 30: 166–176.

Singh, R.J., Kollipara, K.P., and Hymowitz, T. 1990. Backcross derived progeny from soybean and *Glycine tomentella* Hayata intersubgeneric hybrids. *Crop Sci.* 30: 871–874.

Singh, T.P., Singh, K.B., and Brar, J.S. 1974. Diallel analysis in soybean. *Indian J. Genet.* 34(3): 427–432.

Sivkumar, R.J., Patterson, R.P., and Carter, T.E., Jr. 1990. Field drought tolerance of a soybean plant introduction. *Crop. Sci.* 30: 118–123.

Sivkumar, M.V.K., and Shaw, R.H. 1978. Relative evaluation of water stress indicators for soybeans. *Agron. J.* 70: 619–623.

Sloane, R.J., Patterson, R.P., and Carter, T.E., Jr. 1990. Field drought tolerance of a soybean plant introduction. *Crop Sci.* 30: 118–123.

Srivastava, R.L., Sahai, R.N., Saxena, J.K., and Singh, I.P. 1976. Path analysis of yield components in soybean (*Glycine max* (L.) Merrill). *Indian J. Agric. Res.* 10(3): 171–173.

Strong, D.R., Jr. 1979. Biogeographic dynamics of insect-host communities. *Ann. Rev. Entomol.* 24: 89–119.

Sumarno and Fehr, W.R. 1980. Intergenotypic competition between determinate and indeterminate soybean cultivars in blends and alternate rows. *Crop Sci.* 20: 251–254.

Sumarno and Fehr, W.R. 1982. Response to recurrent selection for yield in soybeans. *Crop Sci.* 22: 295–299.

Sutton, J.D., and Weaver, D.B. 1989. Intergenotypic competition between late planted determinate and indeterminate soybean. *Crop Sci.* 29: 1506–1510.

Talwar, M.L., and Singh, S.P. 1983. Genetic analysis of yield and its components is soybean. *Crop Improv.* 10(2): 111–114.

Taware, S.P., Halvankar, G.B., Raut, V.M., and Patil, V.P. 1990. Hybrid vigour in soybean (*Glycine max*). *Indian J. Agric. Sci.* 60(8): 545–546.

Tindale, M.D. 1984. Two new eastern Australian species of *Glycine* Willd (Fabaceae). *Brunonia* 7: 207–213.

Tindale, M.D., and Craven, L.A. 1988. Three new species pf *Glycine* Willd (Fabaceae: Phaseolae) from north-western Australia with notes on amphicary in the genus.

Tinius, C.N., Burton, J.W., and Carter, T.E., Jr. 1991. Recurrent selection for seed size in soybean I. Response to selection in replicate populations. *Crop Sci.* 31: 1137–1141.

Tiwari, S.P., and Bhatnagar, P.S. 1988. Pod-shattering of soybean in India. *J. Oilseeds Res.* 5: 92–93.

Tiwari, S.P., and Bhatnagar, P.S. 1989. Minimising pod-shattering in soybean. *Indian Fmg.* 39: 23–24.

Tiwari, S.P., and Bhatnagar, P.S. 1991a. Pod-shattering as related to other agronomic attributes in soybean. *Trop. Agric.* 68(1): 102–103.

Tiwari, S.P., and Bhatnagar, P.S. 1991b. Genetics of pod shattering in soybean. *Soybean Genet. Newslett.* 18: 150–153.

Tiwari, S.P., and Bhatnagar, P.S. 1993. Consistent resistance for pod-shattering in soybean varieties. *Indian J. Agric. Sci.* 63(3): 173–174.

Tiwari, S.P., and Joshi, H.J. 1989. Correlation and path analysis of seed quality characters in soybean. *J. Oilseeds Res.* 6: 51–57.

Tiwari, S.P., Bhaduria, S.S., Vaddoria, M.A., and Devi Dayal. 1985. Allelopathic effects of weeds in soybean, groundnut and green gram. *Curr. Sci.* 54: 434–435.

Torrie, J.H. 1958. A comparison of the pedigree and bulk methods of breeding soybeans. *Agron. J.* 50: 198–200.

Turnipseed, S.G. 1977. Influence of trichome variations on populations of small phytophagous insects in soybeans. *Environ. Entomol.* 6: 815–817.

Tsuchiya, T., and Sunada, K. 1979. Breeding studies on pod shattering in soybean III. Degree of shattering in the F_1 and F_2. *Bull. of Hokkaido Perfectural Agric. Experiment Stations.* 41: 1–9 (In Japanese). (Plant Breeding Abstracts, 1980, Vol. 50, No. II: 9789).

Upadhyay, H.D., and Singh, B.B. 1979. Induced variability in soybean. *Indian J. Genet.* 39: 207–209.

Van Duyn, J.W., Turnip, S.G., and Maxwell, J.D. 1971. Resistance in soybeans to the Mexican bean beetle I. Source of resistance. *Crop Sci.* 11: 572–573.

Van Schaik, P.H., and Probst, A.H. 1958. The inheritance of inflorescence type, peduncle length, flowers per node and per cent flower shedding in soybean. *Agron. J.* 50: 98–102.

Vello, N.A., Fehr, W.R., and Bahrenfus, J.B. 1984. Genetic variability and agronomic performance of soybean populations developed from plant introductions. *Crop Sci.* 24: 511–514.

Verma, V.D., and Ram, H.H. 1987. Genetics of seed longevity in soybean. *Crop Improv.* 14: 40–42.

Walker, A.K., Cianzio, S.R., Bravo, J.A., and Fehr, W.R. 1979. Comparison of emasculation and nonemasculation for hybridization of soybeans. *Crop Sci.* 19: 285–286.

Watanabe, T., and Kishi, A. 1984. The book of soybeans. Japan Publ. Inc., pp. 1–182.

Weatherspoon, J.H., and Wentz, J.B. 1934. A statistical analysis of yield factors in soybeans. *J. Amer. Soc. Agron.* 26: 524–531.

Weber, C.R., and Moorthy, B.R. 1952. Heritable and nonheritable relationship and variability of oil content and agronomic characters in the F_2 segregation of soybean crosses. *Agron. J.* 44: 202–209.

Weber, C.R., Empig, L.T., and Thorne, J.C. 1970. Heterosis performance and combining ability of two way F_1 soybean hybrids, *Crop Sci.* 10: 159–160.

Werner, B.K., Wilcox, J.R., and Houslex, T.L. 1987. Inheritance of an ethyl methane sulfonate-induced dwarf in soybean and analysis of leaf cell size. *Crop Sci.* 97: 665–668.

Wilcox, J.R., and Cavins, J.F. 1985. Inheritance of low linolenic acid content of the seed oil of a mutant in *Glycine max* (L.) Merr. *Theor. Appl. Genet.* 71: 74–78.

Willmot, D.B., and Nickell, C.D. 1989. Genetic analysis of brown stem rot resistance in soybean. *Crop Sci.* 29: 672–674.

Yamaoto, K., and Nagato, Y. 1984. Variation of DNA content in the genus *Glycine*. *Jap. J. Breed.* 34: 163–170.

Yap, T.C., and Lee, L.F. 1975. Performance and path-coefficient analysis of soybeans under local conditions. *Malaysian Agric. Res.* 4(2): 97–102.

Yaklich, R.W., and Kukik, M.M. 1979. Evaluation of vigour tests in soybean seeds: Relationships of the standard germination test. Vigour classification, seedling length and tetrazolium staining to field performance. *Crop Sci.* 19: 247–252.

Yaklich, R.W., Kukik, M.M., and Garrison, C.S. 1979. Evaluation of vigour in soybean seeds: Influence of date of planting and soil type on emergence, stand and yield. *Crop Sci.* 19: 242–246.

Yee. C.C., and Palmer, R.G. 1984. Summary of locus-to-locus linkage data in soybean. *Soybean Genet. Newslett.* 11: 105–126.

Zheng, H.Y., Chen, R.Y., and Sun, H. 1984. A diploid strain of wild soybean (*Glycine soja*) with four satellited chromosomes. In *World Soybean Research Conf. III*. Abst. 438. R. Shibles, ed. Westview Press, Boulder, Colorado, pp. 96.

6. GROUNDNUT

M.V.R. Prasad

Former, Project-Director, Directorate of Oilseeds Research (ICAR), Rajendra Nagar, Hyderabad, A.P., India

Introduction

Groundnut or peanut (*Arachis hypogaea*) is an annual legume rich in oil and protein and native to South America. The early Spanish and Portuguese explorers found Native Americans cultivating groundnut in the West Indies, Mexico, the northeast coast of Brazil, in all the warm land of Rio de la Plateau basin (Argentina, Paraguay, Bolivia, and southwestern Brazil), and extensively in Peru. From these regions groundnut dispersed to both Africa and Asia and to the Pacific Islands. Peanut plants were carried to Africa and China by the early explorers and missionaries. Groundnut was introduced into India in the 16th century by Portuguese and Spanish explorers through the West Coast. In a period of about 200 years, the acreage under the crop expanded rapidly in more or less contiguous ecological niches to almost 8 million ha in the country. The world's harvest area of groundnut is of the order of 19.4 million ha with a total production of 22.1 million t resulting in productivity of 1.1 t/ha.

Groundnut is one of the principal economic crops of the world, ranking 13th among the food crops (Varnell and McCloud, 1975). It is presently cultivated in over 80 countries from 40°N to 40°S in tropical and warm temperature regions of the world.

The edible oil yielded by groundnut is an important source of fat in the human diet, especially in Asia and Africa. The meal is used for livestock feed. Groundnut is also used directly as food in primary form in Europe, the United States, and Canada. There has been a major increase in food use of groundnut to the tune of 35 per cent in 1980 over the 1970s in the United States.

The groundnut-growing areas in the semi-arid tropics have a short growing season (75–100 days) and are characterized by intermittent drought periods. The studies conducted by Virmani and Piara Singh (1986) showed that the amount of soil moisture in the surface soils is fairly restricted at the time of seed formation and maturity, thereby leading to pod development and harvesting problems. Therefore, the yields are likely to be significantly

Table 1
Area, production and productivity of groundnut in India (1980–81 to 1992–93)

Year	Area ('000 ha)	Production ('000 t)	Yield (kg/ha)
1980–81	6801	5004	736
1981–82	7429	7223	972
1982–83	7215	5282	732
1983–84	7538	7085	940
1984–85	7168	6435	898
1985–86	7124	5121	719
1986–87	6982	5875	841
1987–88	6844	5853	855
1988–89	8528	9658	1132
1989–90	8710	8100	930
1990–91	8309	7514	904
1991–92	8667	7094	818
1992–93	8351	8854	1060

reduced once in every three years due to the failure of seasonal rainfall in the semi-arid tropics. On the other hand, in Southern Africa, where groundnut is grown without any limitations of moisture and nutrients, productivity levels of the order of 9 to 10 t/ha are reported (Smartt *et al.*, 1978b).

A large part of acreage under groundnut in India is in the semi-arid belt comprising Telangana and Rayalaseema regions of Andhra Pradesh and Saurashtra area of Gujarat. The present groundnut area and production in India reflects about 23 and 77 per cent increase, respectively, over acreage and production of 1980–81 (Table 1). South Asia, comprising Bangladesh, Bhutan, India, Malaysia, Nepal, Pakistan, and Sri Lanka accounts for about 43.4 per cent of the world's groundnut area and 35.7 per cent of production. In these countries, the sowing period coincides with the southwest monsoon, which is major growing season. Although the crop is grown in more than one season in India and Sri Lanka, the low average yields of groundnut in this region are due to cultivation of the crop mostly in unirrigated situations on marginal and submarginal lands with low quantum of inputs, use of long-duration varieties, susceptibility of the crop to a plethora of insect pests and diseases, non-availability of efficient farm machinery, and lack of quality seed.

Early Explorations and the Origin of Cultivated Groundnut

Among the various explorers who described groundnut, the significant ones are Bartolome las Casas in 1502, Gonzalo Fernandez de Oviedeo in 1513, Gean de Lery in 1555, Nicholos Monardes in 1574, Portuguese nationalist Gabriel Soares de Suza in 1570, and George Marc Grave in 1648 (Hammons, 1973a). The groundnut plants were found in the prehistoric cemetery of Anion on Peruvian coast near Lima. Ethnobotanical investigations were carried out on the samples of groundnut obtained in Cox-Catlan cave in Tehucaucan Valley in Mexico. Based on the carbon dating, the investigations revealed that the groundnut was cultivated as early as 1200–1500 BC.

The central core of diversity of *Arachis* lies in the Mottogrosso region of Brazil (Krapovickas, 1968). According to Krapovickas (1968), the cultivated groundnut has originated from southern Bolivia and southwestern Argentina. *Arachis hypogaea* was found to exhibit much similarity to *A. villosa*.

As per the available data, the cultivated groundnut originated at the foothills of Andes in southeastern Bolivia, while the genus *Arachis* has originated in the Mottogrosso region of Brazil, which is the main centre of its diversity. The Planalto profile from Corumba to Joazeiro in Brazil is the centre of the area from which the present distribution of genus *Arachis* arose (Krapovickas, 1968; Hammons, 1973a).

Speciation and Cytogenetics

The genus *Arachis* is ascribed to tribe *Aeschynumeneae* and subtribe *Stylosanthinae* on the basis of shared morphological characters of staminal tube with alternately attached basal and dorsal anthers, with flowers on terminal or axillary spikes, pinnate leaves, and leaflets without stipules. The cultivated groundnut belongs to the section *Arachis*. The section is further subdivided into three series. The series *annuae* consists of the species *A. batizocoi, A. duranensis,* and *A. stenosperma.* The series *perennes* consists of *A. helodes, A. villosa, A. diogoi, A. cardenasii* and *A. chachoense.* The third series *amphidiploide* includes two tetraploids —*A. hypogaea* and *A. monticola.*

The species of genus *Arachis* are distributed over a wide range of environments from the south of the Amazon to 34°S latitude and the Atlantic shore to the eastern groups of Andes. The differentiation of these species has generally followed drainage basins and riverbeds. However, the greatest diversity of other species complex is found in the headwaters of Paraguay River (Stalker, 1980). Although 22 species have been described, it is expected that 40–60 others are found in nature. Gregory *et al.* (1973) grouped the species into seven botanical sections. The cultivated species *A. hypogaea* belongs to the section *Arachis* along with another tetraploid species, *A. monticola,* and 10 other diploid species.

Badami (1928) reported somatic chromosome number of $2n = 20$ for the cultivated peanut *A. hypogaea*. Other investigators later confirmed the

Table 2
Species of *Arachis* and corresponding chromosome number

Section/species	Chromosome number	Reference
Arachis		
A. batizocoi Krap. et Greg	20	Smartt and Gregory (1967)
A. pusilla	20	Krapovickas and Rigoni (1957)
A. duranensis) Krap. et Greg. *nom. nud.*		
A. spegazzinii Greg. et Greg. *nom. nud*	20	Gregory and Gregory (1979)
A. stenosperma Greg. et Greg. *nom. nud*	20	Gregory and Gregory (1979)
A. ipaensis Greg. et Greg. *nom. nud*	20	Gregory and Gregory (1979)
A. helodes Martius ex Krap. et Rig	20	Smartt and Gregory (1967)
A. villosa Benth	20	Krapovickas and Rigoni (1949)
A. diogoi Hoehne	20	Mendes (1947)
A. cardenasii Krap. et Gret. *nom. nud.* (10017 GKP)	20	Smartt and Gregory (1967)
A. chacoense Krap. et Greg. *nom. nud* (10602)	20	Smartt and Gregory (1967)
A. hypogaea L.	40	Kawakami (1930)
A. monticola Krap. et Rig.	40	Krapovickas and Rigoni (1957)
Erectoides		
A. guaranitica Chod. et Hassl.	20	
A. tuberosa Benth	20	
A. benthami Handro	20	Smartt and Gregory (1967)
A. martii Handro	20	
A. paraguariensis Chod. et Hassl	20	Smartt and Gregory (1967)
A. oteroi Krap. et Greg. *nom. nud*	20	Smartt and Gregory (1967)
A. rigonii Krap. et Greg	20	Krapovickas and Rigoni (1960)
A. lignosa Chod. et Hassl. (10598)		
Caulorhizae	20	Smartt and Gregory (1967)
A. repens Handro	20	Conagin (1963)
A. pintoi Krap. et Greg. *nom. nud*	20	
Rhizomatosae		
A. burkartii Handro	20	Gregory and Gregory (1979)
A. glabrata Benth	40	Gregory (1946)
A. hagenbeckii Harms	40	Krapovickas and Rigoni (1957)
Extranervosae		
A. marginata Gard	20	Mendes (1947)
A. letescens Krap. et Rig	20	Conagin (1963)
A. villosulicarpa Hoehne	20	Mendes (1947)
A. macedoi Krap. et Greg. (10127)	20	Smartt and Gregory (1967)
A. prostrata Benth	20	Mendes (1947)
Ambinervosae		
None named	20	
Triseminalae		
A. pusilla Benth	20	

Note: *nom. nud.* = species which are not validly described in the botanical gazette.

true number as 2n = 40 (Kawakami, 1930; Husted, 1931). The wild species such as *A. glabrata* was reported to possess 2n = 40 (Gregory, 1946; Mendes, 1947). The chromosome numbers of a wide range of *Arachis* species (Stalker and Dalmacio, 1981) are given in Table 2.

The analysis of the section *Arachis* reveals two genomes with enough variation to identify 10 homologues. The A genome is attributed to most of the species of the section, while B genome is present only in *A. batizocoi*. At least five species can be cytologically identified in the section *Arachis*. Three more cytologically similar species (*A. correntina, A. spegazzinnii,* and *A. villosa*) can also be distinguished. In the section *Arachis, A. batizocoi* hybrids have irregular meiosis and are sterile. Semi-sterility is common in most interspecific hybrids even though meiosis is regular.

The section *Erectoides* consists of plants with taproot system possessing sometimes tuberous roots, tuberiform hypocotyl, horizontal pegs, and medium-large flowers of 16–24 mm. Of the two series included in this section the series *trifoliate* with species *A. guaranatatica* and *A. tuberosa* possesses tuberiform hypocotyl and trifoliate leaves. The second series *tetrafoliate* is characterized with erect or prostrate plants with non-tuberiform hypocotyl, tetrafoliate leaves, and standard orange petals. The species *A. benthami, A. martii,* and *A. paraguariensis* belong to this series.

The series *Rhizomatosae* consists of plants with rhizomatous roots and solid stems in addition to large flowers. The first series of this secction *prorhyzomatosae* represented by *A. burkartii* exhibits delicate plants with large flowers and red veins on both sides of standard petal. The second series, namely *Eurhizomatosae* has robust plants with large flowers. This is a tetraploid group (2n = 4x = 40) consisting of *A. glabrata* and *A. hagenbeckii*.

The section *Extranervosae* consists of basically diploid material (2n = 2x = 20) with thickened lomentiform tuberoid roots, vertical gynophores that sometimes produce adventitious roots. The flowers of this section are small and medium-sized with red veins on the back. The species known to occur in this section are *A. marginata, A. villosulicarpa, A. macedoi, A. prostrata,* and *A. lutescence*. Members of this section exhibit strong crossability barriers with cultivated groundnut.

The section *Ambinervosae* consists of individuals with strong taproot, vertical pegs, and small flowers (8 × 6 mm). The standard petal of the flower has conspicuous red veins on front and back. The members of this section are known to be basically diploids (2n = 2x = 20).

The members of the section *Triaseminale* are also basically diploid (2n = 2x = 20) exhibiting strong crossability barriers with *A. hypogaea*. The plants of the section have a pronounced taproot system and horizontal pegs. The flowers are small (10–12 mm × 8–10 mm) and have a purple mark inside the orange-coloured standard petal. The fruits are trisegmented.

Arachis pusilla is an example of the section.

Kawakami (1930) identified the chromosome number of *A. hypogaea* to be 2n = 4x = 40, while Husted (1933, 1936) reported two distinctive chromosome pairs; one has been termed A chromosomes, which are distinctly similar compared to B chromosomes, which have secondary constriction. All the species of the section *Arachis* have a pair of A chromosomes. However, this pair is absent in the section *Erectoides. Arachis batizocoi*, however, has no identifiable A chromosome pair (Smartt *et al.,* 1978a). The chromosome complement of *A. batizocoi* differs largely from that of the other species of the section *Arachis.* It was proposed earlier (Smartt *et al.,* 1978a) that the cultivated groundnut must be amphidiploid of *A. batizocoi* and *A. cardenasii.* The first interspecific hybrid of diploid species of *Arachis* was produced by Raman and Kesavan (1962) between *A. duranensis* and *A. villosa* var. *correntina.* The hybrid exhibited a regular meiosis.

Intersectional Hybrids

The meiotic behaviour of the F_1 hybrids involving *A. batizocoi* and other members of the section *Arachis* is irregular, leading to sterility due to extensive rearangement of structural elements. The hybrids resulting from the crosses of the section *Erectoides* and *Arachis* were all sterile. Stalker (1978) suggested that the most intersectional hybrids have arisen from the combination of *Erectoides* × *Rhizomatosae.* Very few of the intersectional hybrids are from the crosses involving the *Erectoides, Arachis,* and *Caulorhizae.* The pattern of intersectional cross-compatibility involving the sections *Arachis* and *Erectoides* has some affinity with tetraploid *Rhizomatosae* (Gregory and Gregory, 1979). It is possible that since sections *Arachis* and *Erectoides* are almost competely cross-incompatible, one of the two *Rhizomatosae* genomes confers compatibility with *Erectoides* and the other with species of section *Arachis.* In the section *Arachis,* only members of the series *Annuae* have demonstrated intersectional cross-compatibility. Neither perennials nor tetraploids of this section have produced intersecctional hybrids. The presumed presence of the A genome from the perennial species (Smartt *et al.,* 1978b) may explain the lack of cross-compatibility between *A. hypogaea* and any other section (Gregory and Gregory, 1979).

A considerable extent of diversity exists within the section of *Arachis.* The progenitor species of *A. hypogaea* are yet to be identified. The investigations of Stalker (1980), and Gregory and Gregory (1979) revealed that the section *Arachis* has two genomes. The members with either the A genome (*A. duranensis*) or the B genome (*A. batizocoi*) would hybridize with the section *Rhizomatosae.* However, the other members with A genome, *A. villosa, A. correntina, A. cardenasii,* and *A. chacoense* would not hybridize with the *Rhizomatosae*.

The present indications are that the cultivated groundnut containing A and B genomes is an allotetraploid. Six A genome species and one B genome species have been hypothesized as possible progenitors of *A. hypogaea* (Stalker and Moss, 1987). However, several studies including DNA restriction fragment length polymorphisms (RFLPs) (Kochert *et al.*, 1991), DNA polymerase chain reactions (PCRs) (Halward *et al.*, 1991), isozymes (Stalker, 1991), and seed storage proteins have indicated that *A. batizocoi* is not the donor of the B genome of the cultivated species. A similar observation from chloroplast DNA analyses indicated that *A. batizocoi* cannot be the female parent of *A. hypogaea* (Stalker, 1991). The conclusions drawn from the above molecular studies are that (1) *A. hypogaea* had an allopolyploid origin, (2) a large amount of genomic differentiation between the diploid A and B genomes has occurred as was previously reported (Stalker and Dalmacio, 1981; Singh and Moss, 1982), and (3) definitive identification of the progenitors of *A. hypogaea* has not been completed, although *A. duranensis* is a likely candidate for the female parent. Additionally, the evidence suggests that none of the currently known *A. batizocoi* accessions were involved in the ancestry of *A. hypogaea*. At the best, *A. batizocoi* could have contributed one of the two smallest chromosomes in the *A. hypogaea* genome with high frequency of translocations observed in *A. batizocoi* accessions (Stalker *et al.*, 1991). It was implied that a mechanism for sufficient karyotypic differentiation of a second A genome species could account for the differences in chromosome morphology in *A. hypogaea*. The wild species of *Arachis* represent a valuable genetic repository for several attributes useful in groundnut breeding (Stalker and Dalmacio, 1981).

Although fertilization may be restricted in wide crosses (Sastri and Moss, 1982), it is believed to have limited importance for most crosses. However, the application of mentor pollen or gibberellic acid can increase the percentage of pegging (Sastri and Moss, 1982; Stalker *et al.*, 1987). Lu *et al.* (1990) also showed that several perennial *Arachis* species have smaller stigmas that are surrounded by a protective ring of hairs, as opposed to annuals, which have larger stigmas and lack the ring of hairs. Pegs commonly fail to reach the soil after fertilization (Johansen and Smith, 1956; Sastri and Moss, 1982; Halward and Stalker, 1987), and embryo abortion often occurs within 10 days after pollination. Pattee and Stalker (1991) reported reciprocal differences in *A. hypogaea* crosses where *A. stenosperma* Greg et Greg. *nom. nud.* females had many unfertilized eggs while other crosses resulted in early embryo abortion. Johansen and Smith (1956) observed in other crosses that abortion was due to retarded embryo and endosperm growth and subsequent death occurs prior to embryonic differentiation. Halward and Stalker (1987) and Ozias-Akins and Branch (1990) reported partial seed maturation and subsequent degeneration of interspecific hybrid embryos.

To circumvent the above-mentioned barriers, growth regulators have been applied to reproductive tissues to induce pegging and/or embryo growth with some success (Sastri and Moss, 1982; Mallikarjuna and Sastri, 1985a). To rescue embryos before abortion, cultures of 1- to 4-day-old embryos with accompanying peg tissues have resulted in embryo growth, but not differentiation into heart-shaped embryos (Moss *et al.,* 1988a; Pattee *et al.,* 1988). Johnson (1981) concluded that *in vitro* culture of very small embryos requires a two-step process in which ovules are cultured until they become large enough to remove the embryos, which are then cultured to generate plants. This procedure is now routinely used and plant recovery from some interspecific crosses via embryo culture is now possible (Mallikarjuna and Sastri, 1985b; Stalker and Eweda, 1988; Ozias-Akins and Branch, 1990). The process of embryo development in groundnut is complex and at least under both photo (Zamski and Ziv, 1976; Thompson *et al.,* 1985) and hormonal control (Shushu and Cutter, 1990). Major obstacles to recover interspecific hybrids continue to be the difficulty of first obtaining heart-shaped embryos and then *in vitro* root formation on shoot tissues produced after *in vitro* embryo culture, although significant progress has been made in rooting during recent years (Moss *et al.,* 1988b).

Autotetraploids or amphidiploids with at least eight diploid species have been used to produce hybrids with cultivated groundnut (Gardner and Stalker, 1983; Singh, 1986, 1988). Progenies with resistance to groundnut rust (Singh, 1986), leaf spot (Gardner and Stalker, 1983), and bud necrosis virus (Stalker, 1992) have been developed through cytogenetic approach. With regard to intersectional hybrids, embryo rescue techniques were used to recover at least one hybrid between *A. monticola* and a section *Rhizomatoae* accession (Sastri amd Moss, 1982) and between *A. hypogaea* and *A. villosulicarpa* of *Extranervosae* (Moss *et al.,* 1988a). However, these hybrids exhibit sterility problems and their use in groundnut breeding is thus very much limited.

Botany and Diversity of Cultivated Groundnut

The unique nature of underground fruiting habit and other associated morphological features form the basis for taxonomic classification. The cultivated groundnut is divided into two large botanical groups, (1) virginia and (2) spanish and valenica, on the basis of branching pattern (Richter, 1899). The presence or absence of reproductive nodes on the main axis and the arrangement of reproductive and vegetative nodes on the laterals are the most important criteria for the classification. In the first group i.e., virginia, the main axis is not reproductive. Also, alternating pairs of vegetative and reproductive axes are borne on the cotyledonary laterals and other secondary branches. This system was termed 'alternate branching pattern' by Bunting (1955). The first two branches on the secondary laterals are always vegetative. The alternate branching pattern is repeated in higher order branches.

In the second group, i.e., spanish and valencia, reproductive branches are borne in a continuous series on successive nodes of the cotyledonary and other lateral branches, on which the first branch is always reproductive; this branching is termed 'sequential branching pattern'. Reproductive branches are also borne directly on the main axis at higher nodes. Most of the tertiary and quaternary branches are reproductive. The spanish cultivars produce tertiary vegetative branches irregularly, while valencia cultivars frequently have none. If tertiary branches are found on valencia cultivars, they will be in sequence, distal to the 5th to 8th nodes of secondary branches. The subspecific classification can therefore be as follows:

Arachis hypogaea Linn. (1753)

ssp. *hypogaea* (Krapovickas & Rigoni, 1960)

var. *hypogaea* (= *A. africana* Lour; type 'Braziliano' Dubard, 1906; type 'virginia' Gregory *et al.*, 1951), Bolivia and Amazonia

var. *hirsuta* Kohler (1898) (= *A. asiatica* Lour; type 'Peruano' Dubard, 1906 in part) Peru

ssp. *fastigiata* Waldron (1919)

var. *fastigiata* (= type 'Peruano' Dubard, 1906 in part; type 'valencia' Gregory *et al.,* 1951)

Guarania, Goias and Minas Gerais, Peru and northeastern Brazil.

Var. *vulgaris* Harz (1885) (= type 'Spanish' Gregory *et al.,* 1951)

Guarania, Goias and Minas Gerais, and northeastern Brazil.

A brief description of the subspecies and botanical varieties of *A. hypogaea* is given below (after Gibbons *et al.,* 1972).

Subspecies *hypogaea:* habit procumbent, decumbent or erect; branching alternate; inflorescences simple and never borne directly on the main axis; first branch on the cotyledonary lateral always vegetative; 2 or 2–4 seeds per pod; pod beak pronounced, moderate or absent; pod constriction prominent, moderate or absent; pod very large (> 20 mm) or small (< 10 mm); testa colour commonly tan but red, white, purple, and variegated forms do exist; seed dormancy usually present; foliage dark green.

1) var. *hypogaea*: habit procumbent, decumbent, or erect; main axis in procumbent forms short (not exceeding 40–50 cm); stem usually not very hairy; usually 2 seeded; medium late maturing.

2). var. *hirsuta*: habit procumbent; main axis may exceed 1 m; stem fairly hairy; pods strongly beaked, with 2–4 seeds; very late maturing.

Subspecies *fastigiata*: habit erect to decumbent; branching sequential; inflorescences simple or compound, always present on main axis; first branches on cotyledonary laterals reproductive; seed dormancy usually absent; foliage usually lighter in colour than in ssp. *hypogaea.*

1). var. *fastigiata*: vegetative branches on primaries absent or regularly placed at the distal nodes; inflorescence usually simple; pods with 2 or 2–4 (rarely 5) seeds; beak absent, slight or prominent; size medium to small; testa colour tan, red, white, yellow, purple or variegated.

2) var. *vulgaris*: vegetative branches occasional and irregularly placed; inflorescence compound; pods usually with 2 seeds; beak present or absent size medium to small; testa colour tan, red, white, or purple.

Genetics of Some Important Qualitative Characters

Comprehensive reviews on the genetics of a wide range of characters were published by a number of workers including Hammons (1973b), Wynne and Coffelt (1982), and Reddy (1988). Therefore, no attempt is made now to present the inheritance pattern of different characters. However, the recent investigations pertaining to the genetics of branching pattern, leaf shape, and aerial podding are reviewed in greater detail.

Branching Pattern

Branching pattern is one of the key characters that played a decisive role in the genetic differentiation and evolution of *Arachis hypogaea* L. A number of investigations were carried out with regard to the inheritance pattern of branching attribute by several workers. By and large the conclusions in this regard were that the alternate branching pattern associated with the virginia type was dominant to sequential branching pattern of spanish and valencia types (Dalal, 1962; Balaiah *et al.,* 1977; Jadhav and Shinde, 1979). The F_2 data reported by several workers indicated a monogenic inheritance with 3 : 1 (3 alternate : 1 sequential) segregation pattern. There were, however, certain exceptions (Coffelt, 1974; Varan *et al.,* 1986; Dasale, 1987; Essomba *et al.,* 1988) indicating a deviation from monogenic inheritance.

An analysis of 45 crosses effected by Vanisree (1992) representing alternate × sequential (27), alternate × alternate (8), and sequential × sequential (10) brought out that in F_1 the alternate branching was dominant whenever a parent with alternate branching was involved in the crosses. The F_2 generation, however, presented a varied picture. The F_2 generation of all the sequential × sequential combinations showed all sequential types in F_2. The behaviour of alternate × sequential crosses was variable; the majority of the above (17), of which TMV 2 × TMV 2 NLM was also studied, showed a simple monogenic inheritance pattern with 3 : 1 ratio (3 alternate : 1 sequential), thereby indicating involvement of a pair of a contrasting alleles. One cross, Kadiri-3 × JL 24, revealed F_2 segregation pattern of 9 : 7 ratio (9 alternate : 7 sequential), while 9 such crosses gave 15 alternate : 1 sequential segregation ratio, thereby indicating the involvement of either more than a pair of alleles or the effect of modifier complex on the pair of alleles for branching. It was also observed by Vanisree (1992) that the parents that showed monogenic segregation in certain combinations themselves gave a deviant segregation pattern in certain other combinations, thereby emphasizing the decisive role of modifier gene complex or genetic background in the phenotypic expression of the branching pattern. These conclusions draw support from the investigations of Dalal (1962) and Varan *et al.* (1986).

The studies carried out by Vanisree (1992) indicated that while 7 out of eight crosses involving the parents with alternate branching resulted in no segregation for sequential branching, a lone combination of TMV 2 NLM × GNLM segregated in the F_2 generation in a proportion of 63 alternate to 1 sequential. This observation is of importance in the sense that both the parental types, although they represent alternate branching pattern, were genetically dissimilar and as such divergent. Interestingly, TMV 2 NLM is an alternate branching mutant of a sequential branching variety TMV 2 (Prasad *et al.*, 1984), while GNLM is a mutant of alternate branching variety Punjab-1 (Gopani and Vaishnani, 1970). In other words, the parental types represent two genetically divergent mutational events that could be employed for widening the genetic variability for attaining higher degrees of variances in groundnut breeding. This observation draws support from Coffelt (1974).

Leaf Shape

Leaf shape and leaf orientation are considered to be important attributes of canopy development, which has bearing on the photosynthesis as well as transpiration rate of groundnut plant (Mahapatra, 1966; McCloud *et al.*, 1980). The narrow leaf character in groundnut is supposed to be associated with low levels of transpiration and as such is considered important for developing genotypes with tolerance to moisture stress (Nageshwara Rao, 1992). Gopani and Vaishnani (1970) reported a virginia type Gujarat Narrow Leaf Mutant (GNLM), which has been reported to be one of the genotypes tolerant to drought. Prasad *et al.* (1984) also recovered a narrow leaf virginia mutant from EMS-treated TMV 2 spanish bunch variety, which was found to be genetically different from GNLM. The earlier investigations indicated either a partial (Matlock *et al.*, 1970) or complete dominance (Balaiah *et al.*, 1977) of narrow leaf mutant with monogenic inheritance. The F_2 data of Vanisree (1992) also indicated a typical monogenic inheritance based on two alleles as reported by earlier workers. However, the segregation pattern with regard to crosses involving TMV 2 NLM (induced narrow leaf mutant) on one hand and the normal leaved parents on the other indicated a segregation pattern of 1 narrow : 2 intermediate : 1 normal (Vanisree, 1992). The crosses involving GNLM, on the other hand, exhibited a segregation pattern of 3 narrow : 1 normal, thereby indicating that the TMV 2 NLM and GNLM were genetically different as reported by Prasad *et al.* (1984).

The F_1 of cross involving TMV 2 NLM and GNLM, both narrow leaf types (Vanisree, 1992), showed absolutely narrow leaf type; the F_2 segregated for 1 needle shaped narrow leaves with sterility : 60 narrow leaf types : 3 normal leaf types, suggesting a trigenic model involving recessive sterile due to complementary gene action. Such results were also obtained for chlorophyll attributes by Coffelt and Hammons (1971) and Tai *et al.* (1977). These results amply confirm that the TMV 2 NLM (TMV 2 narrow leaf mutant) reported by Prasad *et al.* (1984) is genetically different from GNLM (Gujarat narrow leaf mutant) of Gopani and Vaishnani (1970).

Aerial Podding

The aerial podding of groundnut was first reported by Prasad (1985). The aerial podding genotypes TAP1 to TAP5 occurred as spontaneous mutations in a population of Brazilian groundnut variety 'Tatu' exhibiting a large number of well-developed aerial pods. The aerial pegs, which otherwise go to waste without developing into pods in traditional varieties, develop into pods with good seed filling in the reported material, perhaps due to a different mechanism of 'Ca' translocation (Prasad and Muralidharudu, 1991) that is absent in the traditional groundnut varieties. The aerial podding genotypes of groundnut offer tremendous potential to enhance productivity of the groundnut plant. An aerial podding segregant was also reported by Nagabhushanam *et al.* (1990) in a progeny of MH2BC28 × ICG(C)8.

The investigations carried out by Vanisree (1992) indicated that the pattern of inheritance of aerial podding character reported by Prasad (1985) varies with the female parent employed in the cross, the male parent being TAP5 (an aerial podding genotype). The crosses involving all the sequential parents on one hand and the aerial podding on the other as well as an alternate branching variety Kadiri-3 × TAP5 have resulted in the F_1 with aerial podding attribute, thereby indicating the dominant nature of this attribute. The F_2 generation of these crosses excepting Kadiri-3 × TAP5 exhibited a monogenic Mendelian inheritance pattern of 3 aerial podding : 1 non-aerial podding. In the case of involvement of the other alternate branching virgina types as female parent, the F_1 did not show any aerial podding, while the F_2 of these crosses segregated in a proportion of 1 aerial podding : 3 non-aerial podding (normal). The same was true in respect of the F_2 of Kadiri-3 × TAP5. These results show that in the F_2s of the crosses of aerial podding involving virginia parents, the heterozygote did not show any aerial podding, thus exhibiting similar phenotypic appearance to the non-aerial podding and hence the above ratio.

The fact that the aerial podding attribute is dominant or partly dominant could be observed from the phenotypic expression of aerial podding nature in respect of the F_1s of all aerial podding crosses involving non-aerial podding and sequential branching traditional genotypes as well as Kadiri 3, a virginia genotype. In the case of the F_2s of crosses involving sequential branching parents, the heterozygote had similar phenotypic expression to that of dominant homozygote with aerial podding, hence the observed proportion of segregation 3 aerial podding : 1 non-aerial podding (normal). It appears that the expression of the aerial podding attribute was very much conditioned by the genetic background. While the genetic background of sequential branching type was conducive to better expression of this trait, the genetic background of alternate branching character tends to modify the expression of aerial podding attribute. Considering the suggestion that the aerial podding attribute could be the result of a different mechanism of 'Ca' translocation to developing gynophore (Prasad and Muralidharudu, 1991), it would be worthwhile to investigate the inheritance pattern of aerial podding attribute with detailed studies on the 'Ca' translocation pattern in all the segregants.

Canopy Compaction

A number of investigations have been undertaken to understand the mode of inheritance of various components of canopy development (Hammons, 1973b; Wynne and Coffelt, 1982). By and large with some exceptions a monogenic type of inheritance has been attributed to branching pattern (Hassan and Srivastava, 1966; Shchori and Ashri, 1970; Balaiah *et al.*, 1977, 1984; Jadhav and Shinde, 1979) and leaf type (Hammons, 1964; Bhide and Dasale, 1970; Balaiah *et al.*, 1977; Branch, 1987; Dasale, 1987). Vanisree (1992) reported a complex type of inheritance for canopy development, with universal dominance of most spreading habit over the rest and trigenic, digenic, and monogenic segregations depending upon the kind of parents involved for canopy development in crosses.

Genetics of Quantitative Characters

The yield in groundnut being a complex character, there is a need to understand the genetics of different component characters contributing to it. Principally, an effort should be made to identify the characters that show strong correlation with the yield.

Correlations

A number of investigations carried out in this context are reported by Hammons (1973b), and Wynne and Coffelt (1982). Seed yield has been shown to be correlated with number of pods and number of seeds per plant, weight of pods per plant, and weight of individual seeds (Coffelt and Hammons, 1972). Balaiah *et al.* (1984) showed that yield was positively correlated with a number of growth parameters, such as number of nodes, number of branches, and leaflet area, and also number of mature pods. Height of main stem was the one vegetative character not correlated with yield in their studies. Nigam *et al.* (1984) calculated correlations between 16 vegetative and reproductive traits in 97 bulks derived from a number of crosses and reported that weight of mature seeds per plant was positively correlated with height of main axis, number of nodes with pegs, total number of pegs, total number of pods, mature pods per plant, mature seeds per pod, and weight per mature seed.

Heterosis

The frequency of heterotic crosses depends more on the cultivars used in the study than on the subspecific groups. Arunachalam *et al.* (1984) analysed two diallels that included spanish bunch (SB) and valencia (VL) of ssp. *fastigiata* and virginia bunch (VB) and virginia runner (VR) of ssp. *hypogaea*. The frequency of heterotic crosses, excluding results where the total number of crosses was less than 10, ranged from 17 per cent for VL × VR to 90 per cent for SB × VB. Both lowest and highest figures were for *fastigiata* × *hypogaea* crosses. These figures were comparable to the range of heterosis of 33 per cent for VL × VL to 83 per cent for VR × VR. Forty per cent of SB × SB crosses were heterotic, while 35 or 70 per cent of VB × VB crosses were heterotic in the two diallels. Although inter-subspecific or intervarietal crosses can give a high degree of heterosis, this can also be achieved from crosses between cultivars within botanical varieties.

In a similar study in North Carolina, Isleib and Wynne (1983) assigned 27 exotic cultivars to groups on the basis of morphological characters. Each cultivar was crossed with a locally adapted, large-seeded virginia breeding line, NC Ac 18000. Heterosis, as a significant deviation from midparent value, was present in some crosses for all characters scored and was positive, except for meat content (shelling percentage). Heterosis was rare in crosses between sequential branching exotics and NC Ac 18000, itself a sequentially branched genotype, but frequent between NC Ac 18000 and exotics with a different branching pattern. Although there was heterosis over midparent value for pod and seed yield, values for these were not greater than that of the adapted parent. Heterosis greater than that of the best parent was expressed for pod size and pod length in intrasubspecific crosses. Heterosis in pod and seed yield was primarily due to dominance, but epistasis was more important in heterosis for pod and seed number and meat content. Even using 27 exotic cultivars from the five subcentres of diversity in South America and from China and Africa, researcher found evidence that the exotic parent contributed favourable alleles only for pod length in crosses with the long-podded, multiple-seeded types.

Genotype × Environment Interaction

Stability over seasons or across environments requires a low genotype × environment interaction. Most studies support the findings of Wynne and Isleib (1978), who grew cultivars adapted to conditions in North Carolina and Virginia (USA), and some breeding lines at different sites for a number of years. Even though the trial was in the area for which the cultivars were bred, there was a large cultivar × location × year interaction for yield, though the average performance over different locations was nearly the same in each year.

Nigam *et al.* (1984) measured 16 characters, including yield (as mature seed weight per plant), and recorded highly significant genotype × year interactions for all characters except number of seeds per pod.

Norden *et al.* (1986) also found large interactions between genotype and environment for pod yield and five quality factors in four peanut multilines and their components, although the differences in yield and market quality between sib-lines were small. However, Schilling *et al.* (1983) studied the stability of yield across four locations and two years for two multilines and their individual components, and found that one multiline was stable with relatively high yields, though the individual components displayed significant variability.

Heritability Studies

The majority of authors conclude that progress can be made by selecting for yield, and that this is the best technique, although theoretically selecting for a number of yield components may help to make better progress. The success of the parctical technique practised by many breeders of early generation selection of plants with many large pods is probably due to the high heritability and correlation with yield of these characters in many crosses.

Genotypic variance was higher than environmental variance for a range of characters in 20 strains each of bunch and spreading groundnuts (Chauhan and Shukla, 1985). The genotypic coefficient of variation, broad-sense heritability, and genetic advance were high for pod yield per plant in both groups.

Reports on heritability of a range of characters associated with yield are conflicting. High heritability for yield, either pod weight or seed weight, was reported by Balaiah and Reddy (1975), Mohammed *et al.* (1978), and Gibori *et al.* (1978), but other authors have reported low heritabilities. However, heritabilities for some yield components are high. Kumar and Yadav (1979) reported a high heritability for 100 kernel weight, indicating that progress could be made in selection for seed size. Similar results were reported by Hammons (1976).

Genetic Improvement

Hybridization Techniques

The largely followed techniques for producing hybrids is mainly emasculation and pollination. The bud at appropriate stage is opened by forceps and anthers are removed. Notwithstanding the variation in the technique, it is essential that the stigma and ovary are not damaged. It is also possible to remove the wing and keel petals to facilitate emasculation. The pollination of the emasculated flowers is carried out early in the next morning using forceps to transfer pollen from the male parent. Care should be taken to effect the pollination before the receptivity of the sigma declines. Whenever the microclimate is dry, it is absolutely necessary to maintain humidity with sprinklers covering individual plants after emasculation. The success rate for pollination could be achieved to the extent of 75 per cent in greenhouses employing trained workers. After pollination, the flower dries up within 10 days. The gynophore can be seen developing. In order to distinguish the peg of the emasculated and pollinated flowers, it would be necessary to use coloured threads for different pegs. The other method is to eliminate or remove the unwanted flowers in the vicinity of the pollinated ones. The F_1 plants should be planted with wider space and the parent lines should also be grown along with F_1s for the purpose of comparison.

Breeding Methodologies for Yield Enhancement

Groundnut breeding, though started as early as 1910, was not pursued with any degree of persistence and therefore made no spectacular headway as in the case of some cereals. This was partly due to the nature of the crop and to the sporadic attempts of workers in different countries. Gregory *et al.* (1951) reviewing the previous work on the subject summed up the position as follows: 'In reporting the published work on peanut breeding and genetics we have been unable to unfold a sequence of events or to build a coordinated body of knowledge, culminating in recent advanced studies logically derived from the results of previous experiments. Instead we have attempted to preserve the mosaic of unrelated patterns.'

Difficulties in hybridization, poor realization of crossed seeds (Venkateswaran, 1980), restricted recombination and hereditary instability (Hammons, 1976), poor percentage of pod set after effecting crosses (Thangavelu *et al.*, 1980), smaller F_2 populations (Reddy, 1980), genetic problems arising out of segmental alloploid nature, and absence of an appropriate selection criterion due to lack of clear-cut relationship between canopy characters and the yield components (Prasad and Kaul, 1980; Prasad *et al.*, 1984) may be cited as some of the reasons for slow progress in the genetic improvement of the crop.

Van der Stock (1910) was the first to start breeding work on groundnut in the Netherlands and the East Indies. He was able to isolate a number of pure-lines from a cultivated variety by selection of single plants. He explained the multiplicity of types of land races as due to hybrid variability caused by chance cross-fertilization. After detailed studies for a number of years, Badami (1928) stated that preliminary selection should be based on plants with the largest number of mature pods and the final selection on the highest weight of seeds per plant.

In America, early breeding work on groundnut was done in Florida and the results were reported by Stokes and Hull (1930). They found that the existing types did not combine the desirable characters such as seed dormancy, erect plant type, large seed, dark green foliage, and light tan seed with high or low oil content and stated that such types could only be produced by hybridization. Umen (1933), on the basis of breeding work conducted in the USSR since 1926, reported that pure-line selection did not yield very favourable results and considered hybridization essential.

Bolhuis (1938) reported that line selection was successful and 'Schwartz 21', a type highly resistant to the slime disease and with high yield, was evolved by application of this method. He enumerated high yield, resistance to slime disease, erect foliage, large seeds, and pods with more than two seeds, only slight or no constriction, and early ripening as the major objectives in groundnut selection. He stated that since further increase in yield seems unlikely, future breeding will probably be concentrated on selection for large pods and early maturity from hybrid material.

Higgins *et al.* (1941) reported that by hybridization and selection, many promising strains were isolated and hybridization between peanut varieties gave rise to a great multiplicity of plant and nut types often quite different in many characteristics from either parent, thereby offering scope for selection. Hull and Carver (1945) were of the opinion that the desired types could not be obtained except by resorting to hybridization and that a very large population of progenies would be required for selection of the desired types.

Bouffil (1947) after years of work in Senegal on the improvement of varieties by breeding stated that selection based on external botanical characters and characters such as number of pods per plant and the mean weight of pod was unsuccessful.

Gregory *et al.* (1951), on the basis of breeding attempts made in North Carolina, concluded that hybridization was necessary for the success of practical groundnut breeding. Seshadri (1956) emphasized that improvement of quality in groundnut could be achieved only through hybridization.

Norden (1973) indicated that the possibilities for groundnut improvement by introduction and selection without hybridization were practically exhausted in the United States. According to him, success in breeding through hybridization depends on the availability of transferable genetic variation and will be more likely if the objectives were clearly defined, the correct parents were selected, and the hybrid populations were managed properly.

Crop Improvement in India

In the crop improvement history of groundnut in India, the years 1905 and 1961 represent important landmarks. Groundnut cultivar Spanish, a pure-line selection from spanish peanut made at Dharwad, was released in 1905 for cultivation in the bunch tract of Bombay-Karnataka region. This was the first official release of a groundnut variety in India. In 1961, C 501, a virginia bunch cultivar derived from a cross between D3 and Ah 477 at Ludhiana, was released for cultivation in Punjab state. This was the first groundnut cultivar in India derived through hybridization in India.

Betwen 1906 and 1960, 12 cultivars were released in different parts of the country. These cultivars were developed either by pure-line or mass selection in local and introduced varieties. Prominent among these were AK 12-24 and TMV-2 (both spanish and released in 1940), Karad 4-11 (Virginia Runner, released in 1957), and T 28 (Virginia Runner, released in 1960). During 1961–1979, 28 more cultivars were released. Of these, 11 originated from hybridization and 17 were mass or pure-line selections.

Groundnut breeding in India received a real impetus in the 1980s when 30 new cultivars were released. These new cultivars included 20 spanish, 7 virginia bunch, and 3 virginia runner types. Contrary to previous decades when heavy reliance in varietal development was placed on mass and pure-line selection in local and introduced material, in the 1980s, 80 per cent of the new cultivars were cross derivatives. This decade also saw a shift in breeding objectives when greater emphasis was laid on stress resistance breeding. Three foliar disease resistant cultivars (Girnar 1, ICGS (FDRS) 10, and ICGV 86590) were released during this period. Similarly, ICGS 11, ICGS 37, and ICGS 44, were specifically released for post-rainy season cultivation in different parts of the country. Till the end of 1992, 72 cultivars were released in India, giving a wide array of choice to farmers.

Unlike in wheat (*Triticum aestivum* L. emend. Fiori & Paol), sorghum (*Sorghum bicolor* (L.) Moench), and rice (*Oryza sativa* L.), the F_2 generations derived from various crosses in groundnut were characterised by small population size with low frequencies of desirable segregants, due to lower degrees of variances for important agronomic attributes (Dutta *et al.*, 1986). The crosses involving different botanical types, however, were also studied for F_1 heterosis (Wynne and Gregory, 1981; Isleib and Wynne, 1983) without adding any information in respect of segregating generations.

The current groundnut breeding activities in India concentrate on developing varieties with high yield, high oil content, early maturity, confectionery traits, and tolerance or resistance to drought, low temperature, iron chlorosis, foliar diseases, and aflatoxin production. It is important that, in future, breeding for high yield *per se* should continue to receive strong emphasis so that the resultant material with high yield potential could form the base for a stress resistance breeding programme.

A number of exotic varieties including those developed with a higher level of productivity in the United States, when evaluated in the coordinated network of varietal trials under the All India Coordinated Research Programme on Oilseeds from 1983 to 1985, were found to yield less than the local adapted cultivars of groundnut (Reddy, 1988), thereby indicating that the possibility of enhancing the productivity through introducing exotic commercial cultivars is very remote. Therefore, there is an urgent need to intensify varietal improvement work to develop genotypes with higher productivity level as well as adaptability to the specific agroecological situations.

During the past decade, significant progress has been made in identification of sources resistant to biotic stresses and developing cultivars resistant to major biotic constraints. However, very little progress has been made in terms of identification and exploitation of genetic traits for physiological factors influencing yield components. There is a need to develop simple screening techniques for such desirable traits. The recent trends in research on groundnut physiology carried out at ICRISAT indicate that substantial variation exists for desirable traits such as water use efficiency, partitioning of dry matter to pods, and efficient rooting system. It is also to be noted that novel techniques have been developed to screen the groundnut germplasm for some of these traits.

The investigations carried out on the stability of different characters indicated that pod number per plant and pod weight per plant, which are commonly employed as selection criteria for higher productivity, are unstable attributes (Madhavi, 1988). This must be one of the reasons for lack of perceptible progress in breeding genotypes with higher levels of productivity. Most of the recently selected and released groundnut cultivars were in no way significantly superior to the locally adapted cultivars.

The work carried out by Nagabhushanam and Prasad (1992a, b) has indicated that pod number and pod weight showed a large degree of environmental interaction as compared to canopy development at peak flowering, measured by canopy dry matter at the peak flowering stage. Therefore, it appears more appropriate to employ some of the stable canopy attributes such as canopy dry matter at peak flowering showing low pooled deviation as criteria for selection, depending upon the feasibility, in addition to pod number and pod weight. This attribute not only expreses adequate stability but also possesses strong positive correlation with kernel yield, apart from the ease and feasibility with which it could be employed. Other characters that merit consideration are number of primary branches at 60 days in secondary branching forms such as spanish bunch and number of secondary branches per plant at 60 days in virginia runner types. Incidentally, these attributes also contribute to the canopy dry matter.

Innovative Breeding Approaches

Using Physiological Attributes as Selection Criteria

The physiological analysis of the wide range of high-yielding varieties of groundnut developed over a period of 40 years carried out by Duncan *et al.* (1978) has indicated that the progressive enhancement of productivity is largely due to partitioning of the assimilates rather than to crop growth rate. Breeders are employing largely pod number and pod weight as principal selection criteria for productivity (Balaiah *et al.,* 1984, Reddy, 1988). However, these attributes are highly unstable (Prasad *et al.*, 1984). Pooled analysis of heritability for four seasons carried out by Nagabhushanam and Prasad (1992b) showed that canopy diameter, canopy circumference, and total dry matter at 60 days had higher heritability values of 91, 85, and 96 per cent respectively. Analysis of the pattern of means and variances in the F_2 generation of a wide range of crosses involving the canopy categories 1, 2, 3, and 4 (Nagabhushanam and Prasad, 1992a) showed that canopy diameter, total number of root nodules, and total root nodule mass, which have bearing on kernel yield, were the least in the F_2s of the crosses involving canopy categories 1×2 and 2×2. In the F_2 generation the parental canopy category combinations 2×4, 3×4 and 1×4 showed high variance with high mean, high variance with medium mean, high variance with low mean, and medium variance with high mean for the characters mentioned above, indicating that such parental cross-combinations should be preferred in groundnut breeding.

According to Prasad (1993), the breeder should use stable characters such as canopy development and peak flowering as the selection criteria and exercise choice for the parental combinations involving intermediate degree of canopy development for generating segregating material in which selection for productivity could be exercised after F_3 generation.

If new variability for genes promoting calcium uptake and its translocation to developing gynophore, leading to assured pod development, could be superimposed over the exploitation of genes for stable attributes such as canopy development, the efficiency of groundnut breeding programmes for higher productivity could be taken to newer heights (Prasad, 1993).

Use of Wild Relatives for Groundnut Improvement

Gregory *et al.* (1973) opined that defects in the genetic constitution of cultivated groundnut in relation to human requirements are many and that there is a need to tap new sources of genetic variability either through induced mutations or transfer of valuable genes from the allied species. Hence, the use of different species of *Arachis* for the improvement of cultivated groundnut in respect of a wide range of agronomic attributes assumes specific importance. Considerable efforts were made to cross-cultivated groundnut with the wild species of *Arachis* since 1938 (Hull and Carvar, 1945). Gregory *et al.* (1973), and Gregory and Gregory (1979) clarified the crossability barriers among the species of *Arachis*, which has helped in the identification of appropriate wild relatives of groundnut improvement.

Ggegory and his coworkers selected 91 parents that represented all seven sections of the genus and attempted 1075 of the 8190 possible crosses between them. In general, intrasectional crosses were successful, but there were few successful intersectional crosses, and often these were restricted to a few species in one series. Thus, series *annuae* of section *Arachis* could be crossed with *Rhizomatosae eurhizomatosae,* and with *Erectoides tetrafoliolatae*, but not with species in other series of section *Arachis*. Caulorhizae can be crossed with *Erectoides tetrafoliolatae*, and *procumbense*, but not with *trifoliolatae*. Other successful intersectional crosses are *ambinervosae* × *extranervosae, Erectoides tetrafoliolatae* × *ambinervosae, Arachis perennes* × *Erectoides procumbense,* and *Erectoides tetrafoliolatae* × *Rhizomatosae, eurhizomatosae*. To date, no other crosses have been authenticated except *A. monticola* × A sp. 276233 (Moss and Sastri, 1986) and *A. hypoagae* × *A.* sp. 276233 (ICRISAT, 1985), though both these crosses were attempted with hormone-aided pollinations, and the latter cross has only proceeded as far as shoots derived from hybrid callus.

These results supported the taxonomic division of the genus based on morphological characters (Gregory *et al.,* 1973). Their practical implication in genetic improvement of groundnuts is that species to be used as sources of genes are currently confined to those in section *Arachis.*

Arachis monticola Krap et Rig. is the only wild tetraploid species in section *Arachis,* and it crosses freely with *A. hypogaea.* It has been used as a parent in the production of Spancross (Hammons, 1970), and no special cytogenetic skills are needed to use it in a breeding programme.

Other wild species in section *Arachis* are diploids, and many hybrids have been produced among them, and between them and *A. hypogaea.* Raman and Kesavan (1962) produced the first hybrid, *A. duranensis* × *A. villosa correntina.* This was a hybrid between two A genome species and was fully fertile with regular meiosis. Although there are differences between hybrids, all AA hybrids are fertile, but hybrids between A genome species and *A. batizocoi,* the only known species with the B genome, have few bivalents at meiosis and low pollen stainability and are usually sterile (Resslar and Gregory, 1979; Smartt *et al.*, 1978a, b; Stalker and Wynne, 1979; Singh and Moss, 1984a). *Arachis hypogaea* has been crossed with many diploid species and many reciprocal crosses succeed. However, *A. helodes* and *A. diogoi* are best used as male parents, and *A. spegazzinii* and *A. ipaensis* can only be used as female parents (Stalker, 1985). Hybrids between A genome or B genome diploid-species and tetraploid *A. hypogaea* are triploid, and therefore expected to be sterile. However, Singh and Moss (1984b) harvested pods from plants of eight triploid hybrid combinations, one plant of *A. hypogaea* × *A. correntina* producing 25 pods. Fertility was not due to somatic chromosome doubling producing fertile hexaploid branches, but due to (1) the formation of restitution nuclei and (2) unequal segregation resulting in haploid to hyperdiploid gametes. Of the 225 F_2 plants derived from triploids, 82 per cent were hexaploid, 80 per cent were tetraploid, and the remainder had chromosome numbers from 20 to 59. The tetraploids are of special interest and mean number of univalents over all hybrids was 8.8 per PMC, indicating that pairing between wild and cultivated chromosomes had taken place. Simpson and Davis (1983) also reported a male fertile complex triploid hybrid. Recently, Stalker (1985) reported that a new diploid accession in section *Arachis, A. spinaclava,* produces sterile hybrids with irregular meiosis when crossed with A or B genome diploids. The chromosomes are subtelocentric, and Stalker designated the new genome as the D genome.

Hybrids between *A. hypogaea* and diploid species of section *Arachis* have been used to transfer pest and diease resistance from the wild species. Initially, the sterility of the triploids was overcome by producing hexaploids by colchicine treatment, and repeatedly backcrossing to *A. hypogaea* to produce pentaploids in BC_1 and tetraploids in later generations (Spielman *et al.*, 1979; Moss, 1980). Pentaploids have reduced fertility and many generations of backcrossing may be needed to restore fertility. Other routes were therefore explored, and the ploidy of the wild species raised to the tetraploid level, either as autotetraploids or by colchicine treatment of diploid interspecific hybrids to produce amphidiploids (ICRISAT, 1980). Fertile, *A. hypogaea*-like plants are produced after a few crosses to *A. hypogaea.*

Few intersectional hybrid combinations have been produced. The hybrids that could be produced by Gregory and Gregory (1979), and other intersectional hybrids have not produced fertile hybrid populations as have the intrasectional hybrids (Stalker, 1985a). Complex crosses have been made in an attempt to produce fertile hybrids, but hybrids were not fertile, and attempts to restore fertility by increasing ploidy level, or to exploit the wild species by developing fertile hybrids with species of section *Arachis* have been unsuccessful (Stalker, 1985a).

Other attempts to produce intersectional hybrids have focussed on crosses where pegs and pods were formed but seeds aborted before maturity. Sequential hormone applications and ovule and embryo culture have been used to produce hybrid embryos, callus, and shoots from *A. hypogaea* × *A.* sp. 276233 *(Rhizomatosae)* crosses (Sastri and Moss, 1982; Nalini and Sastri, 1985a, b). They produced pods by treating flowers with 88 ppm gibberellic acid (GA) at the time of pollination and then treating the developing pegs with 50 or 100 ppm indole acetic acid (IAA) in lanolin 20 days after pollination. Few seeds were produced when pods were left to mature, so ovules or embryos from immature pods were cultured. Callus formation was induced by manipulating hormone concentrations in culture media, and subsequently shoots were produced (ICRISAT, 1985).

Techniques for production of somatic hybrids have not been successful in *Arachis*. Oelck *et al.* (1982) were able to isolate protoplasts that grew and produced callus. The technique has received little attention in *Arachis*, and shoots or plants have not been regenerated from protoplasts, nor are there reports of protoplasts having been successfully fused.

Biotechnology for Groundnut Improvement

Biotechnology, consisting of technologies for genetic analysis and gene transfer, is likely to make a significant impact in groundnut improvement in the future (Weissinger, 1992). Molecular marker technologies are increasing in resolution and declining in cost. Two of the most advanced technologies, restriction fragment length polymorphism (RFLP) and randomly amplified polymorphic DNA (RAPD), have been successfully applied to both the cultivated groundnut and its wild relatives. The current results show that the cultivated types are highly homogeneous, while the diploid species form a reservoir of diverse germplasm. Molecular marker technologies are sufficiently advanced to allow development of genetic maps of both groundnut and of the diploid species. Molecular technologies could also contribute greatly to introgressive gene transfer schemes by allowing alien genes to be followed in backcross progenies.

The application of transformation technology in groundnut is at a relatively primitive level, but offers immense promise. The tissue culture technology is well developed in relation to the wild species. The transient expression and stable transformation of groundnut tissues by microprojectile bombardment have been demonstrated, thereby bringing out the efficacy of microprojectile bombardment for choosing recipient explants and the utility of kanamycin selection of transgenic tissues.

Future Problems and Prospects

Groundnut is different from other grain legumes in the sense that it has aerial flowering and subterranean podding. Due to the positive geotropism of gynophores, the sink moves into the soil where the pod development takes place. Therefore, it becomes difficult for the plant breeder to exercise selection for productivity based on the above-ground characters, which apparently do not indicate the reproductive efficiency. Therefore, groundnut has been termed the 'unpredictable legume' (Gregory *et al.,* 1973). Groundnut breeders heavily depend upon the pod number and pod weight, which are unstable characters, as the principal selection criteria for productivity (Rao, 1976; Reddy, 1988). However, investigations carried out by Prasad *et al.* (1984) and Nagabhushanam and Prasad (1992a) have clearly brought out the relevance of employing more stable character combinations such as canopy development at peak flowering as selection criteria. Investigations carried out by Duncan *et al.* (1978) have brought out that the limited enhancement in productivity achieved so far in groundnut is largely due to partioning of assimilates rather than the higher growth rate of crop. Therefore, there is a need to enhance the growth rate or higher dry matter production of groundnut, which in turn could be partitioned to achieve higher reproductive efficiency and thereby higher kernel yield. In this connection, one should look for induced mutations for higher dry matter production such as TM2-NLM, MH-2, and BC-28 (Prasad, 1988), which could be of immense use in groundnut breeding programmes.

Wynne and Gregory (1981) have observed that there is a need to bring about structural alterations in the groundnut plant so as to enhance number of fruit-bearing sites in order to achieve major advance in productivity. In this connection, the aerial podding attribute reported by Prasad (1985) due to spontaneous mutations in Brazilian groundnut variety Tatu could be of interest. The expression of aerial podding was observed to be due to translocation of calcium through stem to the developing gynophores as has been reported by Prasad and Muralidharudu (1991). Efforts to use the aerial podding genotypes in the breeding programmes have resulted in the development of highly productive groundnut varieties such as DRG-12 and DRG-17 in the All India Coordinated Research Project on Oilseeds (1993).

The groundnut plant offers tremendous potential for higher levels of productivity as it does not suffer from limitations of reduced nitrogen fixation as does soybean, which is incapable of meeting the needs of increased translocation to developing fruit (Sinclair and de Wit, 1975). Hence genetic manipulation of groundnut to achieve novel heights in productivity is feasible. The barriers to interspecific gene transfer in *Arachis* have been successfully overcome and several interspecific derivatives of immense value for groundnut breeding have been developed at ICRISAT. In order to increase the harvest index of groundnut to higher levels and to promote a larger number of fruit-bearing sites in the plant, it would be worthwhile to exploit *Voandzeia subterranea*, which offers a high harvest index. Considering the novel biotechnological tools available for such gene transfer, it should not be very difficult to achieve the recombinants combining the desirable attributes of both groundnut and *Voandzeia.*

Considering the potentials offered by the above possibilites, the future of groundnut breeding appears to be very bright.

REFERENCES

All India Coordinated Research Project on Oilseeds. 1993. Annual Progress Report. National Research Centre for Groundnut, pp. 1–120.

Arunachalam, V., Bandyopadhyay, A., Nigam S.N., and Gibbons, R.W. 1984. Heterosis in relation to genetic divergence and specific combining ability in groundnut (*Arachis hypogaea* L.). *Euphytica* 33: 33–39.

Badami, V.K. 1928. *Arachis hypogaea* (the groundnut). Ph.D. thesis (Vol. II Inheritance studies), Cambridge (England) University Library, pp. 297–374.

Balaiah, C., and Reddy, P.S. 1975. Hertability estimates of some quantitative characters in the F_2 generation of groundnut cross, 'J 11' × 'Gujarat Narrow Leaf Mutant'. *Oilseeds J.* 4: 37–38.

Balaiah, C., Reddy, P.S., and Reddi, M.V. 1977. Genetic analysis in groundnut. I. Inhertiance studies on 18 morphological characters in crosses with 'Gujarat Narrow Leaf Mutant'. *Proc. Indian Acad. of Sci.* 85B: 340–350.

Balaiah, C., Reddy, P.S., and Reddi, M.V. 1984. Studies on linkage between six morphological characters in crosses with 'Gujarat Narrow Leaf Mutant' groundnut (*Arachis hypogaea* L.). *Oleagineux* 39: 325–327.

Bhide, M.V., and Dasale, S.C. 1970. A small leaf mutant in groundnut. *Poona Agri. Coll. Mag.* 59: 113–114.

Bolhuis, G.G. 1938. Cited from *Plant Breed. Abstr.* 10: 235.

Bouffil, F. 1947. Biologie, Ecologie et Selection de L 'Arachide au Senegal. *Paris Bull. Sci.* 1: 1–112. Cited in Seshadri, 1956.

Branch, W.D. 1987. Inheritance of curly-leaf shape in peanut. *J. of Hered.* 78: 125.

Bunting, A.H. 1955. A classification of cultivated groundnut. *Empire J. Exp. Agric.* 23: 158–170.

Chauhan, R.M., and Shukla, P.T. 1985. Variability, heritability and genetic advance in bunch and spreading types of groundnut. *Indian J. Agric. Sci.* 55: 71–74.

Coffelt, T.A. 1974. Inheritance of growth habit in an intraspecific cross population of peanuts. *J. Hered.* 65: 160–162.

Coffelt, T.A. and Hammons, R.O. 1971. Inheritance of an albino seedling character in *Arachis hypogaea* L. *Crop Sci.* 11: 753–755.

Coffelt, T.A., and Hammons, R.O. 1972. Inheritance of sterile brachytic in an intraspecific cross of *Arachis hypogaea* L. *Crop. Sci..* 12: 82–84.

Conagin, C.H.T.M. 1963. Numero do cromossomios das especies selvagens de *Arachis. Bragantia* 22: 125–129.

Dalal, J.L. 1962. Inheritance studies in groundnut crop. *Indian Oilseeds J.* 6: 288–292.

Dasale, S.C. 1987. Genetic studies in groundnut. *J. Maharashtra Agric. Univ.* 12: 125–126.

Duncan, W.G., McCloud, D.W., McGrow, R.L., and Boote, K.J. 1978. Physiological aspects of peanut yield improvement. *Crop. Sci.* 18: 1015–1020.

Dutta, M., Bandyopadhyay, A., Arunachalam, V., Kaul, S.L., and Prasad, M.V.R., 1986. Radiation induced pollen sterility and enhanced outcrossing in groundnut (*Arachis hypogaea* L.) *Environ. Exp. Bot.* 26: 25–29.

Essomba, N.B., Coffelt, T.A., Branch, W.D., and Scoyoc, S.W. 1988 Inheritance of morphological traits in peanut (*Arachis hypogaea* L.). (Abstract) 19: 16 in *Proc. Amer. Peanut Res. Educ. Soc.*

Gardener, M.E.B., and Stalker, H.T. 1983. Cytology and leafspot resistance of section *Arachis* amphidiploids and their hybrids with *Arachis hypogaea. Crop Sci.* 23: 1069–1074.

Gibbons, R.W., Bunting, A.H., and Smartt, J. 1972. The classification of varieties of groundnut (*Arachis hypogaea* L.). *Euphytica* 21: 78–85.

Gibori, A., Hillel, J., Cahaner, A., and Ashri, A. 1978. A 9 × 9 diallel analysis in peanuts (*A. hypogaea* L.): Flowering time, tops weight, pod yield per plant and pod weight. *Theor. Appl. Genet.* 53: 169–179.

Gopani, D.D., and Vaishnani, N.C. 1970. Two mutant forms of groundnut (*Arachis hypogaea* L.). *Indian J. Agric. Sci.* 40: 431–437.

Gregory, W.C. 1946. Peanut breeding program under way. In *Research and Farming,* Sixty-ninth Annual Report, North Carolina Agric. Exp. Stn., pp. 42–44.

Gregory, M.P., and Gregory, W.C. 1979. Exotic germplasm of *Arachis* interspecific hybrids. *J. Hered.* 70: 185–193.

Gregory, W.C., Smith, B.W., and Yarbrough, J.A. 1951. Morphology, genetics and breeding. In *The Peanut—The Unpredictable Legume.* National Fertilizer Association, Washington, DC. pp. 28–85.

Gregory, W.C., Gregory, M.P., Krapovickas, A., Smith, B.W., and Yarbrough, J.A. 1973. Structure and genetic resources of peanuts. In *Peanuts—Culture and Uses.* American Peanut Research and Education Association, Stillwater, Oklahoma, pp. 47–133.

Halward, T.M., and Stalker, H.T. 1987. Incompatibility mechanisms in interspecific peanut hybrids. *Crop Sci.* 27: 465–460.

Halward, T.M., Stalker, H.T., LaRue, E.A., and Kochert, G.D. 1991. The use of single-primer DNA amplifications in genetic studies of peanut. Page 54 in Abstr. of Symp. on Plant

Breeding in the 1990s, 10–14 Mar, Research Report No. 130. North Carolina State University, Raleigh, North Carolina.

Hammons, R.O. 1964. Krinkle, a dominant leaf marker in the peanut, *Arachis hypagaea* L. *Crop Sci.* 4: 22–24.

Hammons, R.O. 1970. Registration of Spancross Peanuts. *Crop Sci.* 10: 459.

Hammons, R.O. 1973a. Early history and origin of the peanuts. In *Peanuts—Culture and Uses.* American Peanut Research and Education Association, Stillwater, Oklahoma, pp. 17–46.

Hammons, R.O. 1973b. Genetics of *Arachis hypogaea* L. In *Peanuts—Culture and Uses.* American Peanut Research and Education Association, Stillwater, Oklahoma, pp. 135–173.

Hammons, R.O. 1976. Peanuts: Genetic vulnerability and breeding strategy. *Crop Sci.* 16: 577–630.

Hassan, M.A., and Srivastava, D.P. 1966 Inheritance of growth habit in groundnut (*Arachis hypogaea* L.) *J. Indian Bot. Soc.* 45: 293–295.

Higgins. B.B., Holley, K.T., Pickett, T.A., and Wheeler, C.D. 1941. Peanut breeding and characteristics of some new strains. Georgia Agric. Exp. Stn. Bull. 213: 3–11.

Hull, F.H., and Carver, W.A. 1945. Peanut improvement. Florida Agric. Exp. Stn. Annual Report., 37 pp.

Husted, L. 1931. Chromosome number in species of peanut, *Arachis. Amer. Naturalist* 65: 476–477.

Husted, L. 1933. Cytological studies on the peanut, *Arachis.* I. Chromosome number and morphology. *Cytologia* 5: 109–117.

Husted, L. 1936. Cytological studies on the peanut, *Archis* II. Chromosome number, morphology and behaviour and their application to the problem of origin of the cultivated forms. *Cytologia* 7: 396–423.

ICRISAT (International Crops Research Institute for the Semi-arid Tropics). 1980. Groundnut: Cytogenetics. Annual Report 1978–79. ICRISAT, Patancheru, India, pp. 145–149.

ICRISAT (International Crops Research Institue for the Semi-arid Tropics). 1985. Groundnut: Barriers to hybridization. Annual Report 1984. ICRISAT, Patancheru, India, pp. 234–236.

Isleib, T.G., and Wynne, J.C. 1983. Heterosis in test crosses of 27 exotic peanut cultivars. *Crop Sci.* 23: 832–841.

Jadhav, G.D., and Shinde, N.N. 1979. Genetic studies in groundnut (*Arachis hypogaea* L.). *Indian J. Agric. Res.* 3: 93–96.

Johansen, E.L., and Smith, B.W. 1956 *Arachis hypogaea* × *Arachis diogoi.* Embryo and seed failure. *Amer. J. Bot.* 43: 250–258.

Johnson, B.B. 1981. Embryo-ovule culture. *Environ. Exp. Bot.* 21: 439 (Abstr.)

Kawakami, J. 1930. Chromosome numbers in Leguminosae. *Bot. Mag.* Tokyo 44: 319–328.

Kohert, G., Halward, T., Branch, W.D., and Simpson, C.E. 1991. RFLP variability in peanut (*Arachis hypogaea* L.) cultivars and wild species.*Theor. Apl. Genet.* 81: 565–570.

Krapovickas, A. 1968. Origen, variabilidad y diffusion del mani (*Arachis hypogaea*). pp. 517–53. In *Acta y Mem. XXXVII Cong. Int. Americanistas,* II. English translation in *The Domestication and Exploitation of Plants and Animals.* P.J. Ucko and G. Dimbleby, eds. Duckworth, London, pp. 427–441, 1969.

Krapovickas, A., and Rigoni, V.A. 1949. Chromosomas de una especies silvestre de *Arachis. Idia* 2: 23–24.

Krapovickas, A., and Rigoni, V.A. 1957. Neuvas especies de *Arachis* vinculaldas al problema del origin del mani. *Darwiniana* 11: 431–455.

Krapovickas, A., and Rigoni, V.A. 1960. La nomenclatura de las subsespecies y variabilidades de *Arachis hypogaea* L. *Revista de investigaciones agricolas* 14: 197–228.

Kumar, P., and Yadava, T.P. 1979. Estimates of variation and heritability in groundnut. *J. Res. Haryana Agric. Univ.* 9: 231–234.

Lu, J., Mayer, A., and Pickersgille, B. 1990. Stigma morphology and pollination in *Arachis* L. (Leguminosae). *Ann. Bot.* 66: 73–82.

Madhavi, K.S. 1988. Studies on some morphological and physiological characters with specific reference to their inter-relationship and heterosis in *Arachis hypogaea* L. M.Sc. thesis, Andhra Pradesh Agricultural University, Hyderabad, India.

Mahapatra, L.N. 1966. Preliminary studies on correlation of yield with other characters of groundnut. *Indian Agric.* 10: 97–106.

Mallikarjuna, N., and Sastri, D.C. 1985a. Utilization of incompatible species in *Arachis:* sequential hormone applications. Pages 147–151. In *Proc. Int. Workshop on Cytogenetics of* Arachis, 31 Oct. to 2 Nov. 1983, ICRISAT Centre, Patancheru, India.

Mallikarjuna, N., and Sastri, D.C. 1985b. *In vitro* culture of ovules and embryos from some incompatible interspecific crosses in the genus *Arachis* L. Pages 153–163 in *Proc. Int. Workshop on Cytogenetics of* Arachis. 31 Oct. to 2 Nov. 1983. ICRISAT Centre, Patancheru, India.

Matlock, R.S., Banks, D.J., Tai, Y.P., and Kriby, J.S., 1970, Inheritance of a narrow-leaflet character in peanuts, *Arachis hypogaea* L. Agronomy Abstracts, American Society of Agronomy, p. 15.

McCloud D., Duncan, W.G., McGraw, R.L., Sibale, P.K., Ingram, K.T., Dreyer., J., and Campbell, I.S. 1980. Physiological basis for increased yield potential in peanuts. *Proc. Int. Workshop on Groundnut,* 13–17 Oct. 1980. ICRISAT Centre, Patancheru, India, pp. 125–132.

Mendes, A.J.T. 1947. Estudos citologicos no genero *Arachis. Braganita* 7: 257–267.

Mohammed, J., Wynne, J.C., and Rawlings, J.P. 1978. Early generation variability and heritability estimates in crosses of virginia and spanish peanuts. *Oleagineux* 33: 81–86.

Moss, J.P., 1980. Wild species in the improvement of groundnuts. pp. 525–535. In *Advances in Legume Science: Proc. Int. Legume Conf.,* 24–29 July 1978, Kew, Surrey. vol. 2. R.J. Summerfield and A.H. Bunting, eds. Royal Botanic Gardens, Kew, Surrey.

Moss, J.P., and Sastri, D.C. 1986. Wide hybridization in *Arachis:* problems and prospects. In *New Genetical Approaches to Crop Improvement.* A.S. Siddiqui and A.M. Faruqui, eds. PIDC Printing Press (Pvt.) Ltd., Karachi, Pakistan, pp. 41–50.

Moss. J.P., Dutt, N.R.G., and Ligamaneni, A. 1988a. Root induction on *in vitro* grown shoots of *Arachis* species and a hybrid. *Int.* Arachis *Newslett.* 4: 25–26.

Moss, J.P., Stalker, H.T., and Pattee, H.E. 1988b. Embryo rescue in wide crosses in *Arachis.* 1. Culture of ovules in pet tips of *Arachis hypogaea. Ann. Bot.* 61: 1–7.

Nagabhushanam, G.V.S., and Prasad, M.V.R. 1992a. Selection criteria for groundnut (*Arachis hypogaea* L.) breeding. *Oleagineux* 47(1): 23–28.

Nagabhushanam, G.V.S., and Prasad, M.V.R. 1992b. Response of canopy and yield attributes of groundnut (*Arachis hypogaea* L.) to chemical mutagen sodium azide. *Indian J. Agric. Sci.* 62(5).

Nagabhushanam, G.V.S., Prasad, M.V.R., and Jagadish, C.A. 1990 A note on the aerial pod bearing variant of groundnut *Arachis hypogaea* L. *J. Oilseeds Res.* 7: 84–85.

Nageshwara Rao, R.C. 1992. Some crop physiological approaches for groundnut improvement. Paper presented at the XL Annual Kharif Oilseed Research Workers Group Meeting, UAS, Dharwad, 21–24 Apr.

Nalini, M., and Sastri, D.C. 1985a. Utilization of incompatible species in *Arachis*: Sequential hormone applications. In *Proc. Int. Workshop on Cytogenetics of Arachis,* 31 Oct. to 2 Nov. 1983. ICRISAT Centre Patancheru, India, pp. 147–151.

Nalini, M., and Sastri, D.C. 1985b. *In vitro* culture of ovules and embroys from some incompatible interspecific crosses in the genus *Arachis* L. In *Proc. Int. Workshop on Cytogenetics of* Arachis, 31 Oct. to 2 Nov. 1983. ICRISAT Centre, Patancheru, India, pp. 153–159.

Nigam, S.N., Dwivedi, S.L., Sigamani, T.S.N., and Gibbons, R.W. 1984. Character association among vegetative and reproductive traits in advanced generation of intersubspecific and intrasubspecific crosses in peanut. *Peanut Sci.* 11: 95–98.

Norden, A.J. 1973. Breeding of the cultivated peanut. In *Peanuts—Culture and Uses.* American Peanut Research and Education Association, Stillwater, Okalahoma, pp. 175–208.

Norden, A.J., Gorbed, D.W., Knauft, D.A., and Martin, F.G. 1986. Genotypes × environment interactions in peanut multiline populations. *Crop Sci.* 26: 46–48.

Oelck, M.M., Bapat, V.A., and Schieder, O. 1982. Protoplast culture of three legumes: *Arachis hypogaea, Melilotus officinalis* and *Trifolium resupinatum. Z. Pflanzenphysiologie* 106: 173–177.

Ozias-Akins, P., and Branch, W.D. 1990. Incompatibility during late embroygeny in some crosses of *Arachis hypogaea* × *A. stenosperma* and the utility of three culture methods for hybrid rescue. *Proc. Amer. Peanut Res. Educ. Soc.* 22: 51 (Abstr.).

Pattee, H.E. and Stalker, H.T. 1991. Observations of reproductive efficiency in crosses between cv NC 6 and two *Arachis* species. Page 74, Abst. Symp. on Plant Breeding in the 1990s. Research Report No. 130. North Carolina State University, Raleigh, North Carolina.

Pattee, H.E., Stalker, H.T., and Moss, J.P. 1988. Embryo rescue in wide crosses in *Arachis.* 2. Embryo development in cultured peg tips of *Arachis hypogaea. Ann. Bot.* 61: 103–112.

Prasad, M.V.R. 1985. Aerial podding in groundnut *Arachis hypogaea* L. *Indian J. Genet.* 45: 89–91.

Prasad, M.V.R. 1988. Genetic restructuring of groundnut plant (*Arachis hypogaea* L.). *Proc. Conf. Cytol. Genet.* 1: 247–254.

Prasad, M.V.R. 1993. Genetic enhancement of groundnut for higher productivity based on canopy development. Nat. Sem. Oilseeds Res. and Dev.: Status and Strategies. Extended Summaries. *Indian Soc. Oilseeds Res.* DOR. 1–5.

Prasad, M.V.R., and Kaul, S. 1980. Problems and prospects in the genetic improvement of groundnut. *Proc. Nat. Sem. Application of Genetics to Improvement of Groundnut,* Tamil Nadu Agric. Univ., Coimbatore, 16–17 July. pp. 127–138.

Prasad, M.V.R. and Muralidharudu, Y. 1991. Calcium uptake in relation to pod development in aerial podding genotype of groundnut (*Arachis hypogaea* L.). Paper for poster session of II International Groundnut Workshop Nov. 25–30. ICRISAT, Patancheru, India.

Prasad, M.V.R., Swarnlatha Kaul, and Jain, H.K. 1984. Induced mutants of peanut (*Archais. hypogaea* L.) for canopy and pod bearing characters. *Indian J. Genet.* 44: 25–34.

Raman, V.S., and Kesavan, P.C. 1962. Studies on a diploid interspecific hybrid in *Arachis Nucleus,* Calcutta 5: 123–126.

Rao, N.G.P. 1976. Groundnut in India. Present status and future strategy. Indian Agric. Res. Inst. Reg. Stn., Hyderabad, Andhra Pradesh, 56 pp.

Reddi, M.V. 1980. Strategy for varietal improvement in groundnut (*Arachis hypogaea* L.). In *Proc. Nat. Sem. Application of Genetics to Improvement of Groundnut,* 16–17 July. Tamil Nadu Agric. Univ., Coimbatore, pp. 191–202.

Reddy, P.S. 1988. Genetics, breeding and varieties. In *Groundnut.* Indian Council of Agricultural Research, New Delhi.

Resslar, P.M., and Gregory, W.C. 1979. A cytological study of three diploids species of the genus *Arachis* l. *J. Hered* 70: 13–16.

Richter, C.G. 1899. Beitrage zur Biologie von *Arachis hypogaea,* p. 39. A. Schrieber, ed. Inaug. Diss., Kgl. Bot. Gart. Breslau.

Sastri, D.C., and Moss, J.P. 1982. Effects of growth regulators on incompatible crosses in the genus *Arachis* L. *J. Exp. Bot.* 53: 1293–1301.

Schilling, T.T., Mozingo, R.W., Wynne, J.C., and Isleib, T.G. 1983. A comparison of peanut multlilines and component lines across environments. *Crop Sci.* 23: 101–105.

Seshadri, C.R. 1956. Future of oilseeds research: groundnut. *Indian Oilseeds J.* 1: 32–37.

Shchori, Y., and Ashri, A. 1970 Inheritance of several macro mutations induced by diethyl sulphate in peanuts *Arachis hypogaea* L. *Radiation Bot.* 10: 551–555.

Shushu, D.D., and Cutter, E.G. 1990. Growth of the gynophore of the peanut *Arachis hypogaea* 2. Regulation of growth. *Can. J. Bot.* 68: 965–978.

Simpson, C.E., and Davis, K.S. 1983. Meiotic behavior of a male-fertile triploid *Arachis* L. hybrid. *Crop Sci.* 23: 581–584.

Sinclair, and De Wit (1975) Photosynthate and nitrogen requirements for seed production by various crops. *Science* 189: 565.

Singh, A.K. 1986. Utilization of wild relatives in the genetic improvement of *Arachis hypogaea* L. 8. Synthetic amphiploids and their importance in interspecific breeding. *Theor. Appl. Genet.* 72: 433–459.

Singh, A.K. 1988. Putative genome donors of *Arachis hypogaea* (Fabaceae), evidence from crosses with synthetic amphiploids. *Plant System. Evol.* 160: 143–151.

Singh, A.K., and Moss, J.P. 1982. Utilization of wild relatives in genetic improvement of *Arachis hypogaea* L. Part 2. Chromosome complements of species in section *Arachis. Theor. Appl. Genet.* 61: 305–314.

Singh, A.K., and Moss, J.P. 1984a. Utillization of wild relatives in genetic improvement of *Arachis hypogaea* L. 5. Genome analysis in section *Arachis* and its implications in gene transfer. *Theor. Appl. Genet.* 68: 355-364.

Singh, A.K., and Moss, J.P. 1984b. Utilization of wild relatives in genetic improvement of *Arachis hypogaea* L. VI. Fertility in triploids: Cytological basis and breeding implications. *Peanut Sci.* 11: 17–21.

Smartt, J., and Gregory, W.C. 1967. Interspecific cross-compatibility between the cultivated peanut *Arachis hypogaea* L. and other members of the genus *Arachis. Oleagineux* 22: 1–5.

Smartt, J., Gregory, W.C., and Gregory, M.P. 1978a. The genomes of *Arachis hypogaea.* 1. Cytogenetic studies of putative genome donors. *Euphytica* 27: 665–675.

Smartt, J., Gregory, W.C., and Gregory, M.P. 1978b. The genomes of *Arachis hypogaea.* 2. The implications in interspecific breeding. *Euphytica* 27: 677–680.

Spielman, I.V., Burge, A.P., and Moss, J.P. 1979. Chromosome loss and meiotic behavior in interspecific bybrids in the genus *Arachis* L. and their implications in breeding for disease resistance. *Z. Pflanzenzuchtung* 83: 236–250.

Stalker, H.T. 1978. Cytological analysis of Erectoides × *Arachis* intersectional hybrids. *Proc. APREA* 10: 62 (Abstr.)

Stalker, H.T. 1980. Cytogenetic investigations in the genus *Arachis.* In *Proc. Int. Workshop on Groundnuts,* ICRISAT Centre, Patancheru, India, pp. 73–81.

Stalker, H.T. 1985. Cytotaxonomy of *Arachis.* In *Proc. Int. Workshop on Cytogenetics of* Arachis, 31 Oct. to 2 Nov. 1983. ICRISAT Centre, Patancheru, India, pp. 65–79.

Stalker, H.T. 1991. A new species in section *Arachis* peanuts with a D genome. *Amer. J. Bot.* 78: 630–637.

Stalker, H.T. 1992. Utilizing *Arachis* germplasm resources: In *Groundnut—a Global Perspective.* S.N. Nigam, ed. Proc. of International Workshop, 25–29 Nov. 1991. ICRISAT Centre, Patancheru, India, pp. 281–295.

Stalker, H.T. and Dalmacio, R.D. 1981. Chromosomes of *Arachis* species section *Arachis. J. Hered.* 72: 403–408.

Stalker, H.T., and Eweda, M.A. 1988. Ovule and embryo culture of *Arachis hypogaea* and interspecific hybrids. *Peanut Sci.* 15: 98–104.

Stalker, H.T., and Moss, J.P. 1987. Speciation, cytogenetics, and utilization of *Arachis* species. *Adv. Agron.* 41: 1–40.

Stalker, H.T., and Wynne, J.C. 1979. Cytology of interspecfic hybrids in section *Arachis* of peanuts. *Peanut Sci.* 6: 110–114.

Stalker, H.T., Seitz, M.H. and Reece, P. 1987. Effect of gibberellic acid on pegging and seed set in *Arachis* species. *Peanut Sci.* 14: 21-25.

Stalker, H.T., Dhesi, J.S., and Parry, D.C. 1991. An analysis of the B genome species *A. batizocoi* (Fabaceae). *Plant System. Evol.* 174: 159–169.

Stokes, W.E., and Hull, F.H. 1930, Peanut breeding. *J. Amer. Soc. Agron.* 22: 1004–1019.

Tai, P.Y.P., Hammons, R.O., and Matlock, R.S. 1977. Genetic relationships among three chlorophyll deficient mutants in peanuts, *Arachis hypogaea* L. *Theor. Appl. Genet.* 50: 35–40.

Thangavelu, S., Radhmanoharan, G., and Vindhya Varman, P. 1980. Studies on pod setting in interspecific crosses of *Arachis*. In *Proc. Nat. Sem. Application of Genetics to Improvement of Groundnut,* 16–17 July. Tamil Nadu Agric. Univ., Coimbatore, pp. 20–26.

Thompson, L.K., Ziv, M., and Deitzer, G.F. 1985. Photocontrol of peanut (*Arachis hypogaea* L.) embryo and ovule development *in vitro. Plant Physiol.* 78: 370–373.

Umen, D.F. 1933. What has been done in groundnut breeding, Lenin Academic Agricultural Institute of Science and Research Oil Curt Krasnodar 5: 9–12 (*Plant Breed. Abst.* 5: 60).

Van der Stok, J.E. 1910. Onderzoekingen omtrend, *Arachis hypogaea* L. (Katang-Tanah). *Mededelingen ven het Department van Landbouw* 12: 176–221.

Vanisree, G. 1992. Analysis of breeding potential of certain induced mutants of groundnut (*Arachis hypogaea* L.) for canopy and reproductive attributes. Ph.D. thesis, Andhra Pradesh Agric. Univ., Hyderabad, India.

Varan, P.V., Rathnasamy, R., Ramalingam, R.S., and Bhat, M.V. 1986 Inheritance of growth habit in groundnut. *Madras Agric. J.* 73: 299–300.

Varnell, R.J., and McCloud, D.E. 1975. Germplasm preservation and genotype evaluation in *Arachis.* International Peanut Program, Gainesville, Florida, pp. 1–19.

Venkateswaran, A.N. 1980. Discriminant function as a tool in groundnut breeding, In *Proc. Nat. Sem. Application of Genetics to Improvement of Groundnut,* 16–17 July. Tamil Nadu Agric. Univ., Coimbatore, pp. 20–26.

Virmani, S.M., and Piara Singh, 1986. Agroclimatological characteristics of the groundnut growing regions in the semi-arid tropics. Proc. Int. Symp. 21–26 Aug. 1985, ICRISAT, Sahelian Center, Niamey, Niger. ICRISAT, Patancheru, India.

Waldron, R.A. 1919. The peanut (*Arachis hypogaea*), its history, histology and utility. *Penn. Univ. Bot. Lab. Contrib.* 4: 301–338.

Weissinger, A.K. 1992. Biotechnology for improvement of groundnut (*Arachis hypogaea*). In *Groundnut—a Global Perspective: Proc. Int. Workshop,* 25–29 Nov. 1991. S.N. Nigam, ed. ICRISAT Centre, Patancheru, India, pp. 317–328.

Wynne, J.C., and Coffelt, T.A. 1982. Genetics of *Arachis hypogaea* L. In *Peanut Science and Technology.* American Peanut Research and Education Society, Yoakum, Texas. pp. 50–94.

Wynne, J.C., and Gregory, W.C. 1981. Peanut breeding. *Adv. Agron.* 34: 39–72.

Wynne, J.C. and Isleib, T.G. 1978. Cultivar × environment interactions in peanut yield tests. *Peanut Sci.* 5: 102–105.

Zamski, E., and Ziv, M. 1976. Pod formation and its geotropic orientation in peanut, *Arachis hypogaea* L., in relation to light and mechanical stimulus. *Ann. Bot.* 40: 631–636.

7. RAPESEED AND MUSTARD

B.Rai
Professor, Plant Breeding, Institute of Agricultural Sciences, Banaras Hindu University, Varanasi, India

Introduction

Rapeseed and mustard are important oil-yielding crops of the Indian sub-continent. They form an important component of the Indian agricultural ecosystem and are largely grown under the hardy conditions of rainfed agriculture with low input management during *rabi* season. Even under such conditions, these crops have a good ability to convert the natural resources into usable biological energy and usually give a higher input : output ratio than many other rainfed *rabi* crops. In India, Uttar Pradesh, Rajasthan, Madhya Pradesh, Punjab, Haryana, West Bengal, Assam, and Bihar are important rapeseed- and mustard-growing states. The important species of oilseed Brassicas that are extensively cultivated commercially are *Brassica campestris, B. juncea,* and *B. napus.* Of these, *B. campestris* and *B. juncea* are largely cultivated in Asia (India, China, Pakistan, Bangladesh, and Nepal), while *B. napus* and *B. campestris* are largely cultivated in Sweden, Germany, France, Canada, and Australia.

Botany and Floral Biology

Rapeseed and mustard belong to the family Cruciferae and genus *Brassica.* There are over 159 species in this genus and the 2n chromosome numbers of 16, 18, 20, 22, 24, 34, 36, 38, and 48 have been reported in different species (Table 1). These species have been used for vegetable, oilseed, or forage production. There are six species of *Brassica* that are generally used for oilseed production. These are *B. campestris, B. juncea, B. napus, B. carinata, B. alba,* and *B. tournefortii. Brassica campestris* var. *toria* or *sarson* is generally termed 'rapeseed' and *B. juncea* is termed 'mustard'. The distingushing features of these two crop species are briefly outlined in Table 2.

The oil-bearing *Brassica* plant is profusely branched and attains a height of 150 to 250 cm at maturity. The leaves are lyrate. Inflorescence is

Table 1
Classification of various *Brassica* species with respect to their somatic chromosome number

Group	Somatic chromosome number	Species	Possible use
I	16	*B. nigra* Koch	Condiment
II	18	*B. obracea* L.	Vegetable
		B. alboglabra L.	Vegetable
		B. sinapistrium	—
III	20	*B. campestris* L.	Oilseeds
		B. rapa L.	Oilseeds
		B. chinensis L.	Oilseeds
		B. japonica Sieb.	Oilseeds
		B. pekinensis	Oilseeds
		B. narinosa Bailey	Oilseeds
		B. tournefortii Gouan	Oilseeds
IV	22	*Eruca sativa*	Oilseeds
V	24	*B. alba* Rab.	Oilseeds
		B. monensis	—
		B. wrightii	—
VI	34	*B. carinata*	Oilseeds
VII	36	*B. juncea* Czern & Coss	Oilseeds
		B. cernua	Fodder
VIII	38	*B. napus*	Oilseeds
		B. napella	—
		B. rugosa	Vegetable
		B. napobrassica	Vegetable
IX	48	*B. cherianthus*	—

Table 2
The major distingushing features of rapeseed (*B. campestris* L.) and mustrad (*B. juncea*)

Characteristic	Rapeseed (*toria, sarson*)	Mustard (*rai* or *raya*)
Leaves	Leaf blade encircles the stem	Leaf blade does not encircle the stem
Plant height	Short to medium (1 to 1.5 m)	Medium to tall (1.5 to 2.5 m)
Days to maturity	90–140	120–150
Surface of siliqua	Uniform and smooth	Beaded or wavy
Seed surface	Smooth	Reticulate
Somatic chromosome number	20	36
Reaction to leaf spot disease	Susceptible	Tolerant

racemose and at the bud stage it looks like a head with the oldest buds at periphery and the youngest ones in the centre. The flowers are yellow, characteristically four-petalled, arranged in the 'cross-form', complete, and bisexual. The flowers in self-incompatible types are bright yellow, widely spread, open, and very attractive. They have four nectaries and the percentage of sugar in the sap of these nectaries varies from 41 to 61 per cent. The flowers in the self-compatible types are light yellow and less open and the sucrose content in their nectaries usually varies from 10 to 12 per cent. The anthers are six in number and are arranged in tetradynamous (4 + 2) condition. The carpel is superior and is bi- or tetralocular and is divided by a false septum. The number of ovules in an ovary varies from 10 to 50. The anthers are at a lower level than the stigmas at the bud stage, but before flower opening, four of them elongate and carry the anthers as high as or above the stigma level.

The plants come to flowering in about 35–75 days after planting. The flowers start opening at 8.00 a.m. and continue to open till 2 p.m. They usually keep open for 2 to 4 days. The stigmas remain receptive 3 days before opening to 3 days after opening of the flower. The pollen viability lasts for about 24 hours. However, under experimental conditions, it has been observed that pollen could remain functional for a year or more in dry storage at –20°C (Brown and Dyer, 1990).

There is a range of cross-pollination in this group of oilseed crops. Toria, *taramira,* and *lotni* brown *sarson* are self-incompatible and cross-pollinated. Mustard (*rai*); yellow *sarson,* and *tora* brown *sarson* are self-compatible. Self-fertile forms are largely (85–95%) self-pollinated. However, owing to insects, specially the honeybees, the extent of cross-pollination in this group varies from 4–16.6 per cent (Singh, 1958; Rai and Singh, 1976; Ram Bhajan *et al.,* 1991). In *toria,* the pollen from the same plant usually requires 24–48 hours to reach the ovule for self-fertilization, whereas the alien pollen grains take usually 5–6 hours to reach the ovules.

The evolution of mating system in genus *Brassica* is interesting. It appears that there is strong intergenomic interaction affecting the mode of pollination. The three primary species, B. *campestris,* B. *nigra,* and *B. oleracea,* with A, B, and C genomes are cross-pollinated, while their amphidiploid products, *B. juncea* (AABB), B. *napus* (AACC), and *B. carinata* (BBCC), are predominantly self-pollinated. The commercially cultivated species *B. campestris,* however, contains both self-compatible and self-incompatible forms. In this species, yellow *sarson* and *tora* brown *sarson* are self-compatible, while *toria* and *lotni* brown *sarson* are self-incompatible. *Lotni* brown *sarson,* which is self-incompatible and cross-pollinated, appears to be the progenitor of all these forms. In this species, the evolution of forms has followed two independent pathways. On one hand *toria* has evolved, retaining the self-incompatibility gene complex. There is very good cross-compatibility, chromosomal homology, and morphological similarity between *lotni* brown *sarson* and *toria.* The only visible difference between them is the days taken to crop maturity. Toria usually takes 90–110 days, while *lotni* brown *sarson* types mature in 125–145 days. On the other hand, *tora* brown *sarson* has evolved from the *lotni* types, primarily because of the preference of the farmers for bold-seeded, uniformly maturing and tall-growing plants for the mixed cropping systems. In the long process of human selection, the self-incompatibility mechanism has been lost. The yellow *sarson* has evolved from *tora* brown *sarson,* as a consequence of macro-mutation for the seed colour.

Selfing and Crossing Techniques

In this group of crops, emasculation is necessary to make crosses in the self-compatible and self-pollinated forms. However, in self-incompatible forms, hand emasculation may not be necessary, if one is sure of the presence of the self-incompatibility gene in the plant. To attempt hand crosses, the open flowers and very immature buds are removed, retaining five to eight well-developed buds for emasculation. Then, two fine slits are gently made in two directions on each of the bud; the sepals and petals are separated and the six immature anthers are removed by their filaments. The buds are closed, labelled, and covered with either a thin muslin cloth bag or a perforated silk bag and supported by a stick. Next morning, ripe dehiscing anthers are collected and pollinations are made by dusting pollen from the ripened and dehiscing anthers over the stigma. One ripened flower can pollinate four to five female flowers. After pollination, flower buds are bagged again. To avoid natural cross-pollination, flowers to be selfed should be bagged before they open. The bags should be large enough to allow proper lengthening of the inflorescence. Cross-pollination of self-incompatible types of *toria* or *lotni* brown *sarson* can be obtained by growing them under the insect-proof cages in which honeybees are kept or by isolating the fields at a safe distance (400 m) from other plants of the same crop.

Genetics of Qualitative Characters

Available information on the genetics of qualitative characters in rapeseed and mustard mostly pertains to the leaf shape, leaf colour, branching pattern, flower colour, blooming time, and seed colour. Unlike maize, wheat, or barley, the precise information on the linkage groups and the gene-chromosome relationship is lacking in these crops. The inheritance patterns have been studied without assigning specific genes to specific chromosomes. The information on the mode of inheritance ralating to qualitative characters in B. juncea an B. campestris has been briefly summarized in Table 3. As could be visualized from this table, a number of morphological attributes of the plant and flower and seed characteristics are reported to be governed by monogenic (3 : 1) inheritance. However, digenic inheritance has also been observed in *B. juncea* for leaf colour, petal colour, and blooming period. Though, in *B. juncea*, hairiness has been observed to be dominant over glabrousness and segregates as monogenic ratio (3 : 1), the degree of hairiness has been observed to be controlled by polygenes.

Table 3

Genetics of some of the qualitative characters in rapeseed and mustard

Characteristic	Genetic ratio in F_2	Reference
B. juncea		
1. *Leaf colour*		
Green vs. purple	3 : 1	Singh *et al.*, 1964
	9 : 7	Yadava *et al.*, 1976
Whitish-green vs. normal green	15 :1	Singh *et al.*, 1964
2. *Leaf surface*		
Hiary vs. glabrous	3 : 1	Singh, 1960 (unpublished)
3. *Branching pattern*		
Basal primary vs. high primary branching	3 : 1	Singh, 1960 (unpublished)
Open vs. adpressed	3 : 1	Nayar and George, 1970
	3 : 1	Rai, 1977 (unpublished)
4. *Petal colour*		
Creamy yellow vs. yellow	12 : 3 : 1 (yellow : pale yellow : white)	Singh *et al.*, 1964; Rawat and Anand, 1986; Brar *et al.*, 1991.
5. *Siliqua arrangement*		
Open vs. adpressed	9 : 7	Brar *et al.*, 1991
6. *Seed colour*		
Yellow vs. brown	3 : 1	Singh, 1960 (unpublished)
Blackish-brown vs. yellow	3 : 1	Nayar and George, 1970
B. campestris var. *toria*		
1. *Plant height*		
Tall vs. dwarf	3 : 1	Rai and Kumar, 1978a, b
	3 : 1	Tyagi *et al.*, 1983
2. *Flower colour*		
Yellow vs. creamy	3 : 1	Rai, 1977 (unpublished)

Cont.

Table 3 Contd.

Characteristic	Genetic ratio in F_2	Reference
3. *Stigma condition*		
Normal vs. protruded	3 : 1	Rai, 1977 (unpublished)
B. campestris var. brown *sarson*		
1. Flower opening		
Normal vs. closed	3 : 1	Singh, 1960 (unpublished)
2. *Apetaly*		
Normal vs. apetalous	3 : 1	Singh, 1960 (unpublished)
3. *Petal merger*		
Free vs. overlapping	3 : 1	Singh, 1960 (unpublished)
B. campestris var. yellow *sarson*		
1. *Plan height*		
Normal vs. dwarf	3 : 1	Hawk and Crowder, 1978
2. *Hypocotyl colour*		
Purple vs. green	3 : 1	Stringam, 1971
3. *Leaf shape*		
Normal vs. acute	3 : 1	Rai, 1977 (unpublished)
4. *Flower colour*		
Yellow vs. white	3 : 1	Richharia, 1957
5. *Petal shape*		
Normal vs. tucked	3 : 1	Hawk and Crowder, 1978
6. *Anther condition*		
Extrorse vs. introrse	3 : 1	Ali Mohammed and Sikka, 1937
7. *Pods*		
Bivalved vs. multivalved	3 : 1	Mohammed *et al.*, 1943
8. *Ovary*		
Normal vs. axial	3 : 1	Singh, 1960 (unpublished)

Brown seed colour is dominant over the yellow in *sarson* and *toria* varieties. However, more than one gene is responsible for the expression of the various grades or shades of brown colour. The various grades of brown colour are controlled by a series of polymorphic factors; the gene for dark reddish-brown colour is designated as Br_1, reddish brown as Br_2, and yellow brown as Br_3. They assort independently (Mohammed *et al.*, 1943). In *sarson,* seed colour is controlled by by two dominant genes, Br_1 and Br_3, but in addition to these, there are some minor genes modifying seed colour (Zaman, 1987). Yellow flower colour has been observed to be dominant over white and the adpressed siliqua arrangement is dominant over the normal. Both the characters are governed by a single gene. The F_2 segregation of multivalved and bivalved siliqua indicated that bivalved condition is dominant over multivalved condition (Singh and Srivastava, 1976). Absence of bloom in two mutants of yellow *sarson* is determined by two different single recessive alleles (Singh and Srivastava, 1976; Stringam, 1976). Straight siliqua condition is dominant over the curved siliqua condition (Singh, 1958).

Genetics of Quantitative Characters

The precise estimation of the nature of gene action for major quantitative characters of economic importance in oilseed Brassicas has now come to stay as an *a priori* requirement for their yield improvement programmes. In these crops, the major emphasis in the past was laid on increasing the seed yield and the oil yield, though due importance is now being given to improving their biochemical attributes from the human nutrition point of view.

Table 4

Genetics of some of the important quantitative characters in rapeseed and mustard

Character	Gene action	Reference
Flowering time	Additive	Singh, 1973; Tewari and Singh, 1973; Asthana and Pandey, 1977; Choudhury and Sharma, 1982; Singh *et al.*, 1982; Duhoon *et al.*, 1982; Singh, 1986
Days to maturity	Additive and non-additive	Yadava *et al.*, 1976
	Non-additive	Yadava *et al.*, 1979
		Yadava *et al.*, 1981
1000-seed weight	Additive	Singh, 1986
Seed yield	Additive	Singh, 1973; Badwal *et al.*, 1976; Paul *et al.*, 1976; Asthana and Pandey, 1977; Singh, 1986
	Additive × non-additive	Yadava *et al.*, 1974; Ram *et al.*, 1976: Asthana and Pandey, 1977; Yadava, 1977; Singh *et al.*, 1982
	Non-additive	Labana *et al.*, 1975; Anand and Rawat, 1978; Yadava *et al.*, 1979; Govil *et al.*, 1981; Singh *et al.*, 1982; Duhoon *et al.*, 1982; Reddy, 1983; Thakral *et al.*, 1986
Oil content	Additive	Rawat, 1975; Pal and Singh, 1986; Singh, 1986; Han, 1990
	Additive × non-additive	Yadava and Gupta, 1975; Yadava *et al.*, 1981; Kumar *et al.*, 1982; Dixit *et al.*, 1983
	Non-additive	Govil *et al.*, 1981; Yadava *et al.*, 1981
Protein content	Additive × non-additive	Dixit *et al.*, 1983

The available information on the genetic architecture of some of the important quantitative characters in rapeseed and mustard has been briefly summarized in Table 4. In these crops, the flowering duration is mostly governed by additive gene action, though its inheritance through non-additive genetic control has also been reported. Varying proportions of additive and non-additive (dominance and epistatic) variances have been reported for days to maturity, number of seeds/siliqua, 1000-seed weight, seed yield, and oil content.

In the genetic control of seed yield, the non-additive gene action has been reported to be more predominant than the additive type of gene action, while the expression of oil content has been reported to be predominantly governed by the additive gene action rather than non-additive gene action.

Genetics of Self-incompatibility

In oilseed Brassicas, sporophytic type of self-incompatibility is present and is under the control of three alleles that show dominance relationship. Therefore, all the pollen obtained from a heterozygote plant react similarly, whether the plant that bears them is heterozygous or homozygous for the self-incompatibiltiy alleles (Rai, 1980). Recently, Lewis *et al.* (1988) have reported the presence of a novel gametophytic-sporophytic system of self-incompatibility in radish variety *radicula* from Poland and it has also been recently observed by Verma *et al.* (1991) in a Japanese radish variety, *Tokinashi.*

The genetic analysis of self-incompatibility in *taramira* has indicated that its control is complex and possibly a three-locus system governs its expression.

Two of these alleles operate like 'S' system. The control is sporophytic and the dominance relationship is present (Verma *et al.,* 1977). Studies conducted on 'Kanpur *lotni* brown *sarson*' have indicated that single locus S with at least four multiple alleles controls the self-incompatibility mechanism. Allele S_3 is dominant over S_2 in the pollen (Sareen and Kakar, 1975).

Genetics of Male Sterility and Fertility Restoration

Cytoplasmic male sterility coupled with fertility restoration system is a workable and convenient tool for the hybrid seed production. The presence of male sterility was initially observed in 1954 in a natural population of *B. oleracea* (broccoli). In the male-sterile plants, the pollen formation ceases at the tetrad stage, after which the young pollen grains degenerate. At the time of anthesis, the anthers look shrivelled, curved, smaller, and comparatively pale. In yellow *sarson,* the tapetum does not seem to function as a nutritive layer. It persists long after the degeneration of the tetrads (Chaudhury and Das, 1966). The non-dehiscence of anthers in the brown *sarson* appeared to be due to the presence of large irregular cells of anther wall in the form of a thick compact layer. The male-sterile plants usually have narrow petals that are generally lighter in shade.

A number of workers have subsequently reported the presence of cytoplasmic male sterility in oilseed Brassicas (Thompson, 1972; Shiga and Baba, 1973; Anand and Rawat, 1978; Hinata and Konno, 1979; Mishra, 1985). Two forms of cytoplasmic male sterility have been described in *B. nigra.* One of these is petaloid, in which the stamens are transformed into petals or carpels with absence of nectaries, and in the other (vestigial anther sterility), the anthers remain rudimentary without in any way affecting the nectar formation, seed setting, and other floral organs. The presence of petaloid or rudimentary anthers without pollen grains coupled with reduced size of floral parts has been reported in mustard by Rawat and Anand (1979). The presence of male sterile plants in M_{11} generation of *B. juncea* (Badwal and Labana, 1983) and a case of functional male sterility in *toria* have also been reported (Tyagi and Verma, 1985).

Shiga (1976) evaluated 131 varieties of *B. napus* for fertility restoration. Of these, 23 varieties restored fertility completely, 79 only partly, and 29 were non-restorers. Based on the type of cytoplasm (S or N) and the number of restorer genes, these varieties were classified as S-O, S-I, S-IIa, S-IIb, S-III, S-IV, N-O, and N-II group. Of the five varieties with S type cytoplasm, two had only one or two fertility-restoring genes, while the remaining three had three such genes (Shiga *et al.,* 1977). Of the 13 varieties with N type cytoplasm, 6 had one, 2 had one or two, and 5 had three fertility-restoring genes.

In *B. napus* variety Bronowski, dominant alleles at two loci were involved in the restoration of male fertility (Rousselle and Renard, 1978). In swede rape, Sernyk (1983) reported that eight loci were involved in the fertility restoration of male-sterile cytoplasm. Mishra (1985), while studying the genetics of fertility restoration, observed a digenic complementary gene action in *B. juncea* that indicated that mechanism of fertility restoration is controlled by two pairs of interacting genes in this species. It has been suggested that it is easy to find male-sterile cytoplasm and their restorers among the varieties in the original centre and maintainers among the varieties far from the original centre of varieties (Fu and Yang, 1991).

Genetics of Fatty Acids

The genetic analysis of fatty acids in *B. juncea* and *B. carinata* has indicated that erucic and ecosenoic acids are controlled by the same gene pair with additive gene action. High linoleic acid was observed to be controlled by a single dominant gene with additive gene action (Rahman, 1976). In the rapeseed varietal crosses Bele × Duro and Bele × Rapido III, the erucic acid content of the oil was observed to be governed by a single gene with incomplete dominance (Jonsson, 1974, 1977). The genetic analysis of fatty acids made between the crosses of Oro and a number of induced mutants for linoleic and linolenic fatty acids indicated that two genes govern these attributes. The gene action was mainly additive but partial dominance was also observed (Nitsch, 1975). Harvey and Downey (1964) have reported a two-gene system working in an additive way to control the expression of erucic acid. Hwe and Chen (1983) have reported that oleic and linoleic acid contents in rapeseed are controlled by the same pair of genes.

Anand and Downey (1981) working with the cross Sangam × Candle observed that the inheritance of erucic acid in *B. campestris* is governed by two alleles at single locus, contributing about 20 to 25 per cent and 30 to 35 per cent respectively towards the synthesis of erucic acid. Kirk and Hudstone (1983) crossed the zero erucic acid *B. juncea* varieties Zem 1 and Zem 2 with four lines each of European and Indian origin. They observed that there are three alleles (E_0, E_1, and E_2) conferring 0, 12, and 20 per cent erucic acid, respectively. Two loci are involved in erucic acid formation, one in each genome (*B. campestris* and *B. nigra*). European lines had E_1E_1 in one and E_0E_0 in the other genome. The two Indian lines had E_1E_1 in one and E_2E_2 in the other genome. The zero erucic acid lines had E_0E_0 genetic constitution in both the genomes in the two species. In *B. napus*, erucic acid has been observed to be controlled by the two major loci each having two alleles with additive effects. The inhertance of linolenic acid content shows a three-locus additive model (Chen and Berersdorf, 1990).

Cytogenetics

Chromosome Number and Karyotype

The number of chromosomes, satellites, and nucleoli in some of the *Brassica* species are given in Table 5. In general, the chromosome length in mustard, *B. juncea* (2n = 36), is less than in *toria*, yellow *sarson,* and *taramira*. Perhaps an increase in the number of chromosomes occurred with a decrease in the size of the chromosomes. The mean number of chiasmata per bivalent in yellow *sarson,* brown *sarson,* and *toria* has been observed to be 1.90, 1.81, and 1.84, respectively. The meiotic behaviour of *B. campestris* (2n = 20) has indicated secondary association of chromosomes, the number ranging from 1 to 4. The cytological basis of this could be secondary balanced polyploidy. Usually 10 bivalents are counted at diakinesis and metaphase stages. In *B. juncea,* 18 bivalents are formed at diakinesis. The frequency of quadrivalent formation is usually less than 3 per cent. This is an amphidiploid species derived from *B. campestris* and *B. nigra* and is expected to form only bivalents. However, the frequency of quadrivalent formation does indicate that the two parental species responsible for its origin are interrelated and have some chromosomes in common. Usually 8 and 11 bivalents are observed in *B. nigra* and *Eruca sativa* respectively at diakinesis stage. Based on the karyotypic analysis, *B. campestris* among the primary species and *B. juncea* and *B. napus* among the amphidiploid species are considered evolved ones (Du Xin and Tang Zejing, 1991).

Table 5

Number of chromosomes, satellites, and nucleoli present and chromosome types in some *Brassica* species

Species	Chromosome number (2n)	Number of different chromosme 'types'						
		Sat.	Nucleoli	Longest	Long	Med.	Small	Smallest
Primary species								
B. campestris L.	20	2	2	2	8	6	2	2
B. nigra	16	4	4	2	8	6	—	—
B. oleracea	18	2	2	2	8	6	2	—
B. rapa	20	2	2	2	8	6	2	2
B. tournefortii	20	2	2	2	8	6	2	2
Amphidiploids								
B. napus	38	4	4	4	16	12	4	2
B. juncea	36	6	6	4	16	12	2	2
B. carinata	34	6	6	4	16	12	2	-

Intra- and Interspecific Hybridization

The crosses between *toria* and yellow *sarson* varieties can be easily made but are more successful when yellow *sarson* is kept as the female parent. The *toria* × yellow *sarson* hybrids usually show pollen sterility ranging from 0.0 to 60.0 per cent. This is because *toria* is self-incompatible and yellow *sarson* is self-compatible. These crosses also lack complete homology of the chromosomes. The coefficient of hybridity between yellow *sarson* and *toria* was observed to be 0.33. The crosses between *toria* and *lotni* brown *sarson* are also easily made. These crosses show 10 bivalents at diakinesis and complete homology of the chromosomes. In certain cases, the presence of chromatin bridges during meiosis indicates inversion of the homologous chromosomes. The coefficient of hybridity between *toria* and brown *sarson* usually varies between 0.18 and 0.20.

Successful crosses involving *toria* (*B. campestris* var. *toria* × *B. rapa*, *toria* × *B. chinensis*, *toria* × *B. pekinensis*, and *toria* × *B. narinosa*) have been reported. However, crosses between *toria* and *B. tournefortii* are not successful. *Sarson* (*B. campastris* var. *sarson*) can be successfully crossed with *B. rapa*, *B. pekinensis*, and *B. narinosa*. There is better success in crosses between *B. campestris* and *B. oleracea* when *B. campestris* is used as the female parent. Crosses involving *B. nigra* × *B. tournefortii* are successful only when *B. nigra* is used as the female parent. Crosses of *B. juncea* with *B. campestris*, *B. nigra*, and *B. carinata* are successful if *B. juncea* is used as a female parent. The F_1 hybrids between *B. juncea* and *B. napus* are either partly or completely sterile and the F_2 plants resemble *B. juncea* very closely.

Amphidiploids in Brassicas have been experimentally produced by crossing two parental species and then inducing polyploidy or by crossing the tetraploids of the two parental species. In this way, *B. juncea*, *B. napus*, and *B. carinata* have been synthesized. The synthetic *B. juncea* closely resembles the natural *B. juncea* forms and crosses readily with them. The synthesized *B. juncea* has larger cell size than natural forms, but this difference is eliminated after it is grown for few generations (Olsson, 1960b). Similarly, synthetic *B. carinata* crosses readily with the naturally occurring *B. carinata*. *Brassica napus* has also been synthesized by crossing *B. campestris* and *B. oleracea* (Prakash and Raut, 1983) and *B. juncea* by crossing *B. campestris* and *B. nigra* (Raut and Prakash, 1985). The reduced fertility reported in the artificial amphidiploids is partly due to the chromosomal differences at meiosis and partly due to the unfavourable genomic interactions between the parental species. The meiotic disturbances and increased gametic and zygotic lethality are also partly responsible for the reduced fertility of the raw amphidiploids (Olsson, 1960a, b).

Genomic Analysis is Brassica

As could be seen in Table 1, the 2n chromosome number of different species of *Brassica* varies from 16 to 48. On the basis of chromosome pairing in interspecific hybrids, it is generally believed that within the genus *Brassica*, there are three primary or the monogenomic species with haploid number of 10, 8 and 9. The genomic constitution of these three species can be regarded as AA, BB, and CC (Morinaga, 1934). The affinity between these three genomes is low. Species with higher hapliod chromosome numbers, namely 17, 18, and 19, are considered to be the amphidiploids or digenomic species of the monogenomic species. This is based on the fact that in any cross between a digenomic species and a monogenomic species, A, B, or C, the chromosomes of the digenomic species pair with the corresponding genome of the monogenomic species. A cross between *B. juncea* (2n = 36, AABB) and *B. campestris* (2n = 20, AA) would produce a hybrid AAB with 28 chromosomes exhibiting, at meiosis, 10 bivalents as a result of primary pairing between A genome of *B. juncea* and A genome of B. *campestris* plus 8 unpaired B genome chromosomes. The pairing relationship in the hybrids of digenomic and monogenomic species have been given in Table 6. Some of the commercially cultivated species with higher chromosome numbers (*B. juncea, B. napus,* and *B. carinata*) have been artificially synthesized (U.N., 1935; Ramanujam and Srinivasachar, 1943; Frandsen, 1943). It provides further support to the hypothesis that species with higher chromosome number are probable amphidiploids of the monogenomic species. Even the monogenomic species with 10, 8, and 9 haploid chromosomes are considered as secondary balanced polyploids. Catcheside (1934), on the basis of secondary pairing in turinp (2n = 20), and Alam (1936) in meiotic studies of *toria, sarson,* and *taramira* regarded them as balanced secondary polyploids with a basic chromosome number of 6. Hussein and Abobakr (1976) as also Chen and Heneen (1991) have hypothesized that the basic chromosome number of Brassica may be x = 3. However, Sikka (1940) from his observations on the secondary association of three species, *B. monensis* (n = 12), *B. sinapistrium* (n = 9) and *B. nigra* (n = 8), concluded that the basic chromosome number for the genus is 5 and not 6. He

Table 6
Pairing relationship in the hybrids of digenomic and monogenomic species

Interspecific hybrid	Pairing behaviour		
	Bivalents	Univalents	Reference
B. juncea × *B. campestris* (n = 18, AB) (n = 10, A)	10	8	Sikka, 1940
B. juncea × *B. nigra* (n = 18, AB) (n = 8, B)	8	10	Morinaga, 1934
B. napus × *B. campestris* (n = 19, AC) (2n = 10, A)	10	9	U.N., 1935
B. napus × *B. oleracea* (n = 19, AC) (n = 9, C)	9–4	10–5	U.N., 1935
B. carinata × *B. oleracea* (n = 17, BC) (n = 9, C)	9	8	U.N., 1935
B. carinata × *B. oleracea* (n = 17, BC) (n = 9, C)	9	8	U.N., 1935
B. napella × *B. campestris* (n = 19, AC) (n = 10, A)	10	9	Morinaga, 1929a
B. cernua × *B. campestris* (n = 18, AB) (n = 10, A)	10	8	Morinaga, 1929b

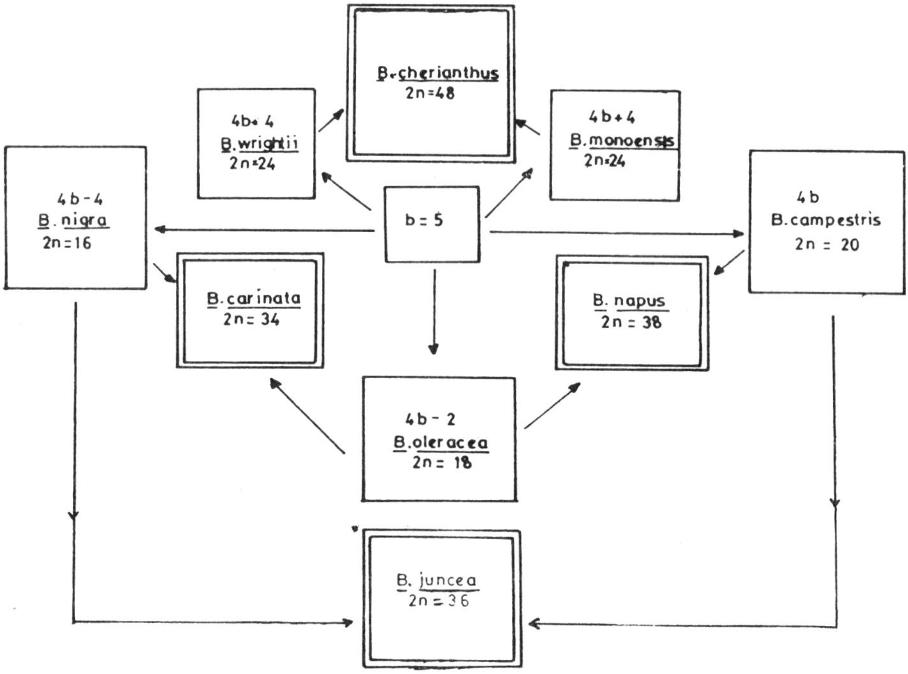

Fig. 1. Genomic relationship between different Brassica species.

suggested that about 50 per cent of the species in *Brassica* have retained the tetraploid 4b on the basis of 5 and the others have resulted from reduplication or loss of certain chromosomes from 4b number (Fig. 1).

The polyploid nature of the primary monogenomic species could also be determined from the number of satellite chromosomes and nucleoli in the somatic complement. Presence of bivalents in the monogenomic haploid species (*B. campestris*) also indicates its polyploid nature.

The three genomes A, B, and C have probably arisen from a common origin and they show weak affinity towards each other in the interspecific crosses. U.N. (1935) observed 0-8 II + 19-3 I at metaphase in F_1 hybrid between *B. campestris* (2n = 20) and *B. oleracea* (2n = 18), which indicates the affinity between A and B genomes. Similarly, in crosses between *B. napus* (2n = 38, AACC) and *B. oleracea* (2n = 18, CC), the F_1 (2n = 28, ACC) shows 0-5 III + 9-4 II + 10-5 I. In this case, the trivalent formation is due to conjugation of certain chromosomes of C *genomes* with those of *A* that must be homologous. When *B. carinata* (2n = 34, BBCC) is crossed with *B. oleracea* (2n = 18, CC), the F_1 (2n = 26, BCC) shows 9-5 III + 6-4 II + 8-12 I, which indicates the affinity between B and C genomes. Sometimes in F_1 crosses between *B. juncea* (2n = 36, AABB) and *B. campestris* (2n = 20, AA) occasionally a quadrivalent is observed at metaphase I that could evidently be due to the pairing between the homologous chromosomes of the genomes A and B.

The study of haploids gives further evidence of the common origin of the various genomes. The haploid of *B. napella* (2n = 38, AACC) showed 0 to 6 and sometimes even 7 bivalents at meiosis. In *B. napus* (2n = 38, AACC), 0 to 9 bivalents at metaphase I of meiosis show the affinity between A and C genomes. Similarly, formation of bivalents in the haploid of *B. juncea* (2n = 36, AABB) shows the affinity between A and B genomes.

It is, thus, evident from the study of cytology of different *Brassica* species as well as from the synthesis of artificial species that hybridization in this genus has played an important role in the evolution of species with higher chromosome numbers. Within the same genome, segmental interchange and inversions also seem to have played an important part in species formation. High multivalent formation in *B. nigra* and *B. wrightii* is considered partly due to the reciprocal translocations. Similarly, the crosses between *B. campestris* and *B. tournefortii* show a ring of four chromosomes. The visualization of the inversions in the hybrids of *B. campestris* × *B. tournefortii* and *B. juncea* × *B. campestris* further substantiates this hypothesis.

Tissue Culture and Gene Transfer

In the past, some attempts have been made in *Brassica* species to explore the possibilities of their genetic improvement through tissue culture technique. Somaclonal variations have been screened for various agronomic attributes. In Brassicas, the monogenomic species differ in their regenerability. *Brassica campestris* (A genome) is highly recalcitrant. *Brassica carinata* with C cytoplasm shows consistently superior performance over B cytoplasm for shoot morphogenesis. B cytoplasm favours rhizogenesis better than the C cytoplasm. The comparative responses of A, AC, and AAC cytoplasms have shown that while C genome can neutralize the inhibitory effect of A in AC, two dosages of A genome in AAC are strong enough to suppress the morphogenetic responses of AAC (Chopra *et al.,* 1986). Anuradha and Chopra (1986, 1989) achieved regeneration from the *in vitro* root culture of two varieties each of *B. juncea* and *B. napus.* Best response was obtained from the 9-day-old seedlings with regeneration and callusing frequencies being 22 and 51 per cent, respectively. *Brassica napus* (AACC) responded better than *B. juncea* (AABB). Within *B. juncea* also, the response of variety Pusa-Bold was better than that of the variety Varuna. In *B. napus,* response of variety BO-15 was better than that of variety BO-54.

Tissue culture technique has also been used to generate low glucosinolate somaclonal mutants in *B. juncea*. In this attempt, cotyledon culture has been observed to be the most efficient regeneration procedure for the Indian lines and leaf culture for the low erucic acid Zem lines (Palmer, 1986). Selection of the salt-tolerant plants from the cultured cotyledons of *B. juncea* has also been tried (Jain *et al.*, 1986; Kirti *et al.*, 1991) to develop salt-tolerant varieties in *B. juncea*. Embryo rescue technique has been used to transfer the desirable genes of *Eruca sativa* (2n = 22) to *B. campestris* (2n = 20). The tetraploids were fertile and were comparable in yield with *B. juncea* varieties and with better disease and pest resistance (Agnihotri *et al.*, 1990). Similarly, this technique has also been used in obtaining hybrids between *B. napus* and *Raphano-Brassica* (Agnihotri *et al.*, 1991).

Ayotte *et al.* (1986) attempted transfer of Atrazine resistance from *B. napus* to *B. oleracea* through tissue culture technique. Guerche *et al.* (1986) attempted direct gene transfer in *B. napus* through electroporation. They used the plasmid PABDI to transform protoplasts. This plasmid contains a chimeric kanamycin-resistant gene cloned in P^{UC9} Promoter and three regions are from the gene VI of cauliflower mosaic virus and the APH(3) II-coding sequence is from Tn5. Leaf protoplasts of *B. napus* were subjected to electroporation and were cultured. Two presumed transformed plants were regenerated and tested for the kanamycin toxicity. Hybridization was detected in the DNA of the plant, PG 20, with no corresponding band in the DNA of the control plant. The regenerated electropored plant PG 20 was observed to be resistant to kanamycin. Thus, they suggested that electroporation could be a suitable technique for the direct transfer of foreign DNA in the genus *Brassica*. The successful application of various biotechnological tools offers great promise to *Brassica* oilseed plant in future (Downey, 1990).

Breeding

Problems and Objectives

In the Indian subcontinent, the major objective of crop improvement programme in case of rapeseed and mustard is to develop high-yielding varieties. This is accomplished through breeding for better plant type, incorporation of disease and insect-pest resistance, fertilizer responsiveness, preventing losses from seed shattering, and other approaches. In addition, tolerance to drought and frost is one of the varietal requirements of these crops in western parts of India, including Punjab, Haryana, Madhya Pradesh, and Rajasthan. In contrast to this, most of the western countries focus major attention on breeding for better oil or seed meal quality, in terms of low or zero erucic acid and low or zero glucosinolates, so that the oil and the meal quality is more suitabe for the human and the livestock consumption, respectively.

High number of siliqua per plant and greater number of seeds per siliqua have been observed to be important and useful selection criteria to select for high yield (Ramanujam and Rai, 1963; Pundir and Rai, 1971; Anand *et al.,* 1975; Nagalakshmi, 1986, 1992; Shabana *et al.,* 1990). In rapeseed, a high expression of oil content is usually associated with bold seed size. Oil yield is highly associated with high seed yield and high oil content (Agrawal and Rai, 1973). Use of proline accumulation as a selection criterion in *B. napus* has been observed to be promising, because of the significant correlation between the proline content at seedling stage and high seed yield at later stages. Dry matter accumulation at rosette stage and leaf area index have also been observed to be positively correlated with the seed yield (Olsson, 1990).

Diseases and insect-pests take a heavy toll of oil-yielding crops. Alternaria leaf blight, downy mildew, and white rust are some of the important diseases of these crops in Asia. Blackleg and clubroot disease (*Plasmodiophora brassicae*) have assumed serious proportions in many western countries. Mustard aphid and mustard sawfly are some of the major insect-pests affecting the commercial production of these crops. *Brassica juncea* varieties possess comparatively better tolerance to mustard aphid than the *B. campestris* varieties (Rai and Sehgal, 1975; Rai *et al.,* 1987). A more or less similar situation has been observed with respect to the resistance to *Alternaria* leaf blight disease in oilseed *Brassica* germplasm (Rai *et al.,* 1976). However, in recent studies *B. carinata* selections have shown comparatively better tolerance than *B. campestris* or *B. juncea* (Bansal *et al.,* 1990).

In *toria,* a source of dwarfing gene has been found and it is suggested that this gene could be used in the development of dwarf varieties in *toria* and *sarson* (Rai and Kumar, 1978a, b; Tyagi *et al.,* 1983).

As indicated earlier, breeding for better oil and meal quality has assumed greater importance in western countries. A beginning has also been made in this direction in India. The mustard oil with zero or low erucic acid content is similar to groundnut or soybean oil in composition. However, consumers in eastern Indian states prefer mustard oil with high pungency for frying fish and for preserving pickles, while those in western parts of the country prefer oil with somewhat sweet smell or low pungency. Mustard oil in most of the Indian states is called *kadwa tel* meaning 'pungent oil', and its purity is usually determined by its pungency (Rai, 1976).

From breeding point of view, Brassicas have an amalgam of the breeding types. In this group, *toria, lotni* brown *sarson,* and *taramira* are highly cross-pollinated because of the presence of self-incompatibility mechanism, entomophily, very high sucrose content (40–61%) in their nectaries to attract honeybees, and the extrorse anther arrangement. In these crops, generation after generation, these mechanisms maintain a very high degree of heterozygosity and frustrate any effort of inbreeding or the fixation of the genotypes. Hence, in such outbreeding populations, breeding for high yield would obviously require adoption of a procedure that maintains the balanced heterozygosity for optimum plant productivity. This could be accomplished through selection (mass selection, recurrent selection, etc.), breeding of synthetic and composite varieties, or breeding of superior-performing commercial hybrids. In other groups of crops, including *B. juncea* and *B. naphus,* with high self-pollination (85–95%) because of the self-compatibility, introrse anther arrangement, and low sucrose content in their nectaries (5–12%), the appropriate breeding procedure would be to develop balanced homozygous populations for high plant productivity. In such crops, pure-line or mass-pedigree selection, hybridization, and mutation breeding would be more appropriate.

Germplasm Resources

The level of success of any crop breeding programme depends upon the amount of favourable genes present in the germplasm collection maintained by the breeder. Presently, over 15,000 entries of oilseed Brassicas are being maintained and evaluated and are available for use in developing new varieties at various centres in India. These genetic stocks are being maintained by sib-mating, bud pollination, or selfing or are grown in long-distance field isolations. Sometimes germplasm lines are also grown in insect-proof cages where they are pollinated by honeybees. Over years, a number of genetic stocks have been identified for desirable agronomic attributes such as earliness, tolerance to diseases, insect-pests, shattering, and frost injury, high oil content, and bold seed size (Table 7). These donor parents are being used in the hybridization programmes. However, a large number of the genetically variable populations are still being used as base populations for simple selection programmes.

Selection

Simple selection in cross-pollinated and self-pollinated Brassicas usually refers to mass selection and pure-line selection, respectively. The pure-line selection involves the principles of progeny testing, while there is no such progeny testing in the mass selection. In mass selection, the desirable plants, selected from a genetically broad-based population, are bulked and this bulked seed is grown cycle after cycle; in the pure-line selection, the selected plants are first progeny-tested in replicated trials and only those whose progeny excel in performance are carried forward. In *toria,* follow-up of one cycle of mass selection in two populations 'MN' and 'EN' of *Bhabhari lahi* land race has given a yield improvement of 8.2 and 15.4 per cent, respectively (Chaubey, 1979).

The commercial varieties Type 9, Type 36, IISA, TL 15, PT 30, BR 23, M 27, and TS 29 in *toria;* BS 2, BS 70, and BSH 1 in brown *sarson* and Type 27 in *taramira* emanated as a result of mass selection from the local germplasm. In yellow *sarson,* the improved varieties Type 151, Benoy, Type

Table 7
Some sources of donors for different agronomic attributes in rapeseed and mustard

Agronomic attribute	*B. juncea* (mustard)	*B. campestris/B. napus* (rapeseed) (*toria*)
Earliness	Seeta, RW 4-c-10, RAURPS, BR 40	TS 29, Shyamgarh local, Bhawani, TL 15
Tolerance to Aphids	T 6342, B85 (glossy), BLM 198, RH 30, RH 7513, RC 558, Kranti, RC 781, RW 3312, RF 25, RC 1329	*B. napus*
Leaf miner	RC 1008	PT 303
Mustard sawfly	Krishna	PT 303
Tolerance to Alternaria leaf blight	RC 781, YRT 3, RC 1401 TM 7, RH 8114, CSR 448, PHR 1	*B. napus:* Karat
White rust	RH 832, RK 803, 8404 RF 32, RH 8113, RC 781, Kranti	—
Downy mildew	RK 803, Kranti	—
Powdery mildew	PR 34	—
Tolerance to drought	RH 7513	—
Tolerance to seed shattering	RH 846, RH 8312, RH 30	—
Tolerance to Orobanche	Durgamani	—
Bold seed size	RH 30, RLM 619, Pusa-Bold, Kranti, Krishna	
High harvest index	RH 30, Krishna, RK 820	—
Resistance to frost injury	RH 781, RW 175, P 26/21A, RSS 3A, RL 18A	—
Sources of zero erucic acid	Zem 1, Zem 2	*B. napus,* Tower, Altex, Global Wester *B. campestris,* Torch Tobin, Span Candle.
High oil content	RLC 1312, RH 786, DIR 226	BSH-1

42, YS 66-197-3, and K 88, and in mustard, the presently grown commercial varieties Varuna, Krishna, Kranti, Shekhar, BR 40, Seeta, RH 30, and Durgamani came out of purposeful pure-line selection programmes. In brown *sarson,* Suphala variety has been developed as a result of the disruptive selection from the local variety, GBS. Following the biparental mating system, a number of BID lines in brown *sarson* have also been developed.

Hybridization

Hybridization aims at recombining a number of desirable attributes from different parents in a single variety. In oilseed Brassicas, it may be effectively used for concentrating favourable genes or gene complexes for characters such as high seed yield, high oil content, and disease and insect-pest resistance. In recent years, the mustard variety Prakash has been developed from the cross RL 18 × T9 and the Pusa-Bold has evolved from the progenies of the cross Varuna × BIC 1780. In case of mustard, Vardan developed as a result of the biparental mating of Varuna, Kesari, CSV 10, IB 1775, 1886, and 786, and Vaibhav originating from biparental crossing of Varuna, Kesari, CSV 10, IB 775, 1886, and 1786 parents, have been released for their cultivation in Uttar Pradesh and Madhya Pradesh.

In *toria,* the first national variety PT 303 has been developed from the advance generation of the intervarietal cross B54 × DS 17 M (Rai, 1987). This is a product of the *toria* × brown *sarson* cross combination, resulting in some introgression of favourable genes from the *sarson* type DS 17 M. In Haryana, hybridization has been used to develop promising entries RH 7846 and RH 8114, which, apart from giving high yield, are also resistant to mustard aphid and the *Alternaria* leaf blight.

In brown *sarson,* a promising selection, BS 34-39, has been developed as a result of biparental mating in F_2 generation of the cross Pusa Kalyani × Assam local. Similarly, another selection, BS 84-65, has been developed from the cross BSH 1 × DS 17 M. Now a number of promising selections from the interspecific hybrids of *B. juncea* × *B. napus*, such as PR 8401, PR 8402, 8404, and 8405, have been developed. These selections combine the desirable genes from both these crop species.

Breeding of Composite and Synthetic Varieties

The usual course of the varietal improvement programme in cross-pollinated crops is the selecion of improved varieties, breeding of composite and synthetic varieties, and ultimately the development of the commercial hybrids. In India, breeding of the commercial composite varieties appears to be more appropriate for stepping up the average yield level of the oilseed Brassicas as these crops are largely (80%) grown in the rainfed conditions of dryland agriculture and are usually subjected to all sorts of biotic and abiotic stresses.

The composite breeding programme in *toria* usually involves the production of a number of intervarietal hybrids and often their blends. This is followed by evaluation of inbreeding depression in seed yield from F_2 onwards and the last step involves the evaluation of the performance of advanced generations of composite against the standard or the regional/national checks. The evaluation of genetic diversity in relation to the expression of heterosis in rapeseed has indicated that crossing between *toria* and *sarson* varieties would be better than crossing between *toria* and *toria* varieties. Therefore, in *toria* composite breeding programmes, meaningful introgression of *sarson* germplasm may be useful for enhancing the yielding potential of the existing *toria* varieties. However, if one has to choose the parental population based on their geographical distribution, then crossing between East Indian *toria* cultures with the exotic (Canadian) brown *sarson* (*B. campestris*) varieties would be comparatively better and more productive than crossing among the Indian *toria* varieties themselves (Agrawal, 1976).

From the *toria* breeding programmes under way at different centres in India, two composites—PT 303 from Pantnagar and *Toria* Ludhiana Composite 1 (TLC-1) from Ludhiana—have been released for commerical cultivation. TLC-1 has been developed by compositing the seeds of 10 *toria* cultures and thereafter by following the simple recurrent selection in its advanced generation.

Development of the synthetic varieties requires selection of the tested inbred lines based on their superior general combining ability, developing their cross combinations, constituting a number of experimental synthetic populations, testing their yield levels over generations and locations, and finally releasing those that excel over the standard checks. Development of the synthetic varieties using the polycross testing programme has also been suggested in these crops (Maini and Ghai,1964). In Haryana, one of the experimental synthetic populations, Syn-65, has been released as a commercial variety named Sangam. Following this breeding approach in case of *lotni* brown *sarson*, Pusa-Kalyani has been developed at IARI station, kanpur.

Development of the Commercial Hybrids

There is a wide range of cross-pollination (5–95%) in oil-yielding rapeseed and mustard crops. Suggestions have often been made to produce commercial hybrids by exploiting the non-additive type of gene action present in case of seed yield and its components. Under experimental conditions, 7–208 per cent heterosis for seed yield has been reported in the various commercially cultivated species of these crops (Singh and Mehta, 1954; Das and Rai, 1972; Asthana and Singh, 1973; Singh, 1973; Yadava *et al.*, 1974; Labana *et al.*, 1975; Agrarwal, 1976; Doloi, 1977; Patnaik and Murty, 1978; Sernyk and Stefansson, 1983; Rai, 1985; Singh, 1986). A range of 20–50 per cent yield heterosis has been considered to be adequate for its commercial exploitation (Rai, 1979). In view of this, it appears that, theoretically, it should be possible to produce the F_1 hybrids in these crops. A theoretical model for producing commercial hybrids in *Brassica* was suggested by Thompson (1964), using the triple-cross technique [(A x B) x C] x [(D x E) x F], based on the sporophytic type of self-incompatibility. However, practically it has so far not met with much success in providing a workable system for the hybrid seed production, as the marker genes tightly linked with the self-incompatibility gene system have not been identified. Presence of genetic male sterility (Choudhury and Das, 1966; Takagi, 1970; Heyn,1973, 1978; Tyagi and Verma, 1985; Chauhan *et al.*, 1986) and the protruded stigma condition (Singh, 1958; Rai, 1983) have not helped much in producing commercial hybrids. This is beacuse the lines containing genes for these characters do not provide 100 per cent male-sterile plants in production plots and, practically, it is very difficult to rogue out the 50 per cent fertile plants at the flowering stage. Therefore, cytoplasmic male sterility appears to be the only practical and workable alternative. Fortunately, the discovery of cytoplasmic male sterility in rapeseed by Thompson (1972), Benerrot *et al.* (1974), Shiga and Baba (1973), Heyn (1978), Hinata and Konno (1979), and Broda and Poplawska (1986) and in mustard by Anand and Rawat (1978), Rawat and Anand (1979), Anand *et al.* (1985), and Singh (1986) has opened up possibilities for the commercial production of hybrids in these crops. Under experimental conditions, a stable source of male sterility in *B. juncea* has been obtained by substituting the genome of this species in the cytoplasm of *B. nigra*. Cytoplasmic male sterility has also been reported by substituting the genome of *B. napus* in the cytoplasm of wild radish and that of the genome of *B. campestris* in the cytoplasm of *Diplotaxis muralis*. At IARI, New Delhi, cytoplasmic male sterility has been transferred in the agronomic background of the mustard varieties Pusa Bold, Varuna, RLM 198, and Prakash through recurrent backcrossing with the respective pollen parents. The male-sterile A lines so produced are being used in producing the heterotic single crosses.

To use cytoplasmic male sterility in seed-producing crops such as rapeseed and mustard, incorporation of male fertility restorer gene in the pollen parents becomes necessary. As indicated before, two to eight loci have been reported to be involved in governing the expression of fertility restoration in these crops (Rouselle and Renard, 1978). Mishra (1985) and Anand *et al.* (1986) have reported a digenic complementary gene action for restoring the fertility in male-sterile cytoplasm. At IARI, New Delhi, two partial fertility restorer (R) lines have been identified from two sources. However, stable and complete pollen fertility restorer lines are still to be identified. In the absence of such fertility restorer lines, it has been suggested that in production plot, one pollinator row of male parent should be grown after every four rows of male-sterile lines and the seeds from both the male and female parents should be harvested separately. The seeds from the two types of lines are then blended together in the proportion of 1 : 4 and used for commercial production. Mixing of seed in this proportion is considered necessary to provide enough pollen for effective seed setting in the hybrid population. Under experimental conditions, when four different proportions of male-fertile : male-sterile lines (1 : 2, 1 : 3, 1 : 4 and 1 : 5) were tried, 1 row of fertile : 4 rows of sterile lines was observed to be optimal (Mishra, 1985). However, in all these combinations, the flower shedding is usually increased with the increase in the number of male-sterile lines, resulting in a lower number of siliqua formed under open pollination conditions. To increase the percentage of seed setting in the field, the keeping of honeybee colonies (5–6 colonies/ha) would be desirable. Information on using 10 : 1 ratio of male-sterile line (A line) to pollen parent has also been reported by Pinnisch and McVetty (1990) from Canada.

Breeding for Improved Plant Type

In Brassicas, seed yield per plant is mainly determined by number of branches, number of siliqua per plant, number of seeds per siliqua, and 1000-seed weight. Main shoot accounts for about 50 per cent of seed yield, of which 10–15 per cent is contributed by first three branches in case of *B. napus* (Sciffert and Boelcke, 1977).

The cross-pollinated rapeseed (*toria* and the *lotni* brown *sarson*) usually possess a pithy and weak stem and spread-over branches. They are prone to lodging on application of even 60 kg N/ha. *Toria* varieties usually have high harvest index (30–40%) but generally lack capacity to produce high amount of dry matter or biomass conducive for producing high seed yield. Mustard (*B. juncea*) varieties have the genetic potential to produce two to three times more dry matter than *toria* varieties. However, harvest index of mustard varieties is poor and varies from 20 to 30 per cent. A number of mustard varieties are very leafly and are only good for fodder or vegetable purposes. A desirable and high-yielding plant type of *toria* requires a strong stem and compact branching habit as in the case of the European or Canadian rapeseed varieties. Fortunately, a new source of dwarfing gene has been obtained in *toria* that could now be profitably incorporated and used in developing a suitable plant type in *toria* as well as *sarson* (Tyagi *et al*., 1983). In brown *sarson*, a compact desirable plant has been developed from the cross BSH 1 × DS 17 M that could accommodate more plants per unit area (Kumar, 1982).

Polyploidy

Several attempts have been made in India to induce polypoidy in rapeseed and mustard through colchicine treatment (Ramanujam and Deshmukh, 1945; Parthasarthy and Rajan, 1953; Rajan, 1955; Rajan and Ahuja, 1956). Autotetraploids were sucessfully produced by the colchicine treatment in *toria* variety T 22. The tetraploid forms of this variety were more leafy and vigorous and produced bolder seeds than the diploid forms. But tetraploids were late in maturity by 10 days and had very poor seed fertility.

The poor seed setting in tetraploids resulted form the failure of fertilization of the embryo sac (Rajan and Ahuja, 1956). The autotetraploids of *B. juncea and B. nigra* usually have thicker leaves and larger buds, flowers, and seeds, and mature later than the corresponding diploid forms. They are also poor in seed fertility. In yellow *sarson*, the seed fertility of autotetraploids came down to 10 seeds/siliqua as compared to 24 seeds/siliqua in the diploid forms. In *B. juncea*, poor seed setting was observed to be due to the poor pollen fertility (71%) of the tetraploids as compared to 95 per cent pollen fertility in case of diploids. Mass pedigree selection was successful in increasing the seed fertility of tetraploids in the initial years, but it deteriorated later and finally this effort failed to produce a commercial veriety in *toria* or in yellow *sarson* (Rajan, 1955; Sikka and Rajan, 1957). Prakash and Raut (1983) synthesized early-maturing and productive amphidiploids of *B. napus* by crossing early indigenous strains of *B. campestris* var. brown *sarson* with *B. oleracea* var. *botrytis.* Polyploidy breeding has been used wih some success in breeding suitable and high-yielding fodder varieties of *Brassica* species in the West European countries.

Mutation Breeding

Mutation breeding has been observed to be useful in developing a number of commerical varieties of the rapeseed in Sweden and other European countries. In India, this approach has been used with varying success for creating new genetic variability for a number of morphological traits in these crops (Verma and Rai, 1980a, b; Mahla *et al.*, 1991). At Bhabha Atomic Research Centre, Trombay, mutagenesis has been used in creating new genetic variability for earliness, compact plant type, and yellow seed colour. Two high-yielding selections, TM 2 and TM 4, have emanated from this programme at BARC. In Punjab, as a result of mutation breeding, high-yielding and early-maturing varities RLM 198, and RLM 514 have been developed and released. Recently, a new bold-seeded variety RLM 619, developed from RL 18 as a result of gamma-ray irradiation, has also been released for its commercial cultivation in Punjab. Induced mutagenesis is now being tried for creating new usable sources of field resistance to *Alternaria* leaf blight disease in mustard.

Breeding for Desirable Oil and Seed Meal Quality

Rapeseed and mustard oil, which is a triglyceride of saturated and unsaturated fatty acids, contains a very high (45–60%) level of erucic acid in its fatty acid profile. The oil also has high glucosinolate content. A high amount of erucic acid is desirable for certain industrial uses. In Sweden, the variety Mustang has been released in 1985 for its high erucic oil content (Persson, 1986), 54 per cent as compared to the 40 per cent of the old Swedish variety Trico. Though the presence of erucic acid is neither harmful nor has an adverse effect on human diet, its reduction or elimination increases the proportion of the oleic and linoleic acids, which are usually considered desirable for human nutrition (Table 8).

Table 8
Fatty acid profile of the normal and zero erucic lines of oilseed Brassicas

		B. napus		*B. campestris*	
Fatty acid	Symbols	Normal variety (Nugget)	Zero erucic acid variety	Normal variety (Arlo)	Zero erucic acid variety
Palmitic	C 16 : 0	3.3	4.7	3.2	4.3
Stearic	C 18 : 0	1.5	1.8	1.1	0.1
Oleic	C 18 : 1	22.8	63.3	26.6	54.8
Linoleic	C 18 : 2	12.2	20.0	17.5	31.1
Linolenic	C 18 : 3	5.4	8.2	8.8	9.7
Ecosenoic	C 20 : 1	14.2	1.3	11.8	0.0
Erucic	C 22 : 1	40.6	0.0	31.0	0.0

Adapted from Downey *et al.* (1969).

Downey and Harvey (1963) have developed a gas liquid chromatographic (GLC) technique that uses only one-half of a cotyledon from a seed to test the erucic acid content. The plant with the embryo could then be grown. With this technique, early segregating population may be analysed and only genotypes with desired oil quality could be selected and carried forward. Using the technique, a number of varieties containing zero or low erucic content have been developed and released for commercial cultivation in Canada, Sweden, France, West Germany, and other European countries.

The rapeseed and mustard seed meal or cake, usually finds little use in poultry and pig feed industries, primarily, because of the presence of a high amount of glucosinolates. These compounds usually break into toxic-sulphur compounds in the intestines of the farm animals and usually impair their growth and reproductive ability. It is thus important that these compounds are either reduced or altogether eliminated through appropriate breeding procedures. With the development of high performance quantitative GLC techniques (McGregor *et al.*, 1983; Sang and Truscott, 1984; Spinks *et al.*, 1984; Brazezinski *et al.*, 1986) it is not only possible now to obtain the profiles of glucosinolates but also feasible to measure their absolute quantitative levels (McGregor and Downey, 1975). It has also been observed that the genes controlling glucosinolate content in rapeseed are either pleiotropic or in linkage with the seed filling stage and have a positive correlation with 1000-seed weight (Olivieri and Parrini, 1986). It has often been observed that genotypes with high seed yield could be lost owing to strict selection for quality characters in early segregating generations. So, it would be better to keep populations heterozygous for quality characters and select plants for these characters in advanced generations (Yongming Zhou and Houli Lin, 1991).

Using the above-mentioned GLC technique, a number of promising double zero varieties (zero erucic and zero glucosinolate) have been developed and released in some of the Western countries and the farmers are taking advantage of them. In India and its neighbouring mustard-growing countries, some useful work has been initiated only recently in this direction. It is hoped that in the years to come, it should be possible to develop double zero varieties for large-scale commercial cultivation in the Indian subcontinent as well.

Future Outlook

The current trends in rapeseed and mustard breeding indicate that to maintain the tempo of progress in yield and quality improvement of oil and seed meal, greatly expanded research efforts would be needed to solve the challenging problems ahead. The breeding of these crops would need accelerated efforts in the areas of collection of new germplasm resources, exploitation of cytoplasmic male sterility for hybrid seed production, application of new breeding approaches and techniques, chromosomal mapping studies through the use of rapid cycling populations of *Brassica*, multidisciplinary cooperation, and ultimately assimilation of many new ideas for tackling the new problems.

A vast potential exists for further collection and use of rapeseed and mustard germplasm. It would be rather desirable to maintain at least one breeding population or gene-pool containing almost the entire collection. Such a population is likely to have more natural recombinations and may provide an opportunity to break some of the undesirable genetic linkages under natural conditions.

Exploitation of heterosis in oilseed Brassicas would require concerted efforts for effective use of cytoplasmic male sterility. Till full fertility restoration mechanism is discovered, it would be wise to exploit the partial use of heterosis through blending the seed of the male-sterile and the pollinator lines in appropriate proportions.

The discovery of rapid cycling populations of *Brassica* indicates their potential use in conducting basic research. For such a study, *B. campestris* variety CERES could be useful. This variety is photoperiodically sensitive to a single long day, as early as 4 days from sowing. As many as 20 plants of this variety could be grown in a pot of 5 cm diameter (Friend, 1985). Use of this variety has provided valuable information on photoperiodic responses, promotion of flowering by sucrose and the changing pattern of enzyme activity during floral initiation (Friend, 1986).

In years to come, an exciting and challenging area of rapeseed and mustard improvement work, perhaps, would be the breeding of commercial varieties with broad-based disease and insect-pest resistance. The development of rapeseed and mustard varieties with in-built genetic resistance to devastating *Alternaria* leaf blight disease and aphids in the Asian countries and to the clubroot (*Plasmodiophora brassicae*) disease in the Western countries should receive more attention.

In future, the oilseed Brassicas should receive more attention with regard to changing their plant types to make them more suitable for growth under different cropping systems. The desired attributes of the new plant types would include increased fertilizer-responsiveness, non-lodging behaviour, compact branching, right type of maturity duration, and a built-in genetic potential to resist the ravages of harsh environments. A plant type with compact arrangement of branches on the main axis may provide more siliqua per unit area, which would ultimately increase the per hectare yield of these crops. Incorporation of dwarfing genes along with other desirable characteristics could also pay rich dividends. Presently, greater importance is being attached in the Indian subcontinent to the genetic improvement of yield *per se.* But the future may see expanded genetical and breeding investigations to improve the fatty acid profile mainly with respect to breeding varieties with low erucic acid and low glucosinolate.

The search for the genes governing thermo- and photoinsensitivity should now receive greater attention from scientists. The incorporation of such genes in rice and wheat has proved useful in expanding their area of adaptation. It is visualized that rapeseed and mustard varieties with such genes could be grown over a wide range of ecological conditions and would show increased overall production.

REFERENCES

Agnihotri, A., Gupta, U., Lakshmi Kumaran, M., Shivanna, K.R., Shyam Prakash and Jagannathan, V. 1990. *Plant Breeding* 104: 281–289.

Agnihotri, A., Lakshmi Kumaran, M., Shyam Prakash and Jagannathan, V. 1991. Embryo rescue technique for intergeneric hybridization between *B. napus* × Raphano-Brassica hybrids. *Cruciferae Newslett.* 14/15: 92–93.

Agrawal, B.L. 1976. Studies on the genetic diversity and heterosis in rapeseed (*Brassica campestris* L.). Unpublished Ph.D. thesis, GBPUAT, Pantnagar, India.

Agrawal, B.L., and Rai, B. 1973. Component analysis of the factors influencing oil production in *toria. Indian J. Genet.* 33: 142–146.

Alam, Z. 1936. Self-sterility in *Eruca sativa. J. Genet.* 32: 257–276.

Ali Mohammed and Sikka, S.M. 1937. Breeding investigations in some of the oleiferous *Brassicae* of Punjab. *Indian J. Agric. Sci.* 7: 849–862.

Anand, I.J., and Downey, R.K. 1981. A study of erucic acid alleles in digenomic rapeseed (*Brassica napus* L.). *Canadian J. Plant Sci.* 61: 199–203.

Anand, I.J., and Rawat, D.S. 1978. Exploiting cytoplasmic male sterility for heterosis breeding in mustard (*B. juncea*) *Proc. Int. Conf. on Cytogenetics and Crop Improvement*, BHU, Varanasi, March 3–6.

Anand, I.J., Mishra, P.K., and Angadi, S.P. 1986. Mechanism of male sterility in *Brassica juncea*. VI. Identification and inheritance of pollen fertility restoration. *Cruciferae Newslett.* 11: 51–53.

Anand, I.J., Mishra, P.K., and Rawat, D.S. 1985. Mechanism of male sterility in *Brassica juncea* I. Manifestation of sterility and fertility restoration. *Cruciferae Newslett.* 10: 44–46.

Anand, I.J., Singh, J.N., and Khanna, P.P. 1975. Interrelationships and diversity in yellow *sarson* (*Brasica campestris* L. var. *sarson* Prain). *Indian J. Agric. Sci.* 45: 253–258.

Anuradha, G., and Chopra, V.L. 1986. *In vitro* differentiation from root cultures of *Brassica juncea* and *B. napus*. *Cruciferae Newslett.* 11: 92–93.

Anuradha, G., and Chopra, V.L. 1989. Genotypic differences in regeneration response of *in vitro* cultured roots of *Brassica juncea* and *B. napus. Genetic Manipulation in Plants* 5: 13–18.

Asthana, A.N., and Pandey, U.K. 1977. Combining ability and rank correlations in a diallel cross of Indian Mustard *B. juncea. Exp. Agric.* 13: 71–79.

Asthana, A.N., and Singh, C.B. 1973. Hybrid vigour in *rai. Indian J. Genet.* 33: 57–63.

Ayotte, R., Harney, P.M., and Souza Machado, V. 1986. The transfer of Atrazine resistance from *B. napus* to *B. oleracea. Cruciferae Newslett.* 11: 95–96.

Badwal, S.S., and Labana, K.S. 1983. Male sterility in Indian mustard (Abstract in Part II). *XV. Int. Genet. Cong.*, New Delhi, 55:514.

Badwal, S.S., Singh, M., Labana, K.S., and Chaurasia, B.D. 1976. General vs. specific combining ability in raya (*B. juncea* L. Czern & Coss). *PAU J. Res.* 13: 319–323.

Bannerot, M., Boulidard, K., Canderon, Y., and Tempe, J. 1974. Transfer of cytoplasmic male sterility from *Raphanus sativa* to *B. oleracea*. Proc. Eucarpia Meeting on Cruciferae. *Crop Sci.* 25: 52–54.

Bansal, V.K., Seguin Swartz, G., Rakow, G.F.W., and Petric, G.A. 1990. Reaction of *Brassica* species to infestation by *Alternaria brassicae. Canadian J. Plant Sci.* 70: 1159–1162.

Brar, B.K., Dhillon, S.S., Singh, K., and Labana, K.S. 1991. Genetics of qualitative traits in Indian mustard. *Cruciferae Newslett.* 14/15: 34–55.

Brazezinski, W., Mendelewski, P., and Munse, B.G. 1986. Comparative study on determination of glucosinolates in rapeseed. *Cruciferae Newslett.* 11: 128–129.

Broda, I.B., and Poplawska, W. 1986. Male sterile forms in summer rapeseed 'Polima'. *Cruciferae Newslett.* 11: 48–49.

Brown, A.P., and Dyer, A.F. 1990. Effect of low temperature storage on the pollen of *Brassica campestris, B. oleracea* and *B. napus. Euphytica* 51: 215–218.

Catcheside, G.D. 1934. The chromosome relationships in the swede and turnip groups of Brassica. *Ann. Bot.* 48: 601–633.

Chaubey, C.N. 1979. Mass selections in *toria. Indian J. Genet.* 39: 194–201.

Chaudhary, J.B., and Das, K. 1966. Male sterility in yellow *sarson. Indian J. Genet.* 26: 374–380.

Chaudhary, S.K., and Sharma, S.K. 1982. Note on the inheritance of some quantitative characters in a cross of Indian mustard. *Indian J. Agric. Sci.* 52: 23–25.

Chauhan, Y.S., Kumar, K., and Maurya, D.M. 1986. Male sterility in *toria* (*Brasica campestris* L. var. *toria). Cruciferae Newslett.* 11: 50.

Chen, B.Y., and Haneen, W.K. 1991. The basic number of Brassica genomes X-3. *Cruciferae Newslett.* 14/15: 20–21.

Chen, J.L., and Berersdorf, W.D. 1990. Fatty acid inheritance in microspore derived populations of spring rapeseed *Brassica napus. Theor. Appl. Genet.* 80: 465–469.

Chopra, V.L., Narsimhulu, S.B., and Shyam Prakash. 1986. Tissue culture studies in Brassica species. *Cruciferae Newslett.* 11: 100–101.

Das, B., and Rai, B. 1972. Heterosis in intervarietal crosses of *toria. Indian J. Genet.* 32: 197–202.

Dixit, R.K., Prasad, K., and Srivastava, A.N. 1983. Combining ability for quality characters in Indian mustard. *Indian J. Agric. Sci.* 53: 776–778.

Doloi, P.C. 1977. Level of self-incompatibility, heterosis and inbreeding depression in *Brassica campestris* L. Unpublished Ph.D. thesis, GBPUAT, Pantnagar.

Downey, R.K. 1990. Brassica oilseed breeding: Achievements and opportunities. *Plant Breed. Abstr.* 60: 1165–1170.

Downey, R.K., and Harvey, B.L. 1963. Methods of breeding for oil quality in rape. *Canadian J. Plant Sci.* 43: 271–275.

Downey, R.K., Craig, B.M., and Young, G.C. 1969. Breeding rapeseed for oil meal quality. *J. Amer. Oil Chemists Soc.* 46(9): 121.

Duhoon, S.S., Chandra, S., Basu, A.K., and Makhija, O.P. 1982. Component of genetic variation for yield and its attributes in a diallel cross of yellow seeded *sarson* (Indian Colza). *Indian J. Agric. Sci.* 52: 154–158.

Du Xin and Tang Zejing. 1991. Karyotypic study as a aid in analysing the evolution and relations of six oilseed species in genus *Brassica. Cruciferae Newslett.* 14/15: 14–15.

Frandsen, K.J. 1943. The experimental formation of *B. juncea* Czern & Coss Dansk. *Botanisk. Archiv.* 11: 1–17.

Friend, D.J.C. 1985. In *Handbook of Flowering.* CRC Press, Boca Raton, Florida, pp. 48–77.

Friend, D.J.C. 1986. Use of rapid cycling Brassicas in photoperiodism research. *Cruciferae Newslett.* 11: 82–83.

Fu, T.D., and Yang Guangsheng. 1991. Relationship between origin and evolution of rapeseed and the development of cytoplasmic male sterile line. *Cruciferae Newslett.* 14/15: 38–39.

Govil, S.K., Chaubey, C.N., and Chauhan, Y.S. 1981. Combining ability studies in Indian mustard. *Indian J. Agric. Sci.* 51: 623–626.

Guerche, P., Charbonnier, M., Jouanin, L., Pelletier, G. 1986. Direct gene transfer in *Brassica napus. Cruciferae Newslett.* 11: 98–99.

Han, J.X. 1990. Genetic analysis of oil content in rape (*Brassica napus*). *Oil Crops of China* 2: 1–6.

Harvey, B.L., and Downey, R.K. 1964. The inheritance of erucic acid in rapeseed (*B. napus*). *Canadian J. Plant Sci.* 44: 104–111.

Hawk, J.A. and Crowder, L.V. 1978. The inheritance of four mutants in early flowering. *Brassica campestris* L. *J. Hered.* 69: 125–127.

Heyn, F.W. 1978. Introgression of restorer genes from *Raphanus sativa* into the cytoplasmic male sterile *B. napus* and the genetic of fertility restoration. *Proc. 5th Int. Rapeseed Conf.* (Malmo, Sweden), vol. 1, pp. 82–83.

Hinata, K., and Konno, N. 1979. Studies on the male sterile strain having *Brassica campestris* nucleus and the *Diplotaxis muralis* cytoplasm. I. On the breeding procedure and some characteristics of the male sterile strain. *Jap. J. Breed.* 29: 305–311.

Hussein, M.M., and Abobakr, M.A. 1976. Secondary associations in *Brassica oleracea*. *Egyptian J. Genet. Cytol.* 5: 174–183.

Hwe, K.K., and Chen, C. 1983. Inheritance of oleic and linoleic acid content in rapeseed oil. *Mem. Coll. Agric. National Taiwan Univ.* 23: 19–26.

Jain, R.K., Sharma, D.R., and Chaudhury, J.B. 1986. Selection of NaCl tolerant plants from cultured cotyledons of *Brassica juncea*. *Cruciferae Newslett.* 11: 96–97.

Jonsson, R. 1974. Breeding for low erucic acid in winter turnip rape (*B. campestris* var. *biennis*). *Zeitschrift fur Pflanzenzuchtung* 73: 259–268.

Jonsson, R. 1977. Erucic acid heredity in rapeseed (*B. napus* and *B. campestris*). *Hereditas* 86: 159–170.

Kirk, J.T.O., and Hudstone, C.J. 1983. Variation and inheritance of erucic acid content in *B. juncea*. *Plant Breed. Abstr.* 53: 8332.

Kirti, P.B., Sarfraz Hadi and Chopra, V.L. 1991. Seed transmission of salt tolerance in regenerants of *Brassica juncea*. *Cruciferae Newslett.* 14/15: 85.

Kumar, P.R. 1982. Strategies for increasing rapeseed and mustard production. *Indian Fmg.* 32(8): 43–47.

Kumar, P.R., Yadava, T.P., and Dhawan, K. 1982. Combining ability analysis for oil content in mustard. *Indian J. Agric. Sci.* 52: 219–221.

Labana, K.S., Badwal, S.S., and Chaurasia, B.D. 1975. Heterosis and combining ability in *B. juncea*. *Crop Improv.* 2: 46–51.

Lewis, D., Verma, S.C., and Zuberi, M.I. 1988. Gametophytic incompatibility in Cruciferae—*Raphanes sativas*. *Heredity* 61: 355–366.

Mahla, S.V.S., Mor, B.R., and Yadava, J.S. 1991. Induced polygenic variation in mustard (*Brassica juncea*). *Cruciferae Newslett.* 14/15: 44–45.

Maini, N.S., and Ghai, B.S. 1964. Relative efficiency of polycross and topcross techniques in determining combining ability *toria* inbreds. *Indian Oilseeds J.* 8: 307–309.

Mc Gregor, D.I., and Downey, R.K. 1975. A rapid and simple assay for identifying low glucosinolate in rapeseed. *Canadian J. Plant Sci.* 55: 181–196.

Mc Gregor, D.I., Mullin, W.J., and Fenwick, G.R. 1983. Review of analysis of glucosinolates: analytical methodology for determining glucosinolate composition and content. *J. Assoc. Anal. Chem.* 66: 825–849.

Mishra, P.K. 1985. Heterosis breeding and exploitation of male sterility in mustard (*B. juncea* Czern & Coss). Ph.D thesis, P.G. School, IARI, New Delhi.

Mohammed, A., Sikka, S.M., and Aziz, M.A. 1943. Inheritance of seed colour in some oleiferous Brassicae. *Indian J. Genet.* 2: 112–127.

Morinaga, T. 1929a. Interspecific hybridization in *Brassica* I. Cytology of F_1 hybrids of *B. napella* and various other species with 10 chromosomes. *Cytologia* 1: 16–27.

Morinaga, T. 1929b. Interspecific hybridization in *Brassica* II. The cytology of F_1 hybrids of *B. cernua* and various other species with 10 chromosomes. *Jap. J. Bot.* 4: 277–289.

Morinaga, T. 1934. Interspecific hybridization in *Brassica* IV. Cytology of F_1 hybrids of *B. juncea* and *B. nigra*. *Cytologia* 6: 62–67.

Nagalakshmi, T.V. 1986. Analysis of yield components in yellow *sarson* (*B. campestris* L. var. *sarson)*. M.Sc. (Ag.) thesis, BHU, Varanasi, India.

Nagalakshmi, T.V. 1992. Analysis of genetic divergence, combining ability and heterosis in Indian mustard. Ph.D. thesis, BHU, Varanasi, India.

Nayar, G.G., and George, K.P. 1970. Inheritance of pod arrangement and seed colour in *B. juncea*. *Indian J. Genet.* 30: 579–580.

Nitsch, A. 1975. Genetical and physiological studies on polygenic fatty acid mutants in Swede rape. Thesis, Georgi August Univ. Gottingen, 58 pp.

Olivieri, A.M., and Parrini, P. 1986. Relationship between glucosinolate content and yield components in rapeseed. *Cruciferae Newslett.* 11: 126–127.

Olsson, G. 1960a. Species crosses within Genus *Brassica* I. Artificial *B. juncea*. *Hereditas* 46: 171–222.

Olsson, G. 1960b. Species crosses within Genus *Brassica* II. Artificial *B. napus* L. *Hereditas* 46: 351–396.

Olsson, G. 1990. Rape yield: Production components. *Svensk Frotidning* 59: 168–169.

Pal, Y., and Singh, H. 1986. Genetic architecture of oil content in Indian mustard (*B. juncea* Czern & Coss) grown in two environments. *Cruciferae Newslett.* 11: 34–35.

Palmer, M.V. 1986. A new approach to the improvement of Indian mustard. *Cruciferae Newslett.* 11: 94.

Parthasarathy, N., and Rajan, S.S. 1953. Studies on the fertility of autotetraploids of *B. campestris* var. *toria*. *Euphytica* 2: 25–36.

Patnaik, M.C., and Murthy, B.R. 1978. Gene action and heterosis in brown *sarson*. *Indian J. Genet.* 38: 119–125.

Paul, N.K., Joarder, O.I., and Enus, A.M. 1976. Analysis of yield and some of its components in mustard. *Indian J. Agric. Sci.* 46(II): 517–524.

Persson, C. 1986. High erucic oil from white mustard (*Sinapis alba*) for technical use. *Cruciferae Newslett.* 11: 134.

Pinnisch, R., and McVetty, P.B.E. 1990. Seed production of hybrid summer rape in field using pol. cytoplasmic male sterility system, a first attempt. *Canadian J. Plant Sci.* 70: 611–618.

Prakash, S., and Raut, R.N. 1983. Artificial synthesis of *B. napus* and its prospects as an oilseed crop in India. *Indian J. Genet.* 43: 282–290.

Pundir, R.P.S., and Rai, B. 1971. Use of multiple selection criteria in the genetic improvement of *toria* populations. *Indian J. Genet.* 31: 377–382.

Rahman, L. 1976. Inheritance of the contents of erucic, ecosenoic, linoleic and oleic acids in crosses between *B. juncea* and *B. carinata*. *Indian J. Agric. Sci.* 46: 192–198.

Rai, B. 1976. Considerations in the genetic improvement of oil quality in rapeseed. *Oilseeds J.* 6: 13–15.

Rai, B. 1979. *Heterosis Breeding*. Agro-Biological Publications, New Delhi, pp. 183.

Rai, B. 1980. Self-incompatibility and its scope for the exploitation of heterosis in *Brassica* species. In: *Breeding Oilseed Crops*. PAU, Ludhiana, pp. 284–295.

Rai, B. 1983. Advances in rapeseed and mustard breeding research. *Indian Fmg.* 33: 3–8.

Rai, B. 1985. Advances in developing commercial hybrids of oilseed crops in India. *Seeds and Farms* 11: 44–47.

Rai, B. 1987. PT 303: The first national variety of *toria*. *Indian Fmg.* 37(3): 16–17.

Rai, B., and Kumar, A. 1978a. Rapeseed and mustard production programme. *Indian Fmg.* 28(5): 27–30.

Rai, B., and Kumar, A. 1978b. New vistas in *toria* production. *Indian Fmg.* 28(11): 28–30.

Rai, B., and Sehgal, V.K. 1975. Field resistance of the *Brassica* germplasm to mustard aphids (*Lipaphis erysimi*). *Sci. and Cult.* 41: 444–445.

Rai, B., and Singh, A. 1976. Commercial seed production in rapeseed. *Indian Fmg.* 26(4): 15–17.

Rai, B., Kolte, S.J., and Tewari, A.N. 1976. Evaluation of the oleiferous *Brassica* germplasm for resistance to *Alternaria* leaf blight. *Indian J. Mycol. Plant Pathol.* 6: 76–77.

Rai, L., Rai, B., Sanghmitra, M., and Rao, B.P. 1987. Control of aphid in mustard and safflower crops. *Indian Fmg.* 37(1): 16–19.

Rajan, S.S. 1955. The effectiveness of mass-pedigree system of selection in improvement of seed setting in autotetraploid *toria*. *Indian J. Genet.* 15: 47–49.

Rajan, S.S., and Ahuja, Y.R. 1956. Seed development in colchicine induced autotetraploids of *toria*. *Indian J. Genet.* 16: 63–76.

Ram Bhajan, Chauhan, Y.S., and Kumar, K. 1991. Natural cross pollination in Indian mustard. *Cruciferae Newslett.* 14/15: 24–25.

Ram, K., Ramakrishnan, Chauhan, Y.S., and Katiyar, R.P. 1976. Partial diallel analysis in the F_3 generation of Indian mustard. *Indian J. Agric. Sci.* 46: 229–232.

Ramanujam, S., and Deshmukh, M.J. 1945. Colchicine induced polyploidy in crop plants. III Oleiferous Brassicae. *Indian J. Genet.* 5: 63–81.

Ramanujam, S., and Rai, B. 1963. Analysis of yield components in yellow *sarson*. *Indian J. Genet.* 23: 312–319.

Ramanujam, S., and Srinivasachar, D. 1943. Cytogenetic investigations in genus *Brassica* and the artificial synthesis of *Brassica juncea*. *Indian J. Genet.* 3: 73–88.

Raut, R.N., and Shyam Prakash. 1985. Synthetic Brassicas: New oilseeds for greater production. In *Genetic Manipulation for Crop Improvement*. Oxford & IBH, New Delhi, p. 326.

Rawat, D.S. 1975. Genetical studies on yield, oil content and characters related to yield in *B. juncea*. Unpublished Ph.D. thesis, P.G. School, IARI, New Delhi.

Rawat, D.S., and Anand, I.J. 1979. Male sterility in Indian mustard. *Indian J. Genet.* 39: 412–419.

Rawat, D.S., and Anand, I.J. 1986. Inheritance of flower colour in Indian mustard. *Indian J. Agric. Sci.* 56: 206–208.

Reddy, W.R. 1983. Genetics of seed coat colour and quantitative characters related to yield in Indian mustard (*B. juncea*) Unpublished M.Sc. thesis, P.G. School, IARI, New Delhi.

Richharia, R.H. 1957. Plant breeding with reference to oilseed crops. *Indian J. Genet.* 17: 218–247.

Rouselle, P., and Renard, M. 1978. Study of cytoplasmic male sterility in rapeseed (*B. napus*). *Cruciferae Newslett.* 3: 40–41.

Sang, J.P., and Truscott, R.J.W. 1984. Liquid chromatographic determination of glucosinolates in rapeseed as desulphoglucosinolates. *J. Assoc. Anal. Chem.* 67: 829–833.

Sareen, P.K., and Kakar, S.N. 1975. Genetics of self-incompatibility in brown *sarson*. *Zeitschrift fur Pflanzenzuchtung* 74: 291–297.

Sciffert, M., and Boelcke, B. 1977. Studies on yield structure in winter swede rape. *Demakrawatischen Republic* No. 149: 185–193.

Sernyk, J.L. 1983. Heterosis and cytoplasmic male sterility in summer rape (*B. napus*). *Diss. Abstr.* (International) 43(8): 2449 (Canada).

Sernyk, J.L., and Stefansson, B.R. 1983. Heterosis in summer rape (*B. napus* L). *Canadian J. Plant Sci.* 63: 407–413.

Shabana, R., Shrief, S.A., Ibrahim, A.F., and Geisler, G. 1990. Correlation and path analysis for some new released double zero spring rapeseed cultivars grown under different competitive systems. *J. Agron. Crop Sci.* 165: 138–143.

Shiga, T. 1976. Studies on heterosis breeding using cytoplasmic male sterility in rapeseed (*B. napus* L.). *Bull. Nat. Inst. Agric. Sci.* (Series D) 27: 1–85.

Shiga, T., and Baba, S. 1973. Cytoplasmic male sterility in oilseed rape (*B. napus*) and its utilization in breeding. *Jap. J. Breed.* 23: 187–197.

Shiga, T., Ohkawa, Y., and Takayanagi, K. 1977. Segregation of fertility restoration in F_2 progenies of crosses between male sterile and European cultivars in rape (*B. napus* L.). Annual Report Nat. Inst. Agric. Sci. Japan, 1977 (II): 30–33.

Sikka, S.M. 1940. Cytogenetics of *Brassica* hybrids and species. *J. Genet.* 4: 441–501.

Sikka, S.M., and Rajan, S.S. 1957. Breeding better Brassicas. *Indian Oilseeds J.* 1: 73–81

Singh, D. 1958. *Rape and Mustard*. Indian Central Oilseeds Committee, Hyderabad, India,

105 pp.
Singh, D., and Mehta, T.R. 1954. Studies on breeding brown *sarson*. I. Comparison of F_1 and their parents. *Indian J. Genet.* 14: 74–77.
Singh, J. 1986. Genetic analysis of oil yield and its component characters in *Brassica napus*. Unpublished Ph.D. thesis, Meerut University, Meerut, India.
Singh, S.P. 1973. Heterosis and combining ability estimates in Indian mustard (*B. juncea*). *Crop Sci.* 13: 497–499.
Singh, R.N., and Srivastava, A.N. 1976. Inheritance of a siliqua mutant in rai (*B. juncea*). *Sci. and Cult.* 42: 524.
Singh, A.P., Chauhan, Y.S., and Singh, P. 1982. Components of genetic variation in Indian mustard. *Indian J. Agric. Sci.* 52: 634–638.
Singh, D.P., Pokhriyal, S.C., and Magnath, K.S. 1964. Inheritance of some characters in *B. juncea*. *Indian J. Genet.* 24: 288–290.
Spinks, E.A., Sones, K., and Fenwick, G.R. 1984. The quantitative analysis of glucosinolates in cruciferous vegetables, oilseeds and forage crops using high performance liquid chromatography. *Fette Seifen Anstrichm.* 86: 228–231.
Stringam, G.R. 1971. Genetics of four hypocotyl mutants in *B. campestris* L. *J. Hered.* 62: 248–250.
Stringam, G.R. 1976. Genetics of two bloomless mutants in *B. campestris*. *Canadian J. Genet. and Cytol.* 18: 225–229.
Takagi, Y. 1970. Monogenic recessive male sterility in oilseed rape (*B. napus*) induced by gamma irradiation. *Z. Pflanzenzuchtung* 64: 242–247.
Tewari, L.P., and Singh, A.B. 1973. Diallel cross analysis of quantitative traits in Indian mustard (*B. juncea* L. Czern & Coss). *Indian J. Agric. Sci.* 43: 1055–1059.
Thakral, S.K. Yadava, A.K., Yadava, T.P., and Kumar, P. 1986. Genetics of seeds yield and some important attributes in Indian mustard. *Cruciferae Newslett.* 11: 31–33.
Thompson, K.F. 1964. Triple cross hybrid kale. *Euphytica* 13: 173–177.
Thompson, K.F. 1972. Cytoplasmic male sterility in oilseed rape. *Heredity* 29: 253–257.
Tyagi, D.V.S., and Verma, R.B. 1985. Functional male sterility in *toria*. *Indian J. Genet.* 45: 219–223.
Tyagi, D.V.S., Rai, B., and Verma, R.B. 1983. A note on bunchy dwarf mutant in *toria*. *Indian J. Genet.* 43: 374–377.
U.N. 1935. Genome analysis in Brassica with special reference to the experimental formation of *B. napus* and particular mode of fertilization. *Jap. J. Bot.* 7: 389–452.
Verma, S.C., Malik, R., and Dhir, I 1977. Genetics of incompatibility system in the crucifer *Eruca sativa*. *Proc. Royal Soc. London* B 196 1123: 131–159.
Verma, S.C., Sareen, T.S., and Jasveer Kaur. 1991. Genetics of self-incompatibility in a Japanese variety of radish (*Raphanus sativus* cv. Tokinashi). *Cruciferae Newsletter* 14/15: 32–33.
Verma, V.D., and Rai, B. 1980a. Note on induced mutagenesis for spotting out usable sources of resistance to Alternaria leaf spot in Indian mustard. *Indian J. Agric. Sci.* 50: 278–280.
Verma, V.D., and Rai, B. 1980b. Mutation in seed coat colour in Indian mustard. *Indian J. Agric. Sci.* 50: 545–548.
Yadava, A.K., Yadava, T.P., and Kumar, P. 1979. Combining ability studies by line × tester analysis in Indian mustard. *Crop Improv.* 6: 141–147.
Yadava, T.P. 1977. Genetics of yield and its components in *toria* (*B. campestris* var. *toria*). *HAU J. Res.* 7: 12–16.
Yadava, T.P., and Gupta, V.P. 1975. Heterosis and genetic structure of oil content in mustard. *Indian J. Genet.* 35: 152–155.
Yadava, T.P., Singh, H., Gupta, V.P., and Rana, R.K. 1974. Heterosis and combining ability in *raya* for yield and its components. *Indian J. Genet.* 34A: 684–695.

Yadava, T.P., Gupta, V.P., and Singh, H. 1976. Inheritances of the days to flowering and maturity in mustard. *SABRAO J.* 8: 81–83.

Yadava, T.P., Kumar, P., and Yadava, A.K. 1981. Genetic architecture for yield and its component traits in Indian mustard. *Indian J. Agric. Sci.* 51: 374–378.

Yongming Zhou and Houli Lin. 1991. Selective strategies on the breeding for quality in *Brassica napus. Cruciferae Newslett.* 14/15: 58–59.

Zaman, M.W. 1987. Inheritance of seed colour in *Brassica campestris. Sveriges Utsades Forenings Tidskriff* 99: 205–207.

8. SUNFLOWER

A. Seetharam
Project Coordinator (Small Millets), Bangalore, India

R.L. Ravikumar
Plant Scientist (Sorghum), University of Agricultural Sciences, Dharwad, India

Introduction

During the last two decades, the cultivated sunflower (*Helianthus annus* L.) has emerged as one of the major edible oil seed crops in the world, ranking second in importance after soybean. Sunflower is planted annually on around 16 million ha producing about 20 million metric t of seed. Sunflower cultivation is largely concentrated in Europe, which accounts for nearly two-thirds of the total production. Among the European countries, the erstwhile USSR is the largest producer of sunflower, followed by Romania, France, and Yugoslavia. The other countries growing sunflower are Argentina, Chile, Uruguay, and Brazil in South America, the United States and Canada in North America, Zimbabwe, South Africa, and Kenya in Africa, China, Turkey and India in Asia, and Australia.

Asia accounts for about 15 per cent of the total sunflower area in the world. The 1980s witnessed significant increases in area and production of sunflower in India. Sunflower area in India has increased more than six times from 1975–77 to 1990–92 (Table 1).

Table 1
Triennial averages for area ('000 ha), production ('000 tonnes) and productivity (kg/ha) of sunflower in India, 1975–77 to 1990–92

	1975–77	1978–80	1981–83	1984–86	1987–89	1990–92
Area	280.2	120.9	470.4	791.9	1323.2	1869.3
Production	165.2	65.6	220.1	367.1	568.5	1036.8
Productivity	583	539	489	467	433	553

Groundnut and rapeseed mustard are the principal oilseed crops in India and their production far outweighs the production of other oilseed crops. Nevertheless, with the passage of time the importance of sunflower to the national edible oil economy is bound to increase further. Till recently, the sunflower production was confined largely to the southern and central peninsula comprising Karnataka, Andhra Pradesh, Maharashtra, and Tamil Nadu. Of late, its cultivation is getting extended to agriculturally advanced states in North India—Punjab, Haryana, and western Uttar Pradesh. It is obvious from the trends being witnessed now in India that the importance of sunflower to Indian agriculture would further increase in the 1990s and the area may touch around 2.5 million ha by 1995–96. The productivity of sunflower in India at present is far lower than the world average. This is to a large extent due to poor quality planting material and several biotic and abiotic stress factors that the crop is invariably subjected to in different regions. The increasing attention being shown in Indian agriculture in the use of quality seeds and inputs should result in increasing production of sunflower. The number of cultivars, particularly hybrids, available for cultivation has phenomenally increased during the last two to three years, which is expected to improve yield and disease control of the crop significantly.

Origin and Domestication

The genus *Helianthus* belongs to the subdivision Tubiflorae and tribe Heliantheae of the family Compositae. Heiser (1957) reported as many as 67 species, all native to South and Central America, under this genus. Two species, *H. annuus*, a diploid (2n = 34), and *H. tuberosus*, a hexaploid (2n = 102) are cultivated, the former as an oilseed crop and the latter as a feed crop for its tubers. The species included are both annuals and perennials; the perenniality arising either from the rhizomatus nature of roots or rarely from tubers. The occurrence of introgression in natural habitats between different species has resulted in variable populations, helping them to occupy a variety of habitats and ecological niches. The basic chromosome number is 17 and both diploid (2n = 2x = 34) and polyploid (2n = 4x = 68 and 2n = 6x = 102) species exist.

The archaeological evidences recorded at many places in North America indicate that sunflower was probably first domesticated in the central part of the present United States by the native Americans (Yarnell, 1972). The domesticated forms in all probability might have originated from the wild and weedy sunflowers growing in that area (Heiser, 1951). The historical accounts in North America subscribe to the view that the native Americans had developed very tall mono head sunflowers as early as the 6th century. This sunflower was introduced most likely from Mexico to Europe through Spain and later on to other parts of Europe, including the USSR.

Evolution of Present-Day High Oil Content Sunflower

The credit of transforming ornamental sunflowers into high-oil forms goes largely to the erstwhile USSR. According to Zhukovski (1950), sunflower was first introduced to Russia in the 18th century and the first attempts to extract oil from the seed were made as early as 1779. The selection work for higher oil content was started in 1860 and persistent work carried out at several agricultural stations resulted in the gradual stepping up of oil content in cultivars from 33 per cent in 1940 to more than 50 per cent in 1965. These high-oil selections later on found their way to all the continents and most of the recently bred varieties owe their origin directly or indirectly to the materials bred in the USSR.

Taxonomy and Floral Biology

Cultivated sunflower is an annual and conspicuously different from the wild and weedy forms in having a single unbranched stem, large capitulum, and medium-sized to large seeds. The cultivated forms vary greatly from one another in height, stem girth, leaf number and size, duration of maturity, test weight, and oil content. The plant height usually varies from 0.5 m to 5.0 m. However, there are reports of types growing up to a height of 12.0 m. The stem may show different degrees of branching, ranging from mono stem nature with single head to highly branched forms with several heads. The stem is supported by a strong, deep central root system with well-developed lateral roots reaching a soil depth of 1.5–2.0 m. The leaves vary in number and are produced in opposite pairs at the lower portion of the stem and in alternate phyllotaxy in whorled form at the above. The leaf morphology varies greatly in size, shape, margin, colour and hairiness of the surface.

In cultivated forms the inflorescence, which is a capitulum or head, is most prominent because of its large size and the conspicuous yellow colour of the ray florets. The capitulum is made up of two distinct flower types—the outer single row of ray florets, which are female and sterile, and the inner disc florets, which are hermaphrodite and fertile. The disc florets are arranged in arcs radiating from the centre of the head. Each disc floret is made up of inferior ovary, two pappus, which are modified sepals, and a tubular corolla formed by the fusion of petals except at the tip. The five anthers are also fused to form a tube with filaments attached independently at the base of corolla tube. The style is inside the anther tube with stigma divided at the tip. When the flower is fully developed, the style elongates and the bifid stigma curls outward.

The unfolding of outer ray florets marks the beginning of flowering in sunflower. This is followed by opening of disc florets from the periphery proceeding gradually towards the centre of the head. The flowering in capitulum is completed in 5–8 days, depending upon the size of the flower and prevailing weather. Anthesis takes place in the morning between 6 and 8 a.m. The sunflower to some extent is protandrous, as anthesis takes place first, accompanied by a time lag of 10–12 hours in the maturation of male and female elements. Because of this, cross pollination is favoured as a rule and insects, particularly bees, play an important role in pollination and fertilization. Putt (1940) showed that pollen is seldom moved by wind. The pollen is spiny and well adapted for transmission by insects.

The fruit is an achene and consists of outer pericarp (hull), thin and papery inner seed coat, and embryo (kernel). In the mature seed, the embryo is the major portion and is made up of mostly cotyledons. The endosperm is largely made up of a single layer of aleurone cells coalesced with the seed coat.

Genetics

Numerous genetic studies have been carried out in sunflower by workers all over the world. The crop is highly amenable to such investigations because of its outbreeding nature, high reproductive ability, availability of cytoplasmic genetic male sterility system, photoperiod insensitivity, wide adaptation, and large diversity in the germplasm. An attempt has been made to present a concise review on quantitative genetic inheritance of a wide range of economic characters. However, reliability of estimates reported is open for debate and re-examination, since genetic analyses are often made using small populations that are subjected to considerable genotype × environment interactions.

Seed Yield

The seed yield shows a wide range of variability and is often substantially influenced by the environment (Schuster, 1970; Reddy and Reddy, 1979; Shinde *et al.,* 1983; Begum *et al.*, 1988). The results are at variance as far as heritability and genetic advance of this character are concerned. However, moderate to high heritability as well as genetic advance has been observed in a number of studies (Kloczowski, 1975; Singh and Yadava, 1986; Chaudhary and Anand, 1987). The predominance of non-additive gene action controlling seed yield has been reported (Pogorletskii, 1974; Miller *et al.*, 1980; Alba *et al.*, 1985; Pathak *et al.*, 1985). However, reports of additive gene action controlling seed yield are not uncommon (Shrivastava and Mishra, 1976; Shinde *et al.*, 1983; Singh *et al.*, 1989).

The preponderance of both types of gene action in different groups of material indicates the suitability of sunflower to both population and heterotic breeding procedures. Most researchers involved in the development of hybrid sunflowers have observed marked differences among inbred lines for combining ability. The heritability of general combining ability (GCA) is high, ranging from 60 to 86 per cent in studies where the relative combining ability of inbred lines was compared to the source populations (Anaschenko, 1974). Hence, the identification of proper cultivars is very important for developing inbred lines with good combining ability. The cultivars with high GCA produced inbred lines with high GCA and early testing of GCA of cultivars is proposed (Gonzalez Garcia and Angeles Espino, 1983). Fick and Zimmer (1976) observed that stability of performance over environments appears to be under genetic control and may have relatively high heritability. In different studies, positive association of seed yield with days to maturity, plant height, head diameter, stem girth, test weight of seed, seed volume weight, total dry matter, and harvest index has been revealed (Putt, 1943; Russel, 1953; Schuster, 1964; Burns, 1970; Anand and Chandra, 1980; Giriraj *et al.*, 1987).

Number of Filled Seeds per Head and 1000-Seed Weight

The number of filled seeds per head and 1000-seed weight show positive association with seed yield (Dhaduk *et al.*, 1985). However, continuous selection for one character often produces a negative effect on the other (Voskoboinik and Marin, 1986). A wide range of variation was observed for number of seeds per capitulum and percentage of self-fertility (Reddy and Reddy, 1979; Sheriff and Appadurai, 1985; Burlov and Krutko, 1986). Petrov and Piskov (1985) found that self-fertility is genetically determined. The seed size varied from 6 to 25 mm in length and 3 to 13 mm in width and most of the cultivated types weigh 40–100 g/1000 seeds (Fick, 1978). Singh and Yadava (1986) observed high genotypic coefficient of variation and high heritability for 1000-seed weight. The results available to date indicate the role of all types of gene action in the control of 1000-seed weight: additive (Singh and Yadava, 1986), dominance (Pathak *et al.*, 1985; Singh and Yadava, 1986), dominance × dominance (Manjunath, 1986), and additive × dominance (Dua and Yadava, 1982). The filling percentage showed a wide variation from 0 per cent seed set to 100 per cent fertility due to presence of self-incompatibility (Fick, 1978).

The number of achenes per head is closely related to number of florets in the inflorescence (Marin, 1986). Reddy and Reddy (1979) found high genotypic and phenotypic coefficient of variation and high genetic advance for seed number compared to other characters. However, heritability values were moderate, indicating the predominance of non-additive gene action (Pathak *et al.*, 1985; Sheriff *et al.*, 1985).

Head Diameter

The direct positive effect of head diameter on seed and oil yield is substantial and hence should be given due weightage while formulating selection indices (Chaudhary and Anand, 1985; Dhaduk *et al.*, 1985). High genotypic coefficient of variation and heritability were reported for this character by Singh and Yadava (1986) while moderate to low heritability estimates were reported by Kloczowski (1975). Fick (1978) reported higher influence of environment on this character. Putt (1966), Pathak *et al.* (1985), Sheriff *et al.* (1985), and Alba *et al.* (1985) have reported both additive and non-additive gene effects controlling head diameter.

Seed Colour

Sunflower seed presents a variety of colours. The presence or absence of pigments in three different layers of pericarp determines the final colour of the achene. Crosses between achenes having different degree of pigmentation gave an intermediate F_1 suggesting a complex inheritance (Mosjidis, 1982). The white colour of the achenes was due to lack of pigmentation in all the three layers and is dominant over dark brown or black colour (Fick, 1978).

Oil Content

Oil content in seeds is dependent upon hull content and oil content in the kernel. Much of the improvement made in oil content till now is through reduction in hull content (Gundaev, 1971). Wide variation has been reported in estimates of heritability (Nikolic-Vig *et al.*, 1971; Kloczowski, 1975; Fick, 1978; Fernendez *et al.*, 1979; Torres *et al.*, 1984). Miller *et al.* (1977) achieved 12.4 per cent improvement in oil content after three cycles of simple recurrent selection. The predominance of additive gene effects controlling this character has been reported by several workers (Carreon and Barreda, 1983; Alba *et al.*, 1985). However, the genetic understanding of oil content is complicated by the observation of complex nature of quantitative inheritance (Vranceanu, 1967; Gundeav, 1971), total maternal parent control of oil percentage (Miller and Zimmerman, 1983), and xenia effect (Seetharam *et al.*, 1977a).

Fatty Acid Composition

The standard sunflower oil comprised primarily palmitic, stearic, oleic, and linoleic acids with oleic and linoleic acids together accounting for about 90 per cent of total fatty acids (Fernandez *et al.*, 1989). Genetic control of fatty acid composition is well established (Putt *et al.*, 1969; Skoric and Bedev, 1978; Fernandez and Knowles, 1982).

Miller and Zimmerman (1983) by crossing high oleic with low oleic parent found lack of dominance and maternal influence, while Miller *et al.* (1977) observed that both oleic and linoleic acids are controlled by the genotype of the maternal parent.

Fick (1984) studied the segregating generations from crosses of high × low oleic plants and reported a single partly dominant gene controlling high oleic character. Urie (1985) concluded that the high oleic trait was controlled by a single dominant gene 'OL' with embryo genotype control. However, in the same cross, there appeared reversal of dominance. Similar results were observed by Hristova and Georgieva-Todorova (1975). A second gene modifying the oleic acid content of the 'OL' gene was reported recently by Miller *et al.* (1987). More recently, Fernandez *et al.* (1989) indicated that the genetic control of high oleic character is more complex than previously reported and suggested the operation of three complementary dominant genes.

Maturity Length

The sunflower shows a wide range of maturity lengths and the genes controlling this character also influence several other important traits such as seed yield, height of plant, leaf number, leaf area and inflorescence size (Ferieys, 1981; Morozov and Raikov, 1983). Correlation studies indicated that seed yied depended to a considerable extend on maturity length (Mishra *et al.*, 1985). The temperature and photoperiod also influences maturity length (Robinson *et al.*, 1967; Schuster, 1970; Goyne, 1985). The heritability of days to maturity is relatively high (Shabana, 1974; Venkateshwaralu *et al.*, 1980). However, the results of a number of studies indicate that days to maturity is controlled by both additive and non-additive types of gene action (Putt, 1966; Dua and Yadava, 1983; Chaudhary and Anand, 1984).

Plant Height

Plant height has direct effect on seed yield. There is large variation for this character ranging from 0.5 m to 7.0 m or even more (Cockerell, 1915; Fick, 1978; Ferieys, 1981). The role of non-additive gene action in the expression of plant height has been reported (Pathak *et al.*, 1985). There are reports indicating the qualitative nature of this character controlled by oligo genes (Enns, 1959; Rodin, 1976).

Branching

Branching is a complex character showing wide variation for the types of branching length and number of branches. Hockett and Knowles (1970) recognized four different branching forms and identified four genes controlling them. Putt (1964) identified a single dominant gene controlling branching on the entire stem and designated it as (Br). Later, Putt (1964b) identified a gene for profuse branching and designated it as (b). This was recessive to non-branching and monogenically inherited. This has been incorporated in the fertility restorer lines for abundant pollen production.

Leaf Shape, Number and Size

Wide variation occurs for leaf shape. The leaves of cultivated sunflower are flat, cordately to subcordately shaped, conspicuously serrated, and usually hispid (Heiser *et al.*, 1969). It is observed that pronounced serration is dominant over finely serrated leaf edge, while finely serrated is dominant over no serration (Skaloud and Kovacik, 1974).

Rostova and Krylova (1987) in a study of wild and cultivated types observed reduction in leaf number and size during the course of domestication. However, leaf number and leaf area show a strong positive correlation with yield and are considered to be very important physiological characters (Fernandez and Orioli, 1984; Baronecelli *et al.*, 1987; Vanishree *et al.*, 1988). The leaf number in a plant is decided by the crop growth period, which in turn determines plant height. The late-maturing types are tall, with larger number of leaves (Ross, 1939; Russell, 1953; Kovacik and Skaloud, 1972; Sheriff *et al.*, 1986).

A wide variation has been observed for leaf number as well as for size (Gundaev, 1971). Genetic control of leaf number and size appears to be complex and both additive and non-additive gene effects play a significant role in its expression (Shabana, 1974; Venkateshwaralu *et al.*, 1980; Chaudhary and Anand, 1984; Sheriff *et al.*, 1986).

Total Dry Matter and Harvest Index

Total dry matter and harvest index are the two important parameters having a significant influence on seed yield (Giriraj *et al.*, 1987). Not much information is available either on the variability existing for these characters or their genetic control.

Self-incompatibility

Sunflower is an obligate outbreeder because of protandrous nature of disc florets and presence of various degrees of self-incompatibility. Self-incompatibility was first reported by Habura (1957) in sunflower and later confirmed by Luciano *et al.* (1965), Asthana (1973), and Oka and Campos (1974). Limited information available on the inheritance of this character indicates that it is sporophytic in nature and its quantitative genetic inheritance is complex (Vranceanu and Pirvu, 1988).

Genetic Male Sterility

The male sterility was first reported by Kuptsov in 1964 (Fick, 1978) but its nature was not known. Subsequently, genetic male sterility was reported by several workers (Leclercq, 1966; Putt and Heiser, 1966; Stoyanova, 1970). The extent of male sterility also varied from complete to partial sterility. The genetic male sterility appears spontaneously and can be induced as well (Anaschenko, 1967; Vulpe, 1967). The vast majority of the data published support the view that a single recessive gene controls this character (Leclercq, 1966; Vranceanu, 1967; Burlov, 1974). However, duplicate and polygenes controlling genetic male sterility character have also been reported (Leclercq, 1968; Vranceanu, 1970; Burlov, 1974). Putt and Heiser (1966) reported recessive control of partial male sterility character.

Cytoplasmic Male Sterility and Fertility Restoration

The discovery by Leclercq (1969) of cytoplasmic male sterility in sunflower and subsequent identification of genes for fertility restoration (Enns *et al.*, 1970; Kinman, 1970; Vranceanu and Stoenescu, 1970; Leclercq, 1971) resulted in the widespread production of sunflower hybrids. The earlier studies indicated a single dominant gene restoring fertility in cytoplasmic male-sterile (CMS) lines. However, further studies revealed the presence of different types of gene action—two independent complementary dominant genes (Fick and Zimmer, 1974b; Reddy and Thammiraju, 1977) in the control of fertility restoration. Dominguez-Gimenez and Fick (1975) indicated that the inheritance of fertility restoration is more complex and up to four non-allelic genes may be involved in this process. The presence of restorer genes is not common among cultivated sunflower but more common among wild sunflower (Dominguez-Gimenez and Fick, 1975).

Recently, several new sources of cytosteriles have been reported in sunflower (Fick, 1978; Whelan, 1981; Heiser, 1982; Serieys and Vincourt, 1987). Attempts are under way to identify suitable restorers to these new sources and understand genetic control of fertility restoration. In the coming years, breeders will have ample choice to diversify the CMS base in the hybrid development programmes.

Disease Resistance

Over 30 diseases are known to cause damage in sunflower. However, in India the important diseases are: *Alternaria* leaf and stem spot caused by *Alternaria helianthi*; rust caused by *Puccinia helianthi*; downy mildew caused by *Plasmopora helstedii*; and various root and stem rots caused by *Sclerotium* sp. and *Rhizoctonia* sp.

Alternaria Leaf and Stem Spot

It is a common disease in many countries all over the world and causes more damage in India than several other diseases (Agrawat *et al.*, 1979; Balasubramanyam and Kolte, 1980). Only field resistance has been reported in cultivated sunflowers and no information is available on the inheritance of resistance to this disease (Morris *et al.*, 1983; Carson, 1985).

Rust

Rust is one of the most destructive diseases of sunflower in the world and can appear throughout the plant gowth. Racial differentiation exists in this fungus (Hoes and Putt, 1962). So far, four North American races and the corresponding resistant genes have been identified (Yang *et al.*, 1989). Resistance to rust is found commonly among wild species of *Helianthus* and successfully incorporated into commercial cultivars (Hennessy and Sackston, 1972). In India, although rust is of a common occurrence, the race pattern is not known. However, resistant sources for local races have been identified and incorporated in the hybrids under cultivation.

Downy Mildew

At present, downy mildew is more serious under temperate conditions than in tropics. However, its occurrence in India was reported in Maharashtra in 1984 (Mayee and Patil, 1986). The pathogen shows wide variation with many races and corresponding resistant genes in the host (Miller and Gulya, 1988). Resistance to each race was found to be controlled by a single dominant gene and the four genes are designated as Pl_1, Pl_2, Pl_3 and Pl_4 (Vranceanu and Stoenescu, 1971; Miller and Gulya, 1987). Apart from major genes, several modifiers are also known to influence resistance (Goossen and Sackston, 1967; Vear, 1974). The resistance to this disease appears to be rare in clutivated annuals but more common in wild perennials (Miller and Gulya, 1988). One of the popular restorer lines, RHA-274, has the Pl_2 gene for resistance to races 2 and 3 (Miller and Gulya, 1984).

Wilt

Resistance to wilt is present in both cultivated and wild sunflowers (Putt, 1964a; Hoes *et al.,* 1973). The available information indicates that resistance is both simple and complex depending upon the material involved in the study (Putt, 1964; Hoes *et al.,* 1973). Fick and Zimmer (1974a) reported a single dominant gene for *Verticillium* wilt resistance.

Sclerotenia Stalk and Head Rot

Extensive studies conducted by Pimakhin *et al.* (1984) on this disease, involving a large number of germplasm have indicated significant variation in susceptibility of the genotypes and variability in pathogen virulence. Resistance to infection of the stem base was not always associated with resistance to infection of the stem and capitulum (Vranceanu *et al.,* 1981). Specific genes for resistance have not been observed and efforts are on to understand the inheritance of resistance.

Wild Species of Sunflower

The basic chromosome number of the genus *Helianthus* is x = 17 (Geisler, 1931; Heiser, 1949) and the genus contains diploid (2n = 34), tetraploid (2n = 68), and hexaploid (2n = 102) species. However, x = 16 in *H. grosseserratus* (Cooper, 1932), x = 15 in *H. mollis* (Kulshreshta and Gupta, 1979), and x = 5, 9. 10, 14, and 15 are also reported in this tribe (Heiser *et al.,* 1962; Turner and David, 1966). These results appear to indicate that x = 17 basic number might be secondarily derived through distant hybridization and polyploidization. Heiser *et al.* (1962) put forth the idea of amphidiploidy in *H. annuus* with basic number as 8 and the idea of a 16-chromosome ancestor. However, this proposition has not been widely accepted in view of the fact that all species of this genus except one or two fall into clear-cut polyploid series of 2n = 34, 68, and 102 indicating 17 as the basic number in the genus. Besides, the high fertility status of all the 34 chromosome species substantiates this group as the primary species of this genus. The different 34-chromosome species might have evolved through deletion and duplication of chromatin resulting in evolution of distinct karyotypes and resultant speciation.

Chromosome Morphology

The chromosome morphology of *H. annuus* is more symmetric. The species possesses 10 pairs of median, 4 pairs of submedian, and 3 pairs of subterminal chromosomes (Ramakanth and Seetharam, 1977). However, Vranceanu *et al.* (1974) grouped the 17 pairs of chromosomes of *H. annuus* into 4 groups. The first group contains two pairs of satellited chromosomes with submedian centromeres; the second group consists of five pairs of chromosomes with median to submedian centromeres; the third group has six pairs with submedian centromeres and the fourth group has four pairs with submedian to subterminal chromosomes. The chromosome length varies from 3.05 to 6.20 μ. The chromosome clumping normally encountered in pachytene appears to have prevented detailed study of pachytene chromosomes of *H. annuus.*

Polyploidy and Speciation

The genus *Helianthus* comprises species with ploidy levels ranging from diploid to hexaploid; the chromosome numbers are mainly multiples of basic number (x = 17). Aneuploids have also been reported in the genus (Leclercq *et al.,* 1970; Gundaev, 1971). The polyploids in general exhibit normal meiotic behaviour, suggesting an allopolyploid nature or an efficient diploidizing system even if they are of autopolyploid origin. However, different diploid species show a high degree of chromosome homology in interspecific hybrids (Georgieva Todorova, 1970; 1974; Kulshreshtha and Gupta, 1979). The cytological studies in interspecific hybrids indicate that the genome of *H. annuus* does not contribute to the genome composition of tetraploid and hexaploid.

Crop Improvement

Breeding Objectives

The ultimate breeding objective is to maximize productivity so that net profit per unit area is increased. Sunflower being an oilseed crop, improvement of both seed yield and oil content is important and deserves equal attention. Early maturity, uniformity, wide adaptation, and resistance to important diseases and pests are other areas requiring breeders' attention.

Breeding Approaches

Sunflower is essentially a cross-pollinated crop and breeding approaches are those that are normally employed in a cross-pollinated species. Sunflower has perfect flowers and insects play an important role in pollination and fertilization. Besides, sunflower shows a wide range of self-incompatibility and many plants do not set seed on enforced self-pollination. In sunflower improvement, two distinct options—either breed superior populations (composites, synthetics, etc.) or hybrids—are available to the breeder. Both approaches have advantages and disadvantages and both find a place in a comprehensive breeding programme. Before discussing the various breeding methods it may be relevant to define the characteristics of the populations on one hand and hybrids on the other.

Characteristics of sunflower populations and hybrids

Material	Genotypic and phenotypic nature	Breeding behaviour
Populations	Homozygous and heterozygous, heterogeneous and highly variable	Polymorphic, stability of population dependent on selection pressure. Erratic behaviour, wide plasticity, imparts certain degree of protection against pests and diseases
Hybrids	Heterozygous, homogeneous, uniform	Degenerates, breaks down but possible to reconstitute, imparts stability to production, can possess wide adaptation but may succumb to epidemics

Population Breeding

Mass selection. Mass selection is an effective method of improving cultivars for certain specific traits. The characters having high heritability are more easily amenable to improvement through mass selection. Mass selection has been adapted in the past for cultivar improvement in the USSR (Gundaev, 1971). The characters improved through mass selection include early maturity, low hull content, resistance to broom rape, higher oil content, and seed size (Morozov, 1947; Robinson, 1967).

Method of reserves (Pustovoit system). As the earlier breeding methods—mass selection and individual family selection—were less effective in improving productivity, academician V.S. Pustovoit developed an effective and simple method for improving sunflower cultivars in the 1970s. This method, which is popularly known as 'Pustovoit's breeding scheme', is essentially a modified recurrent selection laying emphasis on progeny evaluation and subsequent intermating among the selected progenies. The method of reserves breeding scheme (Fig. 1) as described by Pustovoit (1964) comprises the following steps:

(1) A large number of elite plants (up to 15,000) are selected on the basis of phenotype from the base population and harvested separately.

(2) The selected plants are further studied in laboratory for seed yield,

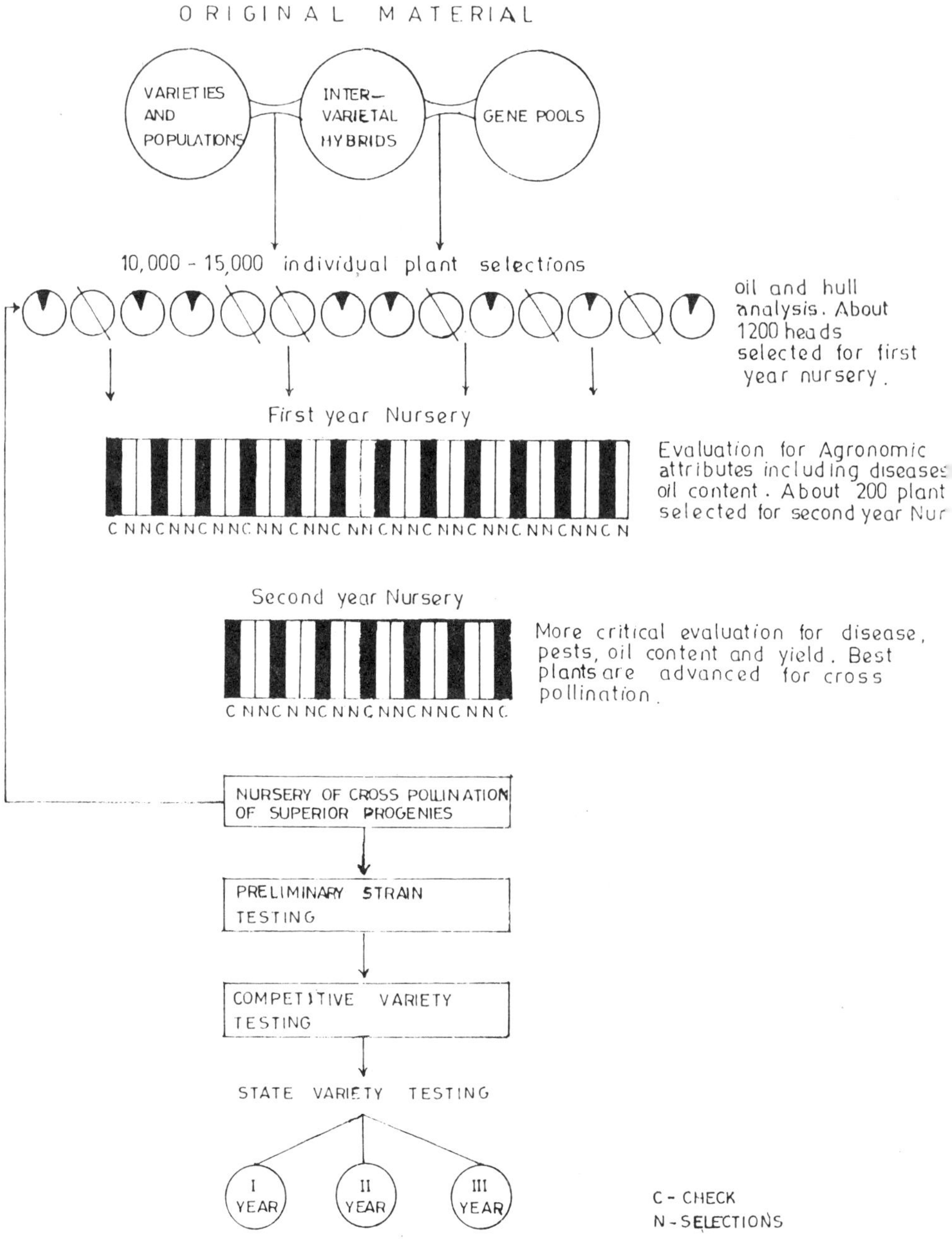

Fig. 1. Scheme of Sunflower breeding according to V.S. Pustovoit

oil content, test weight, and hull content.

(3) On the basis of these analyses, 1200–1400 plants are finally selected for progeny evaluation.

(4) The selected plants are further grouped on the basis of maturity and plant height. The elite plants thus grouped are evaluated in single rows with two replications. For every two progenies a check is included, which is generally the best variety of that maturity group. This procedure of including a control for every two progenies under evaluation is called 'coupled method of comparison' in the USSR. In this nursery the progenies are evaluated for seed yield and other important agronomic attributes. The top-performing plants, usually not exceeding 200, showing superiority over check are selected for further evaluation in the second year.

(5) Reserve seed from the original plants is used for second-year evaluation. In this nursery greater emphasis is laid on disease and pest reaction besides yield and oil content.

(6) The best plants identified from two years of progeny evaluation are taken to a cross-pollination nursery. In this nursery again the original remnant seeds are planted, keeping the identity of each line. The plants are allowed to intermate under strict isolation and roguing. From each progeny, the best plants based on phenotype are harvested separately and again subjected to laboratory analysis. The seed from the best plants thus identified after laboratory analysis is divided into two parts; one part will be carried forward for initiating the next breeding cycle and the other part is mixed to consitute a new strain.

(7) The new strain thus constituted will undergo preliminary evaluation and further testing in yield trials.

The Pustovoit breeding scheme aims at simultaneous improvement of sunflower for several characters. But greater stress is laid on oil content, maturity period and resistance to pests and diseases. The method has been highly rewarding in improving oil content and seed yield of cultivars bred in the USSR.

Heterosis Breeding

There are numerous reports of marked heterosis for both seed yield and oil content. Since the mid-1970s, there has been a marked shift in breeding emphasis from population improvement to heterosis breeding and exploitation of hybrid vigour. This was made possible because of the discovery of cytoplasmic-genetic male sterility by Leclercq (1969), which opened a new era in sunflower breeding. Cultivation of hybrids imparts a few distinct advantages over that of open-pollinated varieties. Hybrids have more production stability, are well suited to high-input agriculture, have high self-fertility resulting in higher seed set even in areas where pollinators are not abundant, and show uniform growth and maturity (Seetharam, 1982).

Development of Inbreds, Cytosteriles and Restorers

Inbred lines must be developed to exploit heterosis. Nevertheless, there are varying degrees of self-fertility and incompatibility that complicate the task of developing inbreds. Marked differences are seen within and between populations for self-fertility. It is therefore natural to expect some failures during the development of inbreds. Most of the open pollinated varieties bred in the USSR show a high degree of self-sterility. So, greater problems and failures are to be expected while attempting development of inbreds from such populations. On the other hand, a variety like Morden (Cernianka 66), which shows relatively higher degree of self-fertility, is more useful in inbred development (Seetharam, 1976).

As in other cross-pollinated crop plants, the procedure of selfing phenotypically desirable plants in open-pollinated cultivars is common in sunflower. Selfing is to be continued for five to six generations till the desired level of uniformity is reached. Nevertheless, selections could be practised both within and among progeny rows looking into desirable plant characters and agronomic attributes. After reaching the desired level of uniformity in the selfed lines sibbing could be practised in subsequent generations. The method of using haploids for development of inbred lines has also been investigated (Gundaev, 1971).

Both genetic and cytoplasmic sterility systems are available for use in heterotic breeding programmes (Leclercq, 1966, 1969; Putt and Heiser, 1966; Vranceanu, 1967). But cytoplasmic male sterility and restorer systems have been more useful in developing hybrids than genetic male steriles of self-incompatibility systems. The inbreds with good GCA could be converted into male steriles by repeated backcrossing with known CMS lines. The most popularly used *H. petiolaris* CMS source has been found to be stable under Indian conditions (Seetharam and Satyanarayana, 1980). The restorer genes apear to be scanty within the group of cultivated sunflowers. However, they are available in abundance in wild and weedy forms of *H. annuus* and in many related species. In most of the studies reported so far, fertility restoration is controlled by one to four dominant genes (Enns *et al.*, 1970; Kinman, 1970; Leclercq, 1971). However, Fick and Zimmer (1974b) and Dominguez and Fick (1975) reported a complex nature of fertility restoration behaviour in certain materials.

Inbred lines, restorers, and MS lines can be evaluated either by line × tester analysis with CMS lines as females or by topcross test using a heterozygous pollen parent as tester (Russel, 1953; Kinman *et al.* 1963; Gundaev, 1971; Sindagi *et al.*, 1979). Chemical induction of male sterility by gibberellic acid to facilitate crossing has proven effective and has been used widely by breeders to produce crossed seeds for combining ability tests (Piquemal, 1970; Seetharam and Kusuma Kumari, 1976).

Production of Hybrids

The lines that nick well, possessing good general and specific combining ability, are converted into CMS and restorer lines. They are subsequently crossed to produce single-cross hybrids (Fig. 2), which are evaluated in multilocation, trials before their release for large-scale cultivation.

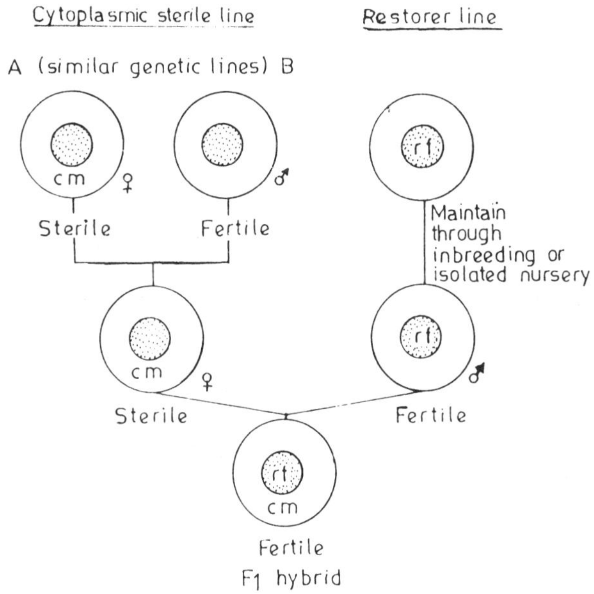

Fig. 2. Schematic presentation of the methods of developing sunflower hybrids.

Mutation Breeding

Induced mutagenesis seems to have limited application in generation of new variability in sunflower. Nevertheless, this method could find application in rectification of simply inherited deficiencies and/or in altering one or two characters in an otherwise agronomically desirable genotype. A range of mutants for earliness, dwarf growth habit, high oil content, thin hull nature, altered fatty acid composition, seed size, and male sterility have been reported through physical and chemical mutagenesis (Gundaev, 1971; Voskoboinik and Soldatov, 1974; Soldatov, 1976; Giriraj *et al.*, 1990).

Interspecific Hybridization

Successful crosses are reported in the genus *Helianthus* within and between polyploid levels involving annual and perennial species (Whelan, 1978). Thus hybridization did not depend on morphological affinities. Such results indicate that morphological discontinuities have arisen due to ecological and geographical barriers besides cross-incompatibilities. There are also instances of asssuming natural hybrids as new species in the genus *Helianthus*. These examples and other information in genus therefore suggest that the origin of species through stabilization of hybrid derivatives might have been the most common pattern of evolution in perennial sunflowers.

The wild species of sunflower are highly variable for many agronomic characters. They are an excellent source of resistance to many major diseases and pests of sunflower (Putt and Sackston, 1957; Fick *et al.*, 1974; Thompson *et al.*, 1981; Morris *et al.*, 1983). The sources of resistance have been successfully transferred to the cultivated types in the USSR (Pustovoit and Gubin, 1974). The boom of hybrid sunflower all over the world during the last two decades is largely due tc CMS source contributed by wiid annual *H. petiolaris* (Leclercq, 1969). This source is now widely used all over the world in commercial hybrid seed production without any problem. Nevertheless, several additional sources of male sterility have been identified in other wild species (Fick, 1978; Heiser, 1982; Whelan, 1981; Serieys and Vincourt, 1987) that could be used in CMS diversification programmes. Virupakshappa *et al.* (1991) have reported a number of maintainer and restorer lines for some of the new CMS sources.

Sunflower Improvement in India

Large-scale cultivation of sunflower in India began in 1972 with the introduction of four varieties from USSR-VNIIMK 8931 (EC 68413), Peredovick (EC 68414), Amoviriskij 3497 (EC 68415), and Armaverts (EC 69874). Of them, two varieties—EC68414 and EC 68415—became popular because of their superior performance and covered large areas in many states. However, sunflcwer production in the country did not increase on anticipated lines because of poor seed set and resulting lower seed yields in the introduced sunflower varieties. There was considerable pessimism in different quarters in the mid and late 1970s about the success of sunflower production in India. The late 1970s were actually the turning point and twc important developments in the research front helped to renew and revive interest in sunflower crop:

(1) identification and release of early-maturing variety Morden sunflower

in 1978 (Seetharam *et al.*, 1978) and

(2) the development and release of the first sunflower hybrid, BSH 1, in 1979 (Seetharam, 1981).

Since then the area under sunflower is on a gradual rise, touching a record 1.6 million ha during 1991–92. The production of the crop is mostly concentrated in southern parts of the country in the states of Karnataka, Tamil Nadu, Maharashtra, and Andhra Pradesh with Karnataka alone accounting for nearly 50 per cent of total area and production in the country. Of late, the crop is spreading to the northern states of Punjab, Uttar Pradesh, and Haryana. If the present trend is any indication, the area under sunflower is likely to touch 2.5 million ha by 1995–96.

Breeding Researches in India since 1972

Sunflower breeding started in India in 1972 under the All India Coordinated Research Project on Oilseeds. During the last 20 years many open-pollinated varieties have been developed and this has widened the varietal base. The high-yielding variety 'Surya' was evolved by mass selection using 'Latur bulk' as base material at Akola in Maharashtra. This variety was released in 1980 and recommended for cultivation in Maharashtra. Another early-maturing variety, CO 1 (selection from Morden), was evolved at Coimbatore for cultivation in Tamil Nadu. This variety matured in 65–70 days as compared to 80–85 days taken by Morden. CO 2 was also released in 1986 for cultivation in Tamil Nadu. It matures in 90–95 days. Another early-maturing variety, SS 56, was evolved at Sholapur for cultivation in Maharashtra. A number of high-yielding open-pollinated populations are in various stages of testing in the coordinated trials. A list of varieties and hybrids released since 1972 for cultivation in different parts of the country is given in Table 2.

The value of hybrids and the importance of heterosis breeding in this crop was recognized quite early. Experimental hybrids were developed during 1974–75 at Bangalore and their performance was compared with Russian varieties then under cultivation (Seetharam *et al.,* 1977b). The hybrids were distinctly superior to the check variety in both seed and oil yield. Multilocation trials conducted in different parts of the country further confirmed the superiority of hybrids over open-pollinated varieties. Based on seed yield, oil content, yield stability, and synchronization of flowering in male and female parents, the hybrid BSH 1 was released for large-scale cultivation in 1980 (Seetharam *et al.,* 1980). It was the first sunflower hybrid to be released in the country. The hybrid base has been further widened through the extension of heterosis breeding work to other centres. During the last five years, four more hybrids have been released for cultivation in different parts of the country (Table 2). Two of the hybrids, LDMRSH 1 and

Table 2
Varieties and hybrids of sunflower recommended for cultivation in different states

Variety/hybrid	Year of release	Adaptation zone	Special features, if any
Varieties			
EC 68414	1972	Maharashtra, Andhra Pradesh, Tamil Nadu, Uttar Pradesh, Punjab, Haryana, West Bengal, Karnataka	Open-pollinated variety of USSR origin
EC 68415	1972	Karnataka	Open-pollinated variety of USSR origin
Morden	1979	Karnataka, Andhra Pradesh, Maharashtra, Tamil Nadu, Uttar Pradesh, Punjab, Haryana	Early maturity, suitable for multiple and mixed cropping
Surya	1983	Maharashtra	Striped seed coat, low oil content
CO 1	1983	Tamil Nadu	Very early maturity
CO 2	1986	Tamil Nadu	—
SS 56	1988	Maharashtra	—
Hybrids			
BSH 1	1980	Karnataka and other states	Resistant to rust and downy mildew, tolerant to *Alternaria* leaf spot
APSH 11	1987	Andhra Pradesh	—
LDMRSH 1	1988	Maharashtra	Resistant to downy mildew
LDMRSH 3	1988	Maharashtra	Resistant to downy mildew
KBSH 1	1992	All states	—

2, released recently possess a high degree of resistance to downy mildew and are recommended for cultivation in Maharashtra, which is endemic to this disease.

Along with the release of hybrids, procedures related to production of hybrid seeds have been worked out (Satyanarayana and Seetharam, 1983; Seetharam and Satyanarayana, 1983). Planting ratio of 1 : 3 of male parent to seed parent is quite satisfactory and 8–9 q of hybrid seed per hectare could be harvested. In seed production plots, natural pollination is to be supplemented by hand pollination to increase the seed set and hybrid seed yield.

Future Thrusts

Production statistics are one of the means of assessing the importance and future trends of the crop. If planting area is any indicator, then progress of sunflower in India appears extremely satisfactory till now. At the same time one should not ignore the fact that India's productivity of 541 kg/ha during 1991–92 is one of the lowest in the world. In spite of lower productivity the Indian farmer is continuing with sunflower because of its highly remunerative price. It is not logical to expect the present situation to continue for ever. Obviously, the sunflower production in the country is bound to be largely influenced by market forces on one hand and productivity level of the crop on the other.

One of the major challenges would be to develop new hybrids and populations superior to the presently grown varieties in seed and oil yield. Imparting stability to production will also be of concern in the coming years. The new varieties bred should combine high seed yield, high oil content, and resistance to pests and diseases. Another important area of breeding research is the development of self-fertile populations and varieties. Sunflower being a self-incompatible crop, production depends on insect pollinators, particularly bees, for pollination and fertilization. In India, poor seed set is one of the serious problems resulting in low yields. Genetic rectification of this problem should aim at developing self-fertile populations.

Intensive and continuous cropping of sunflower has been the major cause for outbreaks of diseases in the recent past. Rust, *Alternaria* leaf spot, and downy mildew are the diseases requiring immediate attention. Root rot caused by *Macrophomina phaseolina* is becoming severe in some parts of Timal Nadu. Hence, development of varieties with multiple resistance is going to be one of the major objectives in the coming years.

REFERENCES

Agrawat, J.M., Chippa, H.P., and Mathur, S.J. 1979. Screening of sunflower germplasm against *Alternaria helianthi. Indian J. Mycol. Pl. Pathol.* 9: 85–86.

Alba, E., Barsanti, P., and Carrozzo, G. 1985, Analysis of combining ability in new cytoplasmically male sterile lines and in fertility restorer lines of sunflower. *Ann. Facolta de Agraria, Univ. di Bari* 31: 363–370.

Anand, I.J., and Chandra, S. 1980. Genetic diversity and interrelationships of oil yielding traits in sunflower. *Sunflower Newslett.* 3: 13–15.

Anaschenko, A.V. 1967. The chemical castration of sunflower plants. *Dokl. Vses. Akad. Sel'skokhoz Nauk im V.I. Lenina* 2: 17–18 (in Russian).

Anaschenko, A.V. 1974. The initial material for sunflower heterosis breeding. In *Proc. 6th Int. Sunflower Conf.*, Bucharest, Rumania, pp. 391–393.

Asthana, A.N. 1973. Selfing studies in sunflower using self and foreign pollen. *Sci. Cult.* 39: 237.

Balasubramanyam, N., and Kolte, S.J. 1980. Effect of Alternaria blight on yield components, oil content and seed quality of sunflower. *Indian J. Agric. Sci.* 50: 701–706.

Baronecelli, S., Cecconi, F., and Pugliesi, C. 1987. Genetic control of some physiological characters in sunflower (*Helianthus annuus* L.). *Genet. Agrar.* 41: 284.

Begum, D., Khan, M.S., and Khaleque, M.A. 1988. Genetic and environmental variability in sunflower. *Bangladesh J. Agric.* 13: 11–15.

Burlov, V.V. 1974. Utilization of male sterility in sunflower breeding for heterosis. In *Proc. 6th Int. Sunflower Conf.*, Bucharest, Rumania, pp. 353–360.

Burlov, V.V., and Krut'ko, D.I. 1986. Expression of the self-compatibility character in sunflower. *Nauchnotekhnicheskii. Bull. Vsesoyuznogo Selektsionno-geneti cheskogo unhita* 2: 45–50.

Burns, R.E. 1970. Head size of sunflower as an indicator of plot yields. *Agron. J.* 62: 112–113.

Carreon, R.J.I., and Barreda, P.D.A. 1983. Recurrent selection for general combining ability (GCA) of S lines in sunflower (*Helianthus annuus* L.). thesis, Universidad Autonoma Chapingo, Mexico.

Carson, M.L. 1985. Epidemiology and yield losses associated with *Alternaria* blight of sunflower. *Phytopathology* 75: 1151–1156.

Chaudhary, S.K., and Anand, I.J. 1984. Heterosis and inbreeding depression in sunflower. *Crop Improv.* 11: 15–19.

Chaudhary, S.K. and Anand, I.J. 1985. Influence of various characters on yield of sunflower. *J. Oilseed Res.* 2: 78–85.

Chaudhary, S.K., and Anand, I.J. 1987. Genetic and morphological variability for quantitative characters in sunflower. *J. Oilseed Res.* 4: 97–102.

Cockerell, T.D.A. 1915. Specific and varietal characters in annual sunflowers. *Am. Nat.* 49. 609–622.

Cooper, D.C., and Mahony, K.L. 1932. Cytological observations on certain Compositae. *American J. Bot.* 22: 843–848.

Dhaduk, L.K., Desai, N.D., Petal, R.H., and Kukadia, M.U. 1985. Correlation and path analysis in sunflower. *Indian J. Agric. Sci.* 55: 52–54.

Dominguez-Gimenez, J., and Fick, G.N. 1975. Fertility restoration of male-sterile cytoplasm in wild sunflowers. *Crop Sci.* 15: 724–726.

Dua, R.P., and Yadava, T.P. 1982. Gene effects for some quantitative characters in sunflower. *Indian J. Agric. Sci.* 52: 439–443.

Dua, R.P., and Yadava, T.P. 1983. Combining ability in sunflower. *Indian J. Genet.* 43: 129–136.

Enns, H. 1959. A study of the inheritance and expressivity of some seedling characters in the sunflower (*Helianthus annuus* L.). M.Sc. Thesis, University of Manitoba, Winnipeg.

Enns, H., Dorrel, D.G., Hoes, J.A., and Chubb, W.O. 1970. Sunflower research, a progress report. In *Proc. 4th Int. Sunflower Conf.* Memphis, Tennessee, pp. 162–167.

Feriey's, M. 1981. Study of the genetic variability of the genus *Helianthus. Bull. CETIOM* 80: 11–13.

Fernadez, L.F., and Orioli, G.A. 1984. Comparative study of leaf structure, light interception and yield in sunflower (*Helianthus annuus* L.). *Ann. Edafologia y Agrobiol.* 42: 2137–2148.

Fernandez, M.N., and Knowles, P.F. 1982. Maternal and embryo effects on the oleic and linoleic and contents of sunflower oil. In *Proc. 10th Int. Sunflower Conf.*, 14–17 March, Australia, pp. 241–243.

Fernandez, M.J., Marquez, F., and Ortiz, J. 1979. Genetics of the seed oil content in sunflower. *Prod. Vegetal* 10: 93–100.

Fernandez, J.M., Jimenez, A., Dominguez, J., Garcia, J.M., Garces, R., and Mancha, M. 1989. Genetic analyses of the high oleic acid content in cultivated sunflower (*Helianthus annuus* L.) . *Euphytica* 41: 39–51.

Fick, G.N. 1978. Breeding and genetics. In *Sunflower Science and Technology.* Jack F. Carter, ed. The American Society of Agronomy, Crop Science Society of America, and Soil Science Society of America, Madison, Wisconsin. pp. 279–338.

Fick, G.N. 1984. Inheritance of high oleic acid in the seed oil of sunflower. In *Proc. Sunflower Res. Workshop,* Bismark, North Dakota, p. 9.

Fick, G.N., and Zimmer, D.E. 1974a. Monogenic resistane to *Verticillium* wilt in sunflowers. *Crop Sci.* 14: 895–896.

Fick, G.N., and Zimmer, D.E. 1974b. Fertility restoration in confectionery sunflowers. *Crop Sci.* 14: 603--605.

Fick, G.N., and Zimmer, D.E. 1976. Yield stability of sunflower hybrids and open-pollnated varieties. In Abstract of papers, *7th Int. Sunflower Conf.* Krasnodar, USSR. pp. 37–78.

Fick, G.N., Zimmer, D.E., Dominguez-Gimenez, and Rehder, D.A. 1974. Fertility restoration and variability for plant and seed characteristics in wild sunflowers. In *Proc. 6th Int. Sunflower Conf.* Bucharest, Rumaina, pp. 333–338.

Geisler, F. 1931. Chromosome numbers in certain species of *Helianthus. Butler Univ. Bot. Standard* II: 53–62.

Georgieva-Todorova, Y. 1970. Zytogenetische Untersuchung der Hybriden *Helianthus argophyllus* × *H. annuus. Z. Pflanzenzuchtung* 64: 357–366.

Georgieva-Todorova, Y. 1974. A new male sterile form of sunflower (*Helianthus annuus* L.). In *Proc. 6th Int. Sunflower Conf.* Bucharest, Rumania, pp. 343–347.

Giriraj, K., Shivaraju, N., Hiremath, S.R., and Prasad, T.G. 1987. Bioproductivity and its relationship to yield attributes and seed yield in inbreds, hybrids and populations of sunflower. *J. Oilseeds Res.* 4: 35–40.

Giriraj, K., Shantha, R., Hiremath, S.R., and Seetharam, A. 1990. Induced variability for flowering, seed weight and oil content in parental lines of sunflower hybrid BSH-1. *Indian J. Genet.* 50: 1–7.

Gonzalez Garica, J., and Angeles Espino, A. 1983. An early test of general combining ability in two varieties of sunflower (*Helianthus annuus* L.). *Revista Chapingo* 42: 42–46.

Goossen, P.G., and Sackston. 1967. Transmission and biology of sunflower downy mildew. *Can. J. Bot.* 46: 5–10.

Goyne, P.J. 1985. Photoperiod and temperature influences on sunflower (*Helianthus annuus* L.) genotypes. *Abstr. International,* B (Sci. Eng.) 45: 36648.

Gundaev. A.I. 1971. Basic principles of sunflower selection. In *Genetic Principles of Plant Selection.* Nauka, Moscow.

Habura, E.C.H. 1957. Parasterilitat bei sonnenblumen. *Z. Pflanzenzuchtung* 37: 280–298.

Heiser, C.B. 1949. Study in the evolution of sunflower species *Helianthus annuus* and *Helianthus bolanderi. Univ. Cali. Pub. Bot.* 23: 157--196.

Heiser, C.B. 1951. The sunflower among the North American Indians. *Proc. Am. Phil.. Soc.* 95: 432–448.

Heiser, C.B. 1957. A revision of the South American species of *Helianthus. Brittonia* 8: 283–295.

Heiser, C.B. 1982. Registration of Indiana-1 CMS sunflower germplasm. *Crop Sci.* 22: 1989.

Heiser, C.B., Martin, W.C., and Smith, D.M. 1962. Species crosses *Helianthus.* I. Diploid species. *Brittonia* 14: 137–147.

Heiser, C.B., Smith, D.M., Clevenger, S., and Martin, W.C. 1969. The North American sunflowers (*Helianthus*). *Mem. Torrey Bot. Club* 22: 1–218.

Hennessy, C.M.R., and Sackston, W.E. 1972. Studies on sunflower rust. X. Specialization of *Puccinia helianthi* on wild sunflowers in Texas. *Can. J. Bot.* 50: 1871–1877.

Hockett, E.A., and Knowles, P.F. 1970. Inheritance of branching in sunflowers (*Helianthus annuus* L.). *Crop Sci.* 10: 432–436.

Hoes, J.A., and Putt, E.D. 1962. Races of *Puccinia helianthi. Phytopathology* 52: 736–737.

Hoes, J.A., Putt, E.D., and Enns, H. 1973. Resistance to *Verticillium* wilt in collections of wild *Helianthus* in North America. *Phytopathology.* 63: 1517–1520.

Hristova, A., and Georgieva-Todorova, Y. 1975. A study of the fatty acid composition and protein content in the seeds of *Helianthus annuus, Helianthus scaberrimus* and their F_1 hybrids. *C.R. Acad. Agric.* G. Dimitrov 8: 55–58.

Kinman, M.L. 1970. New developments in the USDA and state experiment station sunflower breeding programs. In *Proc. 4th Int. Sunflower Conf.* Memphis, Tennessee, pp. 181–183.

Kinman, M.L., Rivers, G.W., Redman, E.J., Wade, J.T., and Flynt, E.L. 1963. *Annual Report of Corporative Sesame, Sunflower and Guar Investigatives.* USDA and Texas Agric. Exp. Stn.

Kloczowski, Z. 1975. Studies on some features of oil sunflower and their significance in breeding that plant in Poland (in Polish). *Hodowla Rosl Akllim, Nasienn* 19: 89–131.

Kovacik, A., and Skaloud, V. 1972. The proportion of the variability component caused by the environment and the correlations of economically important properties and characters of the sunflower *(Helianthus annuus* L.). *Sci. Agric. Bohemoslov* 4: 249--261. (Plant Breed. Abstr. 43: 8100).

Kulshrestha, V.B., and Gupta, P.K. 1979. Cytogenetic studies in the genus *Helianthus* L. *Cytologia* 44: 325–334.

Leclercq, P. 1966. Une sterilité male utilisable pour la production d'hybrides simples de tournesol. *Ann. Amel. Plant* 16: 135–144.

Leclercq, P. 1968. Heredité de quelques caractéres qualitatifs chez le tournesol. *Ann. Amel. Plant* 18: 307–315.

Leclercq, P. 1969. Une sterilité cytoplasmique du tournesol. *Ann. Amel. Plant* 9: 99–106.

Leclercq, P. 1971. La sterilité male cytoplasmique du la fertilité. *Ann. Amel. Plant* 21: 45–54.

Leclercq, P., Cauderon, Y., and Dauge, M. 1970. Selection pour la resistance au mildiour du tournesol a partir d'hybrides topinambour × tournesol. *Ann. Amel. Plant* 20: 363–373.

Luciano, A., Kinman, M.L., and Smith, D. 1965. Heritability of self incompatibilty in the sunflower (*Helianthus annuus* L.). *Crop Sci.* 5: 529–532.

Manjunath, M.E. 1986. Evaluation of male sterile and fertility restorer lines for combining ability in sunflower *(Helianthus annuus* L.). Ph.D. thesis, University of Agricultural Sciences, Bangalore.

Marin, I.V. 1986. Interrelation and inheritance of yield components in sunflower hybrids (abstracts). In *Plant Breed. Abstr.* 58: 10586.

Mayee, C.D., and Patil, M.A. 1986. Downy mildew of sunflower in India. *Trop. Pest Mgmt.* 33: 81–82.

Miller, J.F., and Gulya, T.J. 1984. Sources and inheritance of resistance to race 3 downy mildew in sunflower. *Helia* 7: 17–20.

Miller, J.F., and Gulya, T.J. 1987. Inheritance of resistance to race 3 downy mildew in sunflower. *Crop Sci.* 27: 210–212.

Miller, J.F., and Gulya, T.J. 1988. Registration of six downy mildew resistant sunflower germplasm lines. *Crop Sci.* 28: 1040–1041.

Miller, J.F., and Zimmerman, D.C. 1983. Inheritance of high oleic fatty acid content in sunflower. In *Proc. Sunflower Res. Workshop,* 26 Jan. Fargo, North Dakota. National Sunflower Association, Bismark, North Dakota, 10 pp.

Miller, J.F., Fick, G.N., and Cedeno, J.R. 1977. Improvement of oil content and quality in sunflower (*Helianthus annuus* L.) (abstr.). *Agron. Abstr.,* 64.

Miller, J.F., Hammond, J.J., and Roath, W.E. 1980. Comparison of inbred vs. single cross testers and estimation of genetic effects in sunflower. *Crop Sci.* 20: 703–706.

Miller, J.F., Zimmerman, D.C., and Vick, B.A. 1987. Genetic control of high oleic and acid content in sunflower oil. *Crop Sci.* 25: 923–926.

Mishra, R., Shrivastava, A..N., Chandra, D.R., and Singh, P. 1985. Path analysis for yield components in sunflower (*Helianthus annuus* L.). *Agric. Sci. Digest,* India 5: 38–40.

Morozov, V.K. 1947. *Sunflower Breeding in the USSR.* Piscepromizdat, Moscow, 272 pp.

Morozov, V.K., and Raikov, V.N. 1983. Agronomic and biological feature of early sunflower varieties. *Selskokhozyaistvennaya Biol.* 10: 36–39.

Morris, J.B., Yang, S.M., and Wilson, I. 1983. Reaction of *Helianthus* species to *Alternaria helianthi. Plant Disease* 67: 539–540.

Mosjidis, J.A. 1982. Inheritance of colour in the pericarp and corolla of the disc florets in sunflower. *J. Hered.* 73: 461–464.

Nikolic-Vig, V., Skoric, D., and Bedov, S. 1971. Variability of oil and husk content in the sunflower seed of Peredovik and VNIIMK 8931 varietal populations and their heritability. *Savremena Poljoprivreda* 129: 23–32.

Oka, H.I., and Campos, F.F. 1974. Breeding behaviour and genetic variations of sunflower observed in Central Luzor. *SABRAO J.* 6: 61–67.

Pathak, A.R., Basudeo Singh and Kukadia, M.V. 1985. Genetic architecture of yield and economic characters in sunflower. *Oilseeds J.* 15: 14–18.

Petrov, P.D., and Piskov, A. 1985. Study on the self fertility on some inbred lines, hybrids and varieties of sunflower. *Rasteniv'dni' Nauki* 22: 63–76.

Pimakhin, V.F., Solovov, P.N., and Arkhangelskii, V.N. 1984. Protecting sunflower from *Sclerotinia seclerotiorum* and *Botrytis cinerea* identifying resistant forms. *Maslichnye Kul'tury* 3: 30.

Piquemal, G. 1970. How to produce hybrid sunflower seeds by inducing male sterility with gibberellic acid. In *Proc. 4th Int. Sunflower Conf.*, Memphis, Tennessee, pp. 127–135.

Pogorletskii, B.K. 1974. Yield in single interline hybrids of sunflower (in Russian). *Referativnyi Z.* 2: 187.

Pustovoit, V.S. 1964. Conclusions of work on the selection and seed production of sunflowers. *Agrobiology* 5: 672–697.

Pustovoit, G.V., and Gubin, I.A. 1974. Results and prospects in sunflower breeding for group immunity by using the interspecific hybridization method. In *Proc. 6th Int. Sunflower Conf.*, Bucharest, Rumania, pp. 373–381.

Putt, E.D. 1940. Observations on morphological characters and flowering processes in the sunflower (*Helianthus annuus* L.). *Sci. Agric.* 21:167–179.

Putt, E.D. 1943, Association of seed yield and oil content with other characters in sunflowers. *Sci. Agric.* 21: 377–383.

Putt, E.D. 1964a. Breeding behaviour of resistance to leaf mottle or *Verticillium* wilt in sunflower. *Crop Sci.* 4: 177–179.

Putt, E.D. 1964b. Recessive branching in sunflower. *Crop Sci.* 4: 444–445.

Putt, E.D. 1966. Heterosis, combining ability and predicted synthetics from a diallel cross in sunflower (*Helianthus annuus* L.). *Can. J. Pl. Sci.* 46: 59–67.

Putt, E.D., and Heiser, C.B. 1966. Male sterility and partial male sterility in sunflower. *Crop Sci.* 6: 165–168.

Putt, E.D., and Sackston, E.E. 1957. Studies on sunflower rust. I. Some sources of rust resistance. *Can. J. Pl. Sci.* 37: 43–54.

Putt, E.D., Craig, B.M., and Carson, R.B. 1969. Variation in composition of sunflower oil from composite and single seeds of varieties and inbred lines. *J. Amer. Oil and Chem. Soc.* 46: 126–129.

Ramakanth, R.S., and Seetharam, A. 1977. Cytomorphological studies in the genus *Helianthus.* I. Karyotype studies in the diploid species. *Proc. Indian Acad. Sci.* 86B: 155–158.

Reddy, T.N., and Reddy, P.S. 1979. Studies on heritability and genetic advance in certain hybrids of sunflower. *Indian J. Hered.* 11: 21–23.

Reddy, P.S., and Thammiraju, B. 1977. Inheritance of fertility restoration in sunflower (*Helianthus annuus* L.). *Euphytica* 26: 409–412.

Robinson, R.G. 1967. Registration of Mingren sunflower. *Crop Sci.* 7: 404.

Robinson, R.G., Bernat, L.A., Geise, H.A., Johnson, F.K., Kinman, M.L., Mader, E.L., Oswalt, R.M., Putt, E.D., Swallers, C.M., and Williams, J.H. 1967. Sunflower development at latitudes ranging from 31 to 49 degrees. *Crop Sci.* 7: 134–136.

Rodin, V.F. 1976. Inheritance of short stem in sunflower (abstr.). *Plant Breed. Abst.* 50: 633.

Ross, A.M. 1939. Some morphological characters of *Helianthus annuus* L. and their relations to the yield of seed and oil. *Sci. and Agric.* 19: 372–379.

Rostova, N.S., and Krylova, E.G. 1987. Leaf structure in annual sunflower, its variation and determination. *Plant Breed. Abstr.* 59: 8.

Russel, W.A. 1953. A study of the inter-relation of seed yield, oil content, and other agronomic characters with sunflower inbred lines and their top crosses. *Can. J. Agric. Sci.* 33: 291–314.

Satyanarayana, A.R., and Seetharam, A. 1983. Method of hybrid seed production in sunflower (*Helianthus annuus* L.). II. Effect of planting ratios and method of pollination on seed quality. *Seed. Res.* 11: 8–12.

Schuster, E. 1964. Inbreeding and heterosis in sunflower (*Helianthus annuus* L.) (Plant Breeding Abstract 37: 1207, 1967). In Wilhech Schmitz Verlag Giessen. 135 pp.

Schuster, W. 1970. Die auswirkungen der for tgesetzten inzuchtung van I_0 bis I_{18} verschiedene merkmale der sonnenblume. *Z. Pflanzenzuchtung* 64: 310–334.

Seetharam, A. 1976. Performance of sunflower varieties and the role of various facators affecting seed yield. In *Souvenir of 14th All India Convention of Oilseeds* (Oil trade and industry, pp. 1–8.

Seetharam, A. 1981. Sunflower cultivation. *Indian Fmg.* 25: 1–4.

Seetharam, A. 1982. Breeding strategy for developing high yielding varieties of sunflower. In *Research and Development Strategies for Oil Seed Production in India. Proc. Nat. Symp.* 7–9 Nov. 1979, IARI, New Delhi., 143–152.

Seetharam, A., and Kusuma Kumari, P. 1976. Histological studies in cytoplasmic and GA induced male sterile lines of sunflower. *Indian J. Genet.* 36: 342.

Seetharam, A., and Satyanarayana, A.R. 1980. Stability of cytoplasmic male sterility in lines of sunflower. *SABRAO J.* 12: 109–112.

Seetharam, A., and Satyanarayana, A.R. 1983. Method of hybrid seed production in sunflower (*Helianthus annuus* L.). I. Effect of parental ratios and method of pollination on hybrid seed yield and its attributes. *Seed Res.* 11: 1–7.

Seetharam, A., Kusuma Kumari, P., Patil N.M., and Sindagi, S.S. 1977a. Note on the immediate influence of pollen on seed weight and oil content in sunflower. *Indian J. Agric. Sci.* 47: 212–214.

Seetharam, A., Kusuma Kumari, P., Patil, N.M., and Sindagi, S.S. 1977b. Performance of hybrids of sunflower produced by means of cytoplasmic male sterility. *SABRAO J.* 9: 51–55.

Seetharam, A., Giriraj, K., and Kusuma Kumari, P. 1978. Cernianka-66—A dwarf, early variety of sunflower. *Curr. Res.* 7: 164–165.

Seetharam, A., Giriraj, K., and Kusuma Kumari, P. 1980. Phenotypic stability of seed yield in sunflower hybrids. *Indian J. Genet.* 40: 102–104.

Serieys, H., and Vincourt, P. 1987. Characterisation of some new cytoplasmic male sterility source from *Helianthus* genus. *Helia* 10: 9–13.

Shabana, R. 1974. Genetic variability of sunflower varieties and inbred lines. In *Proc. 6th Int. Sunflower Conf.*, Bucharest, Rumania, pp. 263–269.

Sheriff, N.M., and Appadurai, R. 1985. Genetic variability in some quantitative traits of sunflower. *Madras Agric. J.* 72: 539–540.

Sheriff, N.M., Appadurai, R., and Rangaswamy, M. 1985. Combining ability in sunflower. *Indian J. Agric. Sci.*. 55: 315–318.

Sheriff, N.M., Appadurai, R., and Rangaswamy, M. 1986. Growth pattern, heterosis and correlation in sunflower. *Madras Agric. J.* 73: 308–312.

Shinde, Y.M., Wattamwar, M.J., and Patil, G.D. 1983. Variability and correlation studies in sunflower. *J. Maharashtra Agric. Univ.* 8: 122–123.

Shrivastava, A.N., and Mishra, R. 1976. Genotypic and phenotypic variability in quantitative characters in sunflower. *Madras Agric. J.* 63: 209–210.

Sindagi, S.S., Kulkarni, R.S., and Seetharam, A. 1979. Line × tester analysis of combining ability in sunflower (*Helianthus annuus*). *Sunflower Newslett.* 3(2): 11–12.

Singh, J.V., and Yadava, T.P. 1986. Variability studies of some quantitative characters in sunflower. *J. Oilseeds Res.* 3: 125–127.

Singh, S.B., Labana, K.S., and Virk, D.S. 1989. Detection of epistatic, additive and dominance variation in sunflower. *Indian J. Genet.* 47: 243–247.

Skaloud, V., and Kovacik, A. 1974. Inheritance of some heteromorphic characters in sunflowers (*Helianthus annuus* L.). In *Proc. 6th Int. Sunflower Conf.*, Bucharest, Rumania, pp. 291–295.

Skoric, D., and Bedev, S. 1978. Studies of oil and protein contents and composition in genetically divergent sunflower genotypes. In *Proc. 6th Int. Sunflower Conf.*, Minneapolis, Minnesota. pp. 516–523.

Soldatov, K.I. 1976. Chemical mutagenesis for sunflower breeding. In Abstracts of papers of the *7th Int. Sunflower Conf.*, Krasnodar. USSR, pp. 57–58.

Stoyanova, J. 1970. The inheritance of male sterility in certain sources of this character in sunflowers. *Genetika i Sele k cija* 3: 451–459.

Thompson, T.E., Zimmerman, D.C., and Rogers, C.E. 1981. Wild *Helianthus* as a genetic resource. *Field Crops Res.* 4: 333–343.

Torres, C.E., Ortegon Morales, A.S., and Lopez Perez, E. 1984. Recurrent selection in S_1 lines for yield and oil content in sunflower *Helianthus annuus L. Agric. Tecnica en Mexico* 10: 121–132.

Turner, B.L., and David, F. 1966. Chromosome numbers in the Compositae × North American species. *Amer. J. Bot.* 53: 24.

Urie, L. 1985. Inheritance of high oleic acid in sunflower. *Crop Sci.* 25: 986–989.

Vanishree, G., Ananthasayana, K., Nagabhushanam, G.V.S., and Jagadish, C.A. 1988. Correlation and path coefficient analysis in sunflower (*Helianthus annuus* L.). *J. Oilseeds Res.* 5: 46–51.

Vear, F. 1974. Studies on resistance to downy mildew in sunflowers (*Helianthus annuus*). In *Proc. 6th Int. Sunflower Conf.*, Bucharest, Rumania, pp. 297–302.

Venkateswaralu, V., Reddy, P.S., and Rao, D.V.M. 1980. Heritability studies in sunflower. *Sunflower Newslett.* 4: 10–11.

Virupakshappa, K., Seetharam, A., Jayarame Gowda, and Ravikumar, R.L. 1991. Maintainer and restorer behaviour of some sunflower lines on new cytoplasmic male sterile sources. *J. Oilseeds Res.* 8: 195–198.

Voskoboinik, L.K., and Marin. 1986. Evaluating sunflower for yield components. *Maslichneye Kultury* 4: 35.

Voskoboinik, L.K., and Soldatov, K.I. 1974. The research trends in the field of sunflower breeding for heterosis at the All-Union Research Institute for Oil Crops (VNIIMK). In *Proc. 6th Int. Sunflower Conf.*, Bucharest, Rumania, pp. 383–389.

Vranceanu, A.V. 1967. Ereditatea surselor de androsterilitate la floarea-soarelui. *Problems de Agric.* 12: 4–11.

Vranceanu, A.V. 1970. Advances in sunflower breeding in Rumania. In *Proc. 4th Int. Sunflower Conf.*, Memphis, Tennessee, pp. 136–148.

Vranceanu, A.V., and Pirvu, N. 1988. Genetic advance realised in breeding sunflower in Rumania. *Plant Breed. Abstr.* 60: 629.

Vranceanu, A.V., and Stoenescu, F. 1970. Testing sunflower cytoplasm in relation to genes for male sterility. *Plant Breed. Abstr.* 43: 3837.

Vranceanu, A.V., and Stoenescu, F. 1971. Heritability of immunity to downy mildew of sunflower (abstr.). *Plant Breed. Abstr.* 43: 8106.

Vranceanu, A.V., Stoenescu, F., Ulinici, A., Iliescu, H., and Paulian, F. 1974. *Sunflower. Editura Academiei Republicii* Socialiste, Bucharest, Rumania, 322 pp. (in Romanian).

Vranceanu, A.V., Pirvu, N., Stoenescu, F.M., Iliescu, H., and Lascu, I. 1981. Genetic aspects of the resistance of sunflowers to *Sclerotinia sclerotiorum* (Lib) de By (abstr.). *Plant Breed. Abstr.* 53: 9002.

Vulpe, V.V. 1967. Study of some sunflower hybrids produced through male sterility and fertility restoration system. *Plant Breed. Abstr.* 39: 3.

Whelan, E.D.P. 1978. Cytology and interspecific hybridization. In *Sunflower Science and Technology.* Jack F. Carter, ed. The American Society of Agronomy, Crop Science Society of America, and Soil Science Society of America, Madison, Wisconsin, pp. 339–369.

Whelan. E.D.P. 1981. Cytoplasmic male sterility in *Helianthus giganteus* L. × *H. annuus.* Interspecific hybrids. *Crop Sci.* 21: 855–856.

Yang, S.M., Doweer, W.M., and Luciano, A. 1989. Gene Pu_6: A new gene in sunflower for resistance to *Puccinia helianthi. Phytopathology* 79: 474–477.

Yarnell, R.A. 1972. *Iva annua* var. *macrocarpa.* Extinct American cultigen? *Amer. Anthropol.* 74: 335–341.

Zhukovski, P.M. 1950. *Cultivated Plants and Their Wild Relatives.* (Abstract translated by P.S. Hudson, 1962.) Commonwealth Agricultural Bureau, Farnham Royal, England.

INDEX

Printed in India